Orchid Biochemistry 2.0

Orchid Biochemistry 2.0

Editor

Jen-Tsung Chen

MDPI • Basel • Beijing • Wuhan • Barcelona • Belgrade • Manchester • Tokyo • Cluj • Tianjin

Editor
Jen-Tsung Chen
Department of Life Sciences
National University of
Kaohsiung
Kaohsiung
Taiwan

Editorial Office
MDPI
St. Alban-Anlage 66
4052 Basel, Switzerland

This is a reprint of articles from the Special Issue published online in the open access journal *International Journal of Molecular Sciences* (ISSN 1422-0067) (available at: www.mdpi.com/journal/ijms/special_issues/Orchid_Biochemistry_2).

For citation purposes, cite each article independently as indicated on the article page online and as indicated below:

LastName, A.A.; LastName, B.B.; LastName, C.C. Article Title. *Journal Name* **Year**, *Volume Number*, Page Range.

ISBN 978-3-0365-4652-0 (Hbk)
ISBN 978-3-0365-4651-3 (PDF)

© 2022 by the authors. Articles in this book are Open Access and distributed under the Creative Commons Attribution (CC BY) license, which allows users to download, copy and build upon published articles, as long as the author and publisher are properly credited, which ensures maximum dissemination and a wider impact of our publications.

The book as a whole is distributed by MDPI under the terms and conditions of the Creative Commons license CC BY-NC-ND.

Contents

Jen-Tsung Chen
Orchid Biochemistry 2.0
Reprinted from: *Int. J. Mol. Sci.* **2022**, *23*, 6823, doi:10.3390/ijms23126823 1

Emilia Brzosko, Andrzej Bajguz, Justyna Burzyńska and Magdalena Chmur
Nectar Chemistry or Flower Morphology—What Is More Important for the Reproductive Success of Generalist Orchid *Epipactis palustris* in Natural and Anthropogenic Populations?
Reprinted from: *Int. J. Mol. Sci.* **2021**, *22*, 12164, doi:10.3390/ijms222212164 5

Francesca Lucibelli, Maria Carmen Valoroso, Günter Theißen, Susanne Nolden, Mariana Mondragon-Palomino and Serena Aceto
Extending the Toolkit for Beauty: Differential Co-Expression of *DROOPING LEAF*-like and Class B MADS-box Genes during *Phalaenopsis* Flower Development
Reprinted from: *Int. J. Mol. Sci.* **2021**, *22*, 7025, doi:10.3390/ijms22137025 35

Li-Min Huang, Hsin Huang, Yu-Chen Chuang, Wen-Huei Chen, Chun-Neng Wang and Hong-Hwa Chen
Evolution of Terpene Synthases in Orchidaceae
Reprinted from: *Int. J. Mol. Sci.* **2021**, *22*, 6947, doi:10.3390/ijms22136947 53

Danqi Zeng, Jaime A. Teixeira da Silva, Mingze Zhang, Zhenming Yu, Can Si and Conghui Zhao et al.
Genome-Wide Identification and Analysis of the APETALA2 (AP2) Transcription Factor in *Dendrobium officinale*
Reprinted from: *Int. J. Mol. Sci.* **2021**, *22*, 5221, doi:10.3390/ijms22105221 73

Emilia Brzosko, Andrzej Bajguz, Magdalena Chmur, Justyna Burzyńska, Edyta Jermakowicz and Paweł Mirski et al.
How Are the Flower Structure and Nectar Composition of the Generalistic Orchid *Neottia ovata* Adapted to a Wide Range of Pollinators?
Reprinted from: *Int. J. Mol. Sci.* **2021**, *22*, 2214, doi:10.3390/ijms22042214 95

Conghui Zhao, Zhenming Yu, Jaime A. Teixeira da Silva, Chunmei He, Haobin Wang and Can Si et al.
Functional Characterization of a *Dendrobium officinale* Geraniol Synthase DoGES1 Involved in Floral Scent Formation
Reprinted from: *Int. J. Mol. Sci.* **2020**, *21*, 7005, doi:10.3390/ijms21197005 123

Can Si, Jaime A. Teixeira da Silva, Chunmei He, Zhenming Yu, Conghui Zhao and Haobin Wang et al.
DoRWA3 from *Dendrobium officinale* Plays an Essential Role in Acetylation of Polysaccharides
Reprinted from: *Int. J. Mol. Sci.* **2020**, *21*, 6250, doi:10.3390/ijms21176250 139

Juan Chen, Bo Yan, Yanjing Tang, Yongmei Xing, Yang Li and Dongyu Zhou et al.
Symbiotic and Asymbiotic Germination of *Dendrobium officinale* (Orchidaceae) Respond Differently to Exogenous Gibberellins
Reprinted from: *Int. J. Mol. Sci.* **2020**, *21*, 6104, doi:10.3390/ijms21176104 157

Gah-Hyun Lim, Se Won Kim, Jaihyunk Ryu, Si-Yong Kang, Jin-Baek Kim and Sang Hoon Kim
Upregulation of the MYB2 Transcription Factor is Associated with Increased Accumulation of Anthocyanin in the Leaves of *Dendrobium bigibbum*
Reprinted from: *Int. J. Mol. Sci.* **2020**, *21*, 5653, doi:10.3390/ijms21165653 181

Zhenming Yu, Conghui Zhao, Guihua Zhang, Jaime A. Teixeira da Silva and Jun Duan
Genome-Wide Identification and Expression Profile of *TPS* Gene Family in *Dendrobium officinale* and the Role of *DoTPS10* in Linalool Biosynthesis
Reprinted from: *Int. J. Mol. Sci.* **2020**, *21*, 5419, doi:10.3390/ijms21155419 . . . . **199**

Editorial
Orchid Biochemistry 2.0

Jen-Tsung Chen

Department of Life Sciences, National University of Kaohsiung, Kaohsiung 811, Taiwan; jentsung@nuk.edu.tw

In the Special Issue entitled "Orchid Biochemistry", researchers explored the biochemistry and molecular mechanisms of pigment formation, flower scent, bioactive compounds, plant–microbial interaction, as well as aspects of biotechnology, and these studies have greatly enriched the understanding in the field of orchid biology [1]. In the second volume of this Special Issue, entitled "Orchid Biochemistry 2.0", one literature review and nine original research articles were published, and the Special Issue provides further insight into several critical subtopics, including reproduction biology, functional genomics in secondary metabolites, as well as polysaccharides and orchid mycorrhizae.

1. Pollination and Flowering Biology

Orchids are ideal models for the study of pollination biology due to their diverse flowers that adapted structurally and chemically during evolution. Brzosko et al. studied plant–pollinator interactions using *Neottia ovata* (L.) Bluff & Fingerh., a generalist orchid, as a model to explore the impact of soil parameters on flower structure and nectar chemistry [2]. The authors found that carbon and the ratio of carbon to nitrogen might be the most important factor affecting flower structure and nectar composition. Furthermore, Brzosko et al. investigated the impact of the flower structure and nectar chemistry in *Epipactis palustris* (L.) Crantz, another generalist orchid that is pollinated by over 100 species of pollinators, on reproductive success [3]. The authors concluded that there are significant differences in nectar chemical properties between natural and anthropogenic populations of *E. palustris* and pointed out that future study is needed to clarify the most critical factor between pollinator differentiation and soil characteristics.

The molecular basis of flowering in orchids is yet to be fully understood, and one of the principle questions is the function of MADS-box genes. Lucibelli et al. conducted an in silico differential expression analysis to identify two YABBY *DL/CRC* homologs in *Phalaenopsis equestris* (Schauer) Rchb.f., namely *PeDL1* and *PeDL2* [4]. It was found that *PeDL2* regulates the differentiation of labella, and this finding enriches our knowledge of the regulatory network for orchid flower development. *Apetala 2* (*AP2*) is a gene that codes for a transcription factor, and it belongs the *AP2/EREBP* gene family, which plays key roles in regulating growth and development in plants. Zeng et al. identified 14 homologs of *AP2* in a popular medicinal orchid, *Dendrobium officinale* Kimura & Migo, namely *DoAP2-1* to *DoAP2-14* [5]. After subcellular localization and functional analysis of these genes, it was found that *DoAP2* may encode transcriptional repressors and be involved in flower development, stress response and other biological activities.

2. Functional Genomics in Secondary Metabolites and Polysaccharides

Terpenes, the largest family of plant secondary metabolites, possess a range of vital roles in plant growth and development. Yu et al. studied the gene family of terpene synthase (TPS) in *D. officinale*. The authors identified 34 TPS genes (*TPSs*) and analyzed their expression patterns, and it was found that the predominantly expressed organ is flowers. Among these genes, *DoTPS10* was selected for further investigation under abiotic stress, and it was found that the targeted subcellular localization of DoTPS10 is in chloroplasts, and in in vitro test, it was shown to convert geranyl pyrophosphate into linalool specifically [6]. Huang et al. contributed a review article for the comprehensive analysis of the evolution pathway of *TPS*s in orchids [7]. The authors refined the phylogeny of *TPS*s and suggested that the driving force of evolution in each sub-*TPS* gene family might be different and chiefly depend on pollinator attraction, stress tolerance, and/or genotype-specific characteristics. Among terpenes, geraniol is commercially important and is involved in plant–pollinator interaction and stress biology in nature. Zhao et al. identified *DoGES1*, a gene which encodes geraniol synthase, using genomic annotation data of *D. officinale* [8]. The authors studied the subcellular localization and functions of DoGES1 and finally concluded that *DoGES1* was highly expressed in the petals of semi-open flowers and effectively controlled geraniol biosynthesis in *D. officinale*. The MYB (myeloblastosis) family of transcription factors can be found in animals and plants. Among them, MYB2 acts as a transcriptional repressor in anthocyanin biosynthesis. Lim et al. studied the role of MYB2 in *Dendrobium bigibbum* Lindl., and it was demonstrated that the accumulation of purple color in leaves is associated with the increased expression of *MYB2* [9].

Polysaccharides are critical constituents in medicinal orchids, and theoretically, the chemical modification of acetylation or deacetylation could certainly affect their bioactivities. Si et al. investigated the homologs of *REDUCED WALL ACETYLATION* (*RWA*), which encode acetyltransferases, in *D. officinale*, and three *DoRWA* were identified. Eventually, DoRWA3 was demonstrated to be involved in transferring acetyl groups to polysaccharides [10].

3. Orchid–Fungus Symbiosis

Orchid mycorrhizae are symbiotic relationships between orchids and fungi, particularly in the early stage of seed germination and the subsequent development of seedlings. Chen et al. investigated the effect of an exogenous gibberellic acid (GA_3) on the symbiotic germination of *D. officinale* in vitro [11]. The results indicated that exogenous GA_3 had a dose-dependent effect on the establishment of the symbiotic relationship, and it was shown that it might act on the complicated signaling networks or biosynthetic pathways of hormones.

4. Conclusions and Perspectives

Altogether, this present Special Issue makes certain progress in revealing the secrets of orchid biology, and interestingly, five out of ten articles chose to conduct their research on a medicinal orchid, *Dendrobium officinale*. It is worth mentioning that the field of functional genomics for the exploration of biosynthesis and signaling networks in secondary metabolites is also currently receiving attention. Additionally, in this Special Issue, our knowledge regarding pollination biology, flowering mechanisms and symbiosis in orchids was expanded; however, there are still numerous questions that have yet to be answered. Shortly, the picture of orchid biology will be more complete due to the efforts of orchid researchers in the application of advanced and high-throughput technologies such as genome editing and integrative multi-omics.

Funding: This research received no external funding.

Institutional Review Board Statement: Not applicable.

Informed Consent Statement: Not applicable.

Conflicts of Interest: The author declares no conflict of interest.

References

1. Chen, J.-T. Orchid Biochemistry. *Int. J. Mol. Sci.* **2020**, *21*, 2338. [CrossRef] [PubMed]
2. Brzosko, E.; Bajguz, A.; Chmur, M.; Burzyńska, J.; Jermakowicz, E.; Mirski, P.; Zieliński, P. How Are the Flower Structure and Nectar Composition of the Generalistic Orchid *Neottia ovata* Adapted to a Wide Range of Pollinators? *Int. J. Mol. Sci.* **2021**, *22*, 2214. [CrossRef] [PubMed]
3. Brzosko, E.; Bajguz, A.; Burzyńska, J.; Chmur, M. Nectar Chemistry or Flower Morphology—What Is More Important for the Reproductive Success of Generalist Orchid *Epipactis palustris* in Natural and Anthropogenic Populations? *Int. J. Mol. Sci.* **2021**, *22*, 12164. [CrossRef] [PubMed]

4. Lucibelli, F.; Valoroso, M.C.; Theißen, G.; Nolden, S.; Mondragon-Palomino, M.; Aceto, S. Extending the Toolkit for Beauty: Differential Co-Expression of DROOPING LEAF-Like and Class B MADS-Box Genes during *Phalaenopsis* Flower Development. *Int. J. Mol. Sci.* **2021**, *22*, 7025. [CrossRef] [PubMed]
5. Zeng, D.; Teixeira da Silva, J.A.; Zhang, M.; Yu, Z.; Si, C.; Zhao, C.; Dai, G.; He, C.; Duan, J. Genome-Wide Identification and Analysis of the APETALA2 (AP2) Transcription Factor in *Dendrobium officinale*. *Int. J. Mol. Sci.* **2021**, *22*, 5221. [CrossRef] [PubMed]
6. Yu, Z.; Zhao, C.; Zhang, G.; Teixeira da Silva, J.A.; Duan, J. Genome-Wide Identification and Expression Profile of TPS Gene Family in *Dendrobium officinale* and the Role of DoTPS10 in Linalool Biosynthesis. *Int. J. Mol. Sci.* **2020**, *21*, 5419. [CrossRef] [PubMed]
7. Huang, L.-M.; Huang, H.; Chuang, Y.-C.; Chen, W.-H.; Wang, C.-N.; Chen, H.-H. Evolution of Terpene Synthases in Orchidaceae. *Int. J. Mol. Sci.* **2021**, *22*, 6947. [CrossRef] [PubMed]
8. Zhao, C.; Yu, Z.; Silva, J.A.T.d.; He, C.; Wang, H.; Si, C.; Zhang, M.; Zeng, D.; Duan, J. Functional Characterization of a *Dendrobium officinale* Geraniol Synthase DoGES1 Involved in Floral Scent Formation. *Int. J. Mol. Sci.* **2020**, *21*, 7005. [CrossRef] [PubMed]
9. Lim, G.-H.; Kim, S.W.; Ryu, J.; Kang, S.-Y.; Kim, J.-B.; Kim, S.H. Upregulation of the MYB2 Transcription Factor is Associated with Increased Accumulation of Anthocyanin in the Leaves of *Dendrobium bigibbum*. *Int. J. Mol. Sci.* **2020**, *21*, 5653. [CrossRef] [PubMed]
10. Si, C.; Teixeira da Silva, J.A.; He, C.; Yu, Z.; Zhao, C.; Wang, H.; Zhang, M.; Duan, J. DoRWA3 from *Dendrobium officinale* Plays an Essential Role in Acetylation of Polysaccharides. *Int. J. Mol. Sci.* **2020**, *21*, 6250. [CrossRef] [PubMed]
11. Chen, J.; Yan, B.; Tang, Y.; Xing, Y.; Li, Y.; Zhou, D.; Guo, S. Symbiotic and Asymbiotic Germination of *Dendrobium officinale* (Orchidaceae) Respond Differently to Exogenous Gibberellins. *Int. J. Mol. Sci.* **2020**, *21*, 6104. [CrossRef] [PubMed]

Article

Nectar Chemistry or Flower Morphology—What Is More Important for the Reproductive Success of Generalist Orchid *Epipactis palustris* in Natural and Anthropogenic Populations?

Emilia Brzosko *, Andrzej Bajguz *, Justyna Burzyńska and Magdalena Chmur

Faculty of Biology, University of Bialystok, Ciolkowskiego 1J, 15-245 Bialystok, Poland; j.burzynska@uwb.edu.pl (J.B.); m.chmur@uwb.edu.pl (M.C.)
* Correspondence: emilka@uwb.edu.pl (E.B.); abajguz@uwb.edu.pl (A.B.); Tel.: +48-85-7388424 (E.B.); +48-85-7388361 (A.B.)

Abstract: The aim of this study was to determine the level of reproductive success (RS) in natural and anthropogenic populations of generalist orchid *Epipactis palustris* and its dependence on flower structure and nectar composition, i.e., amino acids and sugars. We found that both pollinaria removal and female reproductive success were high and similar in all populations, despite differences in flower traits and nectar chemistry. Flower structures were weakly correlated with parameters of RS. Nectar traits were more important in shaping RS; although, we noted differentiated selection on nectar components in distinct populations. Individuals in natural populations produced nectar with a larger amount of sugars and amino acids. The sucrose to (fructose and glucose) ratio in natural populations was close to 1, while in anthropogenic ones, a clear domination of fructose and glucose was noted. Our results indicate that the flower traits and nectar composition of *E. palustris* reflect its generalist character and meet the requirements of a wide range of pollinators, differing according to body sizes, mouth apparatus, and dietary needs. Simultaneously, differentiation of nectar chemistry suggests a variation of pollinator assemblages in particular populations or domination of their some groups. To our knowledge, a comparison of nectar chemistry between natural and anthropogenic populations of orchids is reported for the first time in this paper.

Keywords: floral display; fruiting; marsh helleborine; nectar amino acids; nectar sugars; pollinaria removal

1. Introduction

To achieve the highest possible reproductive success, plants have evolved different strategies. In animal pollinated plants, the strategies are directed at relations with pollinators. The masters in building the most specialized interaction with their pollinating partners are representatives of Orchidaceae. The majority of them are specialists connected to only one pollinator species (67% of all orchids) or a single functional group [1–4]. On the opposite point of the continuum of the specialization–generalization scale are generalists, pollinated by a wide range of animals from different systematic and ecological groups. An example of the last group is the object of the present study of *Epipactis palustris*, which is pollinated by more than 100 species [5,6].

To attract pollinators, orchids adapted their flowers structurally and chemically. Many of them (30–40% species) have developed deceptive tactics (mainly food or sexual deception) [7–11]. The important part of Orchidaceae constitutes rewarding species, which reward pollinators through different attractants, such as nectar, fragrances, oils, resin, and wax [12]. The first of them is the most effective for pollination success in orchids [13]. Although the role of the presence of nectar for the reproductive success (RS) of orchids is unquestionable [9,11,13,14], its quantity and quality for pollination effectiveness are documented only for some species [15–19]. Most studies on nectar in orchids, although valuable, only reported about the presence of sugars without ratios between them, or even did not distinguish between the sugars in floral and extrafloral nectar [20,21]. Nevertheless, studies on other plants well document the great variation of nectar properties in different species, distinct populations of a given species, dependence on habitat, flower position on inflorescence, flower age, and other factors. One of the most important findings, due to an evolutionary point of view, is that nectar produced by a given plant species meets the requirements of their pollinators. Relationships between nectar properties and pollinator types confirm many studies [22–26]. Pollinators' requirements of nectar properties are connected with their body size and behavior (energetic needs), the possibility for them to acquire nectar (mouth apparatus), and gustatory (taste caused by some amino acids (AAs)) [26–31]. Preferences of pollinators concern both nectar concentration, sugar proportion, and amino acids composition. For example, bats and hawkmoths feed on the nectar of lower concentrations of sugars, while bees prefer a higher concentration [22,32,33]. The concentration of sugars in orchid nectar sits within a wide range, from a low percentage to 90% [34]. Different pollinators also show distinct preferences to the ratio of main sugars, i.e., saccharose, glucose, and fructose. The extreme example of pollinators preferences to sugar components are some nectarivorous birds and ants, which prefer sucrose-free nectar due to their physiological constrains—the lack of invertase prevents them from attaining sucrose assimilation [35]. Pollinators also select nectar depending on its AA composition. Butterflies choose nectar with high AA concentration, while birds or flies prefer those with lower concentration [28]. Moreover, in the nectar of different species, distinct compositions of AAs were noted—with the domination of some AAs combined with lower concentrations or even absence of others [24,36,37]. The use of nectar by pollinators depends not only on its composition but also on its availability. In orchids, nectar is accumulated in shallow, cup-like structures, at the base of the labellum, in long spurs, in the base of the flower alongside the ovary, and on the side-lobes or along the central groove of the labellum [38]. Nectar located inside the corolla or in the spur is available for specific, restricted groups of pollinators with longer mouth apparatus, while exposed nectar is available for a wide range of pollinators, differing with respect to body sizes and dietary requirements. Exposed nectar is more vulnerable for evaporation and robbery than nectar located in deeper parts of flowers [24,37,39]. Moreover, nectar accumulated in open nectaries is often dominant in hexoses (e.g., fructose and glucose), while in concealed nectaries is most often sucrose dominant [21,24,36,40].

Flower structure plays an important role in shaping relationships between plants and their pollinators; therefore, structure shows adaptation to pollinating animals. The mutual match between pollinator and flower traits is the result of phenotypic selection [41,42]. Despite the fact that the general architecture of orchid flowers has the same scheme, details are differentiated in particular representatives of the family [12] that is strictly connected with the pollinator's properties. One of the best examples of a match between flowers and pollinators are spurred orchids, for which pollinator-mediated selection on flower traits—especially nectar spur length and corolla tube width—are well documented [41–50]. Studies on such species show that increasing the mechanical fit between flower and pollinator increases the precision of pollen transfer, thus affecting plant fitness [51]. Pollinator-mediated selection on flower traits is also documented by studies on deceptive orchids [52–56].

The application of orchids' distinct reproductive strategies translates into their level of reproductive success (RS). For example, rewarded species achieve higher RS than deceptive ones, and among rewarding species, those which produce nectar have the best effectiveness of pollination [7,11,13]. Many data document that reproductive success in orchids is strictly related to an important component of reproductive strategies—the flower's properties [42,46,49,56]. Generally, orchids are known as a group with a relatively low fruit set, especially non-autogamous species, mainly due to their limited pollinators [11,13,57]. Pollinator deficiency is often noted in anthropogenic populations [58]. Under such circumstances, increasing competition for pollinators may cause intensification of selection on floral traits by increasing pollen limitation [51,59–61]. Anthropogenic habitats also offer distinct soil resources, which can shape plant traits such as their size, flower production, or nectar quantity and quality. The dependence of these traits on soil parameters is well documented in orchids [16–19]. Differentiation of the above-mentioned factors causes spatial and temporary variation of reproductive success [11,13,57,62–64]. Some orchids, including *E. palustris*, may colonize different types of secondary habitats [65–68]. This presents the opportunity for maintaining orchids' diversity, since they are one of the plant groups that are the most sensitive to habitat loss and destruction due to human activity; therefore, they belong to the most endangered plant groups [69,70] with different threat levels of particular species [71]. The extinction risk of all known orchid species is c.a. 47% [72], and, as an example, 25% of globally extinct orchids are Australian [73]. Kull and Hutchings [74] compared changes in orchids distribution in the United Kingdom and Estonia and found that the mean decline in distribution range for 49 species in the United Kingdom was 50% and the mean decline for 33 orchid species in Estonia was 25%. Similar trends were observed by Jacquemyn, et al. [75] in Flanders and the Netherlands, where during 70 and 50 years, respectively, 81% species decreased distribution range in Flanders, and 78% species decreased distribution area in the Netherlands. Moreover, few species in each area went extinct. Reduction of distribution area and population number in Europe is noted for the object of our studies, *E. palustris* [67]. The threat to orchids is strengthened by the global decline of many plant pollinators, including those crucial for the pollination of orchids [58,76]. For example, the 25% loss of honey bee colonies in Central Europe between 1985 and 2005 has been observed [58].

The main aim of our study was to evaluate the level of pollinaria removal and fruiting and to determine the role of flower structure and nectar composition in shaping RS in natural and anthropogenic *E. palustris* populations. We supposed that RS should be high, due to the following traits of this orchid: (a) as a generalist, it is pollinated by a wide range of pollinators; (b) self-compatible properties enable autogamous and geitonogamous pollination; (c) the presence of nectar enhances the probability of pollination. We also hypothesized the differentiation of nectar characteristics and flower properties between populations, especially between natural and anthropogenic ones.

The answer to the question "what is more important for the reproductive success of generalist orchid *E. palustris* in natural and anthropogenic populations—nectar components or flower morphology?" can help elucidate the evolutionary pathways of different floral traits. Moreover, although the importance of nectar for RS is unquestionable, only a few studies document the role of nectar composition for RS in orchids. Additionally, to our knowledge, this is the first paper where a comparison of nectar chemistry between natural and anthropogenic populations of orchids has been reported.

2. Results

2.1. Floral Display and Flower Structure

E. palustris populations differ significantly in all floral display parameters, i.e., shoot height, inflorescence length, and flower number (Table 1 and Table S1, Figure S1). The highest shoots were observed in the natural ZAB population (62.6 ± 16.1 cm) and the lowest in the anthropogenic SIL population (42.46 ± 7.4 cm). Inflorescence length and number of flowers per shoot was the highest in the anthropogenic SOP population. In ZAB and SIL populations, all floral display traits were monotonically correlated to each other ($r_s = [0.38, 0.74]$; i.e., $r_s = 0.74$ for length of inflorescence vs. shoot height, $r_s = 0.58$ for length of inflorescence vs. shoot height, respectively), while in SOP, the length of inflorescence and number of flowers depended on shoot height ($r_s = 0.65$ and $r_s = 0.41$, respectively). In ROS, statistically significant correlation was found between inflorescence length and number of flowers ($r_s = 0.53$) (Table S2).

Table 1. Variation of floral display and flower structure in *Epipactis palustris* natural (Nat.) and anthropogenic (Ant.) populations. Data ($n = 30$) represent the mean ($\bar{x}$) ± standard error (SE), lower quartile (Q_1), median (Q_2), upper quartile (Q_3), and interquartile range (IQR). Different lowercase letters indicate statistically significant differences according to Tukey's post-hoc test ($p < 0.05$). Different uppercase letters indicate statistically significant differences according to the pairwise Wilcoxon Rank Sum test with Benjamini–Hochberg adjustment ($p < 0.05$). Additional comparisons were shown only when populations within Nat. or Ant. (or both) do not differ significantly.

Parameter	Statistic/Comparison	Natural (Nat.) Populations			Anthropogenic (Ant.) Populations	
		ZAB	ROS	SIL	SOP
Shoot height (cm)	$\bar{x}$ ± SE	62.59 ± 3.05 [a]	51.69 ± 1.61 [b]	42.47 ± 1.36 [c]	54.62 ± 2.14 [b]
	Q_1	51.50	47.01	37.5	43.25
	Q_2 (IQR)	64.75 (18.75)	53 (10)	42.75 (9)	56.75 (16.75)
	Q_3	70.25	57.00	46.5	60.00
Inflorescence length (cm)	$\bar{x}$ ± SE	9.43 ± 0.58 [b]	10.43 ± 0.53 [b]	10.26 ± 0.33 [b]	13.46 ± 0.79 [a]
	Q_1	7.25	9.00	9.00	10.62
	Q_2 (IQR)	9.25 (3.75)	10 (3.00)	10.00 (3.00)	13.50 (5.38)
	Q_3	11.00	12.00	12.00	16.00
	Nat. vs. SIL vs. SOP		b		a
Flower number	$\bar{x}$ ± SE	9.00 ± 0.54	10.66 ± 0.46	9.53 ± 0.43	11.85 ± 0.81
	Q_1	7.00	9.00	9.00	9.25
	Q_2 (IQR)	8.00 (4.50) [B]	11.00 (3.00) [A]	9.00 (1.75) [B]	11.00 (5.75) [A]
	Q_3	11.50	12.00	10.75	15.00
Length of dorsal sepal (LDS) (mm)	$\bar{x}$ ± SE	11.22 ± 0.13 [a]	10.01 ± 0.09 [c]	10.70 ± 0.10 [b]	11.55 ± 0.18 [a]
	Q_1	10.64	9.56	10.33	10.99
	Q_2 (IQR)	11.37 (1.04)	9.99 (0.81)	10.69 (0.66)	11.34 (1.07)
	Q_3	11.68	10.37	10.99	12.06
Width of dorsal sepal (WDS) (mm)	$\bar{x}$ ± SE	4.48 ± 0.06 [a]	3.83 ± 0.05 [a]	3.99 ± 0.04 [b]	4.37 ± 0.11 [b]
	Q_1	4.34	3.64	3.87	4.02
	Q_2 (IQR)	4.53 (0.28)	3.84 (0.31)	4.02 (0.27)	4.34 (0.68)
	Q_3	4.61	3.95	4.15	4.70
	ZAB vs. ROS vs. Ant.	a		b	
	Nat. vs. SIL vs. SOP		ab	b	a
	Nat. vs. Ant.		do not differ significantly		
Length of petal (LP) (mm)	$\bar{x}$ ± SE	9.38 ± 0.12 [a]	8.18 ± 0.11 [b]	9.27 ± 0.10 [a]	9.36 ± 0.23 [a]
	Q_1	9.06	7.74	8.77	8.63
	Q_2 (IQR)	9.41 (0.74)	8.03 (0.95)	9.29 (0.86)	9.31 (1.34)
	Q_3	9.80	8.70	9.63	9.97
	ZAB vs. ROS vs. Ant.	a	b		a

Table 1. *Cont.*

Parameter	Statistic/Comparison	Natural (Nat.) Populations		Anthropogenic (Ant.) Populations	
		ZAB	ROS	SIL	SOP
Width of petal (WP) (mm)	$\bar{x} \pm SE$	4.01 ± 0.05 [a]	3.44 ± 0.06 [b]	3.83 ± 0.06 [a]	3.78 ± 0.11 [a]
	Q_1	3.89	3.18	3.69	3.48
	Q_2 (IQR)	4.06 (0.29)	3.38 (0.42)	3.86 (0.34)	3.67 (0.53)
	Q_3	4.17	3.60	4.03	4.01
	ZAB vs. ROS vs. Ant.	a	c	b	
Length of lateral sepal (LLS) (mm)	$\bar{x} \pm SE$	11.34 ± 0.13	11.01 ± 0.12	11.35 ± 0.10	12.80 ± 0.19
	Q_1	10.82	10.53	11.05	11.93
	Q_2 (IQR)	11.35 (1.01) BC	10.87 (0.91) C	11.40 (0.65) B	12.63 (1.36) A
	Q_3	11.81	11.44	11.70	13.29
	Nat. vs. SIL vs. SOP	B		B	A
Width of lateral sepal (WLS) (mm)	$\bar{x} \pm SE$	4.17 ± 0.04	3.76 ± 0.04	3.83 ± 0.04	4.33 ± 0.10
	Q_1	4.03	3.63	3.69	3.96
	Q_2 (IQR)	4.15 (0.29) B	3.71 (0.34) A	3.88 (0.32) A	4.18 (0.56) B
	Q_3	4.32	3.96	4.01	4.52
Width of flowers (FW) (mm)	$\bar{x} \pm SE$	22.70 ± 0.35	20.83 ± 0.24	22.59 ± 0.16	24.31 ± 0.38
	Q_1	22.26	20.22	21.87	22.66
	Q_2 (IQR)	22.94 (1.64) B	20.89 (1.43) A	22.47 (1.40) B	24.01 (2.70) C
	Q_3	23.90	21.65	23.26	25.37
Length of flowers (FH) (mm)	$\bar{x} \pm SE$	11.47 ± 0.14 [ab]	10.36 ± 0.10 [c]	11.01 ± 0.12 [b]	11.94 ± 0.22 [a]
	Q_1	11.04	10.02	10.69	11.33
	Q_2 (IQR)	11.45 (0.83)	10.35 (0.76)	10.97 (0.77)	11.82 (1.30)
	Q_3	11.86	10.78	11.46	12.63
Length of labellum (LL) (mm)	$\bar{x} \pm SE$	11.91 ± 0.16	11.46 ± 0.15	11.45 ± 0.14	11.93 ± 0.26
	Q_1	11.56	11.01	11.23	11.42
	Q_2 (IQR)	12.06 (0.81) B	11.47 (0.89) A	11.54 (0.64) A	11.99 (1.20) B
	Q_3	12.37	11.88	11.87	12.62
	ZAB vs. ROS vs. Ant.	A	B	AB	
Width of hypochile (HW) (mm)	$\bar{x} \pm SE$	5.81 ± 0.13 [a]	4.90 ± 0.10 [b]	4.84 ± 0.14 [b]	5.54 ± 0.12 [a]
	Q_1	5.25	4.64	4.31	5.06
	Q_2 (IQR)	5.79 (1.15)	4.95 (0.61)	4.69 (0.98)	5.53 (0.83)
	Q_3	6.40	5.25	5.29	5.9

Table 1. *Cont.*

Parameter	Statistic/Comparison	Natural (Nat.) Populations			Anthropogenic (Ant.) Populations	
		ZAB	ROS	SIL	SOP
Length of hypochile (HL) (mm)	$\bar{x} \pm SE$	5.03 ± 0.13	4.58 ± 0.07	4.44 ± 0.06	4.80 ± 0.10
	Q_1	4.62	4.38	4.20	4.36
	Q_2 (IQR)	$4.89 (0.61)$ [C]	$4.65 (0.36)$ [AB]	$4.47 (0.38)$ [A]	$4.76 (0.77)$ [BC]
	Q_3	5.23	4.74	4.58	5.12
Length of epichile (LE) (mm)	$\bar{x} \pm SE$	7.11 ± 0.08	6.74 ± 0.11	7.03 ± 0.09	7.44 ± 0.11
	Q_1	6.87	6.38	6.82	7.14
	Q_2 (IQR)	$7.10 (0.51)$ [B]	$6.88 (0.74)$ [A]	$7.05 (0.44)$ [B]	$7.51 (0.64)$ [C]
	Q_3	7.38	7.12	7.26	7.78
Length of isthmus (LI) (mm)	$\bar{x} \pm SE$	2.07 ± 0.04 [ab]	1.79 ± 0.03 [ab]	1.94 ± 0.03 [b]	2.24 ± 0.06 [a]
	Q_1	1.92	1.66	1.82	2.09
	Q_2 (IQR)	$2.08 (0.33)$	$1.79 (0.29)$	$1.95 (0.26)$	$2.31 (0.41)$
	Q_3	2.25	1.94	2.08	2.50
	Nat. vs. SIL vs. SOP	b		b	a
Width of epichile (WE) (mm)	$\bar{x} \pm SE$	7.98 ± 0.11 [ab]	7.71 ± 0.09 [ab]	7.56 ± 0.09 [b]	8.07 ± 0.16 [a]
	Q_1	7.52	7.46	7.27	7.47
	Q_2 (IQR)	$7.89 (0.84)$	$7.66 (0.49)$	$7.51 (0.54)$	$8.11 (0.94)$
	Q_3	8.36	7.95	7.81	8.42
	Nat. vs. SIL vs. SOP		ab	b	a
Width of isthmus (WI) (mm)	$\bar{x} \pm SE$	0.84 ± 0.01	0.90 ± 0.02	0.82 ± 0.01	1.02 ± 0.03
	Q_1	0.82	0.81	0.75	0.94
	Q_2 (IQR)	$0.85 (0.06)$ [B]	$0.91 (0.15)$ [A]	$0.82 (0.11)$ [B]	$1.01 (0.13)$ [C]
	Q_3	0.88	0.96	0.86	1.07
Isthmus area (AI) (mm^2)	$\bar{x} \pm SE$	1.75 ± 0.05 [b]	1.60 ± 0.05 [b]	1.59 ± 0.05 [b]	2.32 ± 0.11 [a]
	Q_1	1.65	1.42	1.41	1.94
	Q_2 (IQR)	$1.75 (0.29)$	$1.58 (0.37)$	$1.54 (0.34)$	$2.28 (0.71)$
	Q_3	1.94	1.79	1.75	2.65
	Nat. vs. SIL vs. SOP	b		b	a

All measured flower traits differed between populations (Table 1, Figure S1). The smallest flowers (both their length and width) were noted in natural ROS populations (length of flowers (FH): 10.34 ± 0.56 mm; width of flowers (FW): 20.75 ± 1.34 mm), while the largest in the SOP population (FH: 11.9 ± 1.2 mm and FW: 24.3 ± 2.0 mm). Values of other traits most often shaped according to the same pattern as FH and FW. It should be highlighted that the isthmus area (AI), on which surface nectar is secreted, was larger in anthropogenic (especially in SOP) populations than in natural populations. Spearman's correlation analysis revealed that almost all flower traits in SOP correlated positively with each other strongly or very strongly (Table S2). In the remaining 3 populations, FH was always correlated with LDS and LP (r_s = [0.72, 0.96]), and AI with LI and WI (r_s = [0.71, 0.85]).

Furthermore, the flower structure dataset was subjected to principal component analysis (PCA) and its preliminary tests. The *p*-value from Bartlett's test of sphericity was approximately equal to 0, while the calculated overall measure of sampling adequacy (MSA) from the Kaiser–Meyer–Olkin test was equal to 0.84. MSA for individual parameters ranged from 0.48 (for width of isthmus (WI)) to 0.96 for the width of flowers (FW) (Table S3). Thus, according to Kaiser [77], the MSA value is high enough to perform PCA. According to Cattell's rule, one or two components should be selected (Figure S2) [78], while Kaiser's rule indicates that three components should be retained [79]. On the basis of the first axis (Dim1), which accounts for 53.3% of the variation, a separation of all four populations is visible—the following pattern: SOP > ZAB > SIL > ROS usually occurs for all floral structure parameters. SOP and ZAB are mostly associated with positive values of Dim1 (thus higher than average values of floral parameters), while SIL and ROS– are mostly associated with negative values of Dim1. Thus, a sign-based distinction between natural and anthropogenic populations is not possible (Figure 1 and Figure S3).

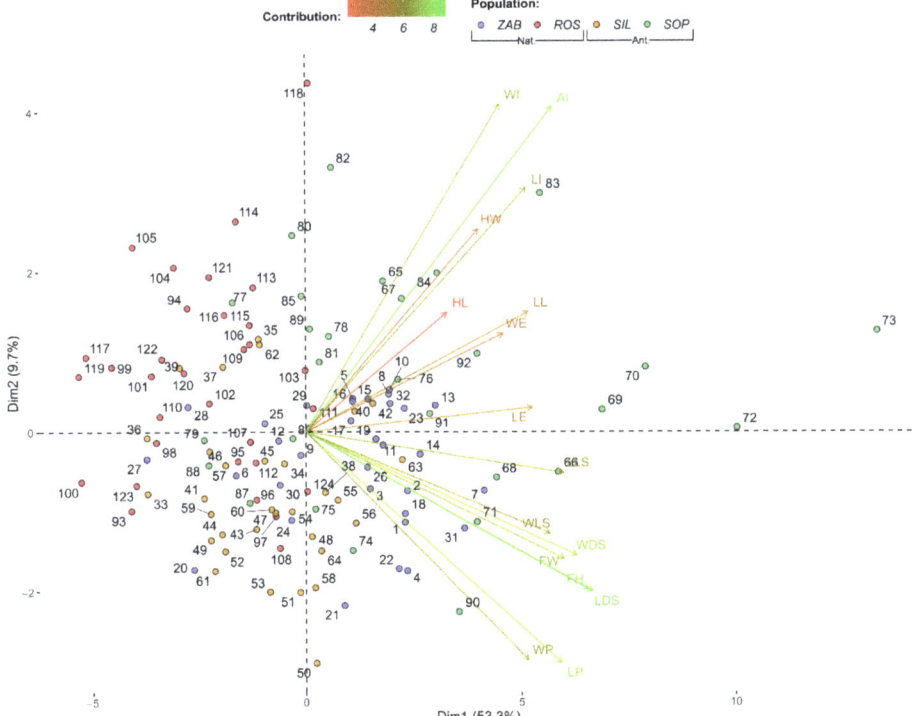

Figure 1. Biplot of flower structure profiles for *Epipactis palustris* natural (Nat.) and anthropogenic (Ant.) populations, showing the first two dimensions or factors (Dim1-2) of PCA that, together, explain 63% of the variance. Biplot vectors indicate the strength and direction of factor loading for the first two factors. Individuals (populations) are color-coded by population.

2.2. Nectar Chemistry

2.2.1. Sugars

Our analyses document very low *E. palustris* nectar amounts of three common sugars, i.e., sucrose, fructose, and glucose. We found statistically significant differences between populations in sugars quantity (sum of sugars), excluding glucose content. The total amount of sugars was significantly lower in anthropogenic than in natural populations (SIL: 34.05 mg/mL and SOP: 35.0 mg/mL vs. ZAB: 48.09 mg/mL and ROS: 40.68 mg/mL) (Table S4). Participation of sucrose in nectar was also significantly lower in anthropogenic than in natural populations (Figure 2, Table S5). On the other hand, the sucrose to (fructose and glucose) ratio was more balanced in natural ZAB and ROS populations (0.93 and 0.86, respectively), while in SIL and SOP anthropogenic ones, a clear domination of fructose and glucose was found (0.57 and 0.58, respectively). No statistically significant differences in fructose to glucose ratios were found among populations (Table S4).

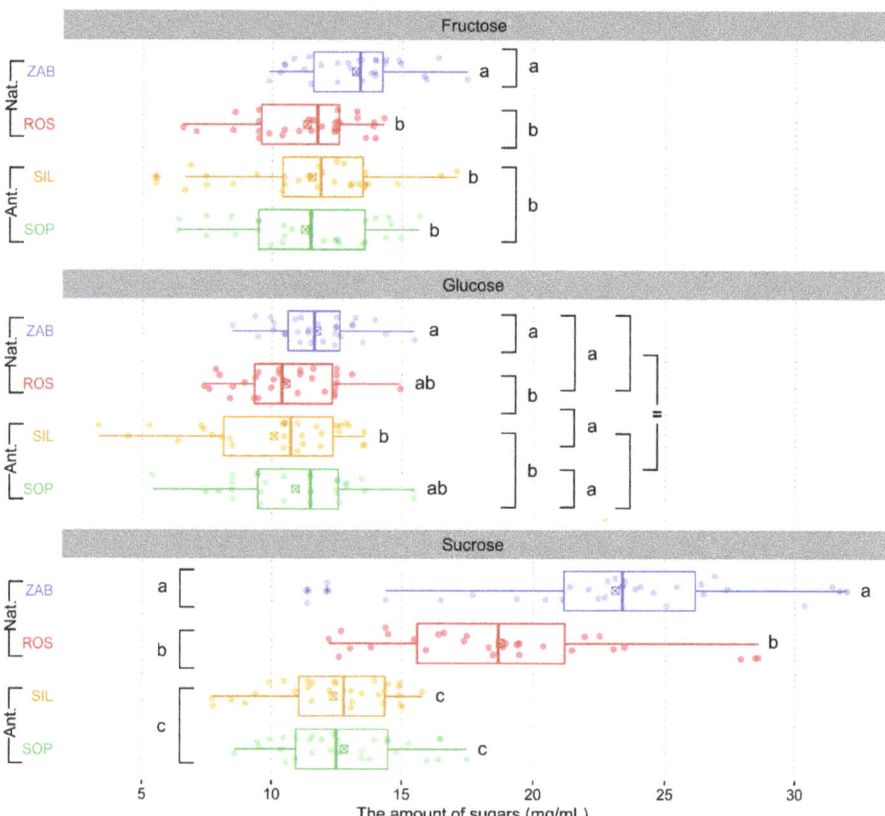

Figure 2. Boxplots of sugar amounts for *Epipactis palustris* natural (Nat.) and anthropogenic (Ant.) populations. Colored dots are individual samples. The crossed square shows the mean. The lower and upper hinges correspond to the lower (Q_1) and upper (Q_3) quartiles. Thus, box length shows the interquartile range (IQR). The thicker lines inside the boxes corresponds to the median. The lower whisker extends from the hinge to the smallest value at most $Q_1 - 1.5 \times$ IQR of the hinge. The upper whisker extends from the hinge to the largest value no further than $Q_3 + 1.5 \times$ IQR. Data beyond the end of the whiskers, indicated with an asterisk symbol, are outliers. Different lowercase letters indicate statistically significant differences according to Tukey's post-hoc test ($p < 0.05$). Symbol "=" means they did not differ significantly. Additional comparisons on the left or right side were shown only when the Nat. or Ant. (or both) populations did not differ significantly.

2.2.2. Amino Acids

The amount of AAs in *E. palustris* nectar ranged from 0.39 ± 0.002 mg/mL in SOP to 0.52 ± 0.002 mg/mL in ZAB. Statistically significant differences between populations were noted in the sum of AAs, and the largest differences were observed between natural and anthropogenic populations (Table 2 and Table S6, Figure S4,). In total, 27 distinct AAs were detected in *E. palustris* nectar (20 proteogenic and 7 non-proteogenic) with their different participation in particular populations. Nevertheless, some of them dominated in all populations (glutamic acid (Glu) and glutamine (Gln)—always above 10%). Glu, tyrosine (Tyr), arginine (Arg), and β-alanine (β-Ala) had a significantly higher percentage in natural than in anthropogenic populations. On the other hand, in anthropogenic populations, participation of proline (Pro), alanine (Ala), and phenylalanine (Phe) in nectar was higher than in natural places. It should be noted that β-Ala was observed only in natural populations (ZAB and ROS), citrulline (Cit) only in one natural population (ZAB) (but only in some individuals), and Tau was absent in one anthropogenic population (SOP). It is interesting that anthropogenic populations were characterized by a higher percentage of proteogenic AAs in nectar than natural ones, while non-proteogenic AAs had higher participation in natural populations.

In natural populations, strong monotonic correlations were found, i.e., leucine (Leu) vs. isoleucine (Ile) (r_s = {ZAB: 0.75, ROS: 0.74}), ornithine (Orn) vs. glutamine (Gln) (r_s = {ZAB: −0.78, ROS: 0.60}), taurine (Tau) vs. Orn (r_s = {ZAB: 0.55, ROS: 0.63}), and Tau vs. Gln (r_s = {ZAB: −0.60, ROS: 0.68}). Additionally, in ZAB, a correlation between methionine (Met) and lysine (Lys) was noted (r_s = −0.64), while in ROS, r_s = 0.67 was reported for tryptophan (Trp) vs. threonine (Thr) and Orn vs. glycine (Gly). In anthropogenic populations, no common strong or very strong correlations were reported. However, in the case of SIL, strong monotonic correlations (r_s = [0.60, 0.79]) were noted between the following: asparagine (Asn) vs. serine (Ser), histidine (His) vs Asn, Ile vs. alanine (Ala), valine (Val) vs. Leu and Trp, as well as Orn vs His. While, in case of SOP, strong monotonic correlations (r_s = [0.60, 0.79]) were noted between the following: Arg vs. Ala, Trp vs. Leu and Met, as well as Orn vs. Lys. It should be also highlighted that, between natural and anthropogenic populations, no intersection of strongly or very strongly correlated AA pairs exist (Table S7).

Different relations between production of sugars and AAs in particular populations was noted. In both natural populations, the sum of AAs positively correlated with the sum of sugars (ZAB: r_s = 0.43; $p < 0.05$ and ROS: r_s = 0.40; $p < 0.05$), in ZAB with fructose and sucrose amount (r_s = 0.44; $p < 0.05$ and r_s = 0.38; $p < 0.05$), and in ROS with sucrose amount (r_s = 0.44; $p < 0.05$). In anthropogenic populations, positive correlations between AAs amount and percentage of hexoses (i.e., sum of fructose and glucose) were observed (SIL: r_s = 0.44; $p < 0.05$ and SOP: r_s = 0.38; $p < 0.05$) and negative correlations were observed between AAs amount and percentage of sucrose (SIL: r_s = −0.44; $p < 0.05$ and SOP: r_s = 0.38; $p < 0.05$).

We found a notable difference between natural and anthropogenic populations in participation of AA from distinct taste classes (Figure 3). The percentage share of class II AAs was approximately 35–48% for natural populations and 48–56% for anthropogenic populations, while class IV was 36–42% for SOP and 42–48% for SIL. The class II of AAs had higher participation in natural populations. On the other hand, the class III group, represented in *E. palustris* nectar only by Pro, had about five times higher participation in anthropogenic populations than in natural populations (Table 2).

Table 2. The concentration of amino acids (μM) and total amount of amino acids (mg/mL) in *Epipactis palustris* nectar. The number of classes represents the effect of amino acids on insect chemoreceptors: I—no effect; II—inhibition of chemoreceptors; III—stimulate the salt cell; IV—the ability to stimulate the sugar cell. Data ($n = 30$) represent the mean ($\bar{x}$) ± standard error (SE), lower quartile (Q_1), median (Q_2), upper quartile (Q_3), and interquartile range (IQR). Different lowercase letters indicate statistically significant differences, according to Tukey's post-hoc test ($p < 0.05$). Different uppercase letters indicate statistically significant differences according to the pairwise Wilcoxon Rank Sum test with Benjamini-Hochberg adjustment ($p < 0.05$). ND—not detected. Additional comparisons were shown only when populations within Nat. or Ant. (or both) do not differ significantly.

Amino Acid	Class	Statistic	Natural (Nat.) Populations		Anthropogenic (Ant.) Populations	
			ZAB	ROS	SIL	SOP
			Proteogenic amino acids (μM)			
Aspartic acid (Asp)	I	$\bar{x}$ ± SE	357.55 ± 6.47	377.50 ± 6.73	200.25 ± 6.81	259.51 ± 5.06
		Q_1	329.36	351.40	178.80	228.94
		Q_2 (IQR)	359.31 (56.82) [B]	380.49 (50.09) [A]	198.98 (48.50) [D]	267.79 (49.53) [C]
		Q_3	386.17	401.49	227.30	278.48
Glutamic acid (Glu)	I	$\bar{x}$ ± SE	740.90 ± 8.71 [b]	884.43 ± 7.46 [d]	430.24 ± 6.87 [d]	525.11 ± 4.78 [c]
		Q_1	706.06	858.27	400.67	509.22
		Q_2 (IQR)	730.92 (71.49)	885.59 (48.87)	421.47 (64.14)	522.14 (24.38)
		Q_3	777.55	907.13	464.81	533.60
Alanine (Ala)	I	$\bar{x}$ ± SE	92.59 ± 2.48	116.35 ± 1.74	84.42 ± 1.10	89.31 ± 2.40
		Q_1	80.82	111.75	79.64	81.50
		Q_2 (IQR)	89.79 (23.47) [B]	117.99 (9.50) [A]	84.07 (9.36) [C]	83.99 (12.90) [C]
		Q_3	104.29	121.25	89.01	94.40
		ZAB vs. ROS vs. Ant.	B	A		B
Cysteine (Cys)	I	$\bar{x}$ ± SE	163.35 ± 3.07	229.32 ± 3.52	149.33 ± 3.41	93.10 ± 1.10
		Q_1	152.41	217.42	136.75	89.49
		Q_2 (IQR)	168.28 (20.05) [B]	229.17 (17.12) [A]	143.13 (26.49) [B]	92.88 (7.34) [C]
		Q_3	172.46	234.54	163.24	96.83
Glycine (Gly)	I	$\bar{x}$ ± SE	68.81 ± 1.92	88.60 ± 1.03	69.36 ± 1.60	57.25 ± 0.69
		Q_1	61.34	83.57	63.75	54.38
		Q_2 (IQR)	71.38 (16.50) [B]	88.39 (9.56) [A]	66.38 (7.20) [B]	56.96 (6.03) [C]
		Q_3	77.83	93.13	70.96	60.41

Table 2. Cont.

Amino Acid	Class	Statistic	Natural (Nat.) Populations		Anthropogenic (Ant.) Populations	
			ZAB	ROS	SIL	SOP
Serine (Ser)	I	$\bar{x} \pm SE$	273.61 ± 3.77	326.20 ± 2.31	225.79 ± 2.77	184.75 ± 1.71
		Q_1	263.55	318.96	214.73	180.13
		Q_2 (IQR)	275.68 (23.06) [B]	326.89 (12.68) [A]	227.17 (16.76) [C]	185.01 (9.16) [D]
		Q_3	286.61	331.64	231.50	189.29
Threonine (Thr)	I	$\bar{x} \pm SE$	141.01 ± 4.28 [a]	139.20 ± 1.34 [a]	123.17 ± 1.01 [b]	80.10 ± 0.99 [c]
		Q_1	119.21	133.71	118.35	76.39
		Q_2 (IQR)	139.94 (41.64)	139.5 (9.77)	122.99 (9.15)	80.38 (7.99)
		Q_3	160.85	143.48	127.49	84.38
		Nat. vs. SIL vs. SOP		a	b	c
Tyrosine (Tyr)	I	$\bar{x} \pm SE$	12.29 ± 0.43	15.35 ± 0.34	6.91 ± 0.40	6.67 ± 0.41
		Q_1	10.49	13.64	6.41	5.64
		Q_2 (IQR)	12.89 (2.81) [B]	15.39 (3.48) [A]	7.41 (1.91) [C]	7.38 (2.63) [C]
		Q_3	13.30	17.12	8.32	8.28
		ZAB vs. ROS vs. Ant.	B	A	C	
Arginine (Arg)	II	$\bar{x} \pm SE$	24.84 ± 0.79 [b]	34.94 ± 0.42 [a]	15.65 ± 0.29 [c]	16.15 ± 0.42 [c]
		Q_1	21.95	32.70	14.42	14.39
		Q_2 (IQR)	24.59 (5.29)	34.39 (3.80)	16.13 (2.08)	16.39 (3.10)
		Q_3	27.25	36.50	16.49	17.49
		ZAB vs. ROS vs. Ant.	b	a	c	
Asparagine (Asn)	II	$\bar{x} \pm SE$	313.16 ± 4.99	432.35 ± 2.86	246.77 ± 3.26	237.69 ± 1.99
		Q_1	302.5	421.59	232.62	228.42
		Q_2 (IQR)	317.31 (25.06) [B]	430.94 (20.08) [A]	241.48 (31.30) [C]	238.41 (15.70) [C]
		Q_3	327.56	441.67	263.92	244.12
		ZAB vs. ROS vs. Ant.	B	A	C	

Table 2. *Cont.*

Amino Acid	Class	Statistic	Natural (Nat.) Populations			Anthropogenic (Ant.) Populations	
			ZAB	ROS	SIL	SOP
Glutamine (Gln)	II	$\bar{x} \pm SE$	529.06 ± 6.84 [b]	623.95 ± 5.04 [a]	414.67 ± 2.64 [c]	351.19 ± 3.72 [d]
		Q_1	509.06	600.73	406.49	332.55
		Q_2 (IQR)	521.50 (30.51)	623.48 (41.83)	412.54 (15.84)	349.53 (31.96)
		Q_3	539.57	642.56	422.33	364.51
Histidine (His)	II	$\bar{x} \pm SE$	133.04 ± 3.02 [a]	101.81 ± 1.01 [b]	71.05 ± 1.10 [c]	68.62 ± 0.71 [c]
		Q_1	128.42	97.40	66.89	65.64
		Q_2 (IQR)	132.50 (15.12)	102.48 (8.75)	69.65 (8.37)	68.49 (5.53)
		Q_3	143.54	106.15	75.27	71.17
		ZAB vs. ROS vs. Ant.	a	b	c	c
Lysine (Lys)	II	$\bar{x} \pm SE$	167.59 ± 3.61	231.48 ± 2.75	128.46 ± 1.29	115.63 ± 0.74
		Q_1	152.75	219.45	124.67	113.68
		Q_2 (IQR)	162.49 (25.32) [B]	230.45 (18.80) [A]	128.49 (7.92) [C]	116.39 (4.53) [D]
		Q_3	178.07	238.24	132.60	118.21
Proline (Pro)	III	$\bar{x} \pm SE$	87.53 ± 2.92 [b]	82.45 ± 2.56 [b]	362.10 ± 7.85 [a]	373.20 ± 8.13 [a]
		Q_1	76.60	71.53	331.58	335.57
		Q_2 (IQR)	89.14 (25.72)	83.13 (18.10)	374.09 (69.02)	380.04 (69.61)
		Q_3	102.32	89.62	400.60	405.18
		ZAB vs. ROS vs. Ant.	b	b	a	a
		Nat. vs. SIL vs. SOP				
		Nat. vs. Ant.	differ significantly			
Isoleucine (Ile)	IV	$\bar{x} \pm SE$	162.45 ± 3.79 [a]	143.93 ± 1.61 [b]	117.30 ± 1.60 [c]	100.93 ± 0.62 [d]
		Q_1	148.25	138.68	111.08	98.77
		Q_2 (IQR)	162.48 (30.83)	144.89 (12.13)	118.29 (10.70)	101.39 (4.52)
		Q_3	179.08	150.80	121.78	103.28

Table 2. Cont.

Amino Acid	Class	Statistic	Natural (Nat.) Populations		Anthropogenic (Ant.) Populations	
			ZAB	ROS	SIL	SOP
Leucine (Leu)	IV	$\bar{x} \pm SE$	333.40 ± 4.77 [a]	290.65 ± 3.82 [b]	240.59 ± 3.13 [c]	242.72 ± 1.78 [c]
		Q_1	312.63	278.40	224.67	237.50
		Q_2 (IQR)	331.99 (31.35)	291.74 (24.45)	239.15 (28.87)	243.39 (11.74)
		Q_3	343.97	302.85	253.54	249.23
		ZAB vs. ROS vs. Ant.	a	b	c	c
Methionine (Met)	IV	$\bar{x} \pm SE$	92.10 ± 3.11 [a]	87.79 ± 0.79 [a]	67.46 ± 0.66 [b]	50.42 ± 0.79 [c]
		Q_1	79.01	84.48	66.03	47.66
		Q_2 (IQR)	89.19 (24.54)	87.33 (5.79)	67.43 (3.47)	50.20 (5.55)
		Q_3	103.55	90.27	69.50	53.21
		Nat. vs. SIL vs. SOP	a		b	c
Phenylalanine (Phe)	IV	$\bar{x} \pm SE$	44.87 ± 1.66 [ab]	46.20 ± 0.61 [ab]	48.72 ± 1.17 [a]	42.50 ± 0.89 [b]
		Q_1	37.49	43.47	44.35	39.93
		Q_2 (IQR)	44.95 (14.70)	46.42 (4.90)	49.4 (8.94)	43.38 (6.41)
		Q_3	52.19	48.37	53.29	46.33
		Nat. vs. SIL vs. SOP		ab	a	b
Tryptophan (Trp)	IV	$\bar{x} \pm SE$	189.82 ± 3.71	255.90 ± 2.97	145.15 ± 1.20	128.45 ± 1.22
		Q_1	176.42	244.04	140.58	124.39
		Q_2 (IQR)	188.14 (25.84) [B]	254.37 (22.42) [A]	145.04 (9.16) [C]	127.39 (5.84) [D]
		Q_3	202.26	266.46	149.74	130.23
Valine (Val)	IV	$\bar{x} \pm SE$	76.77 ± 1.51 [a]	69.89 ± 0.94 [b]	49.85 ± 0.82 [c]	47.51 ± 0.83 [c]
		Q_1	71.41	66.29	46.37	44.62
		Q_2 (IQR)	77.69 (12.03)	68.52 (8.12)	50.05 (6.13)	47.39 (6.46)
		Q_3	83.44	74.40	52.50	51.08
		ZAB vs. ROS vs. Ant.	a	b	c	c

Table 2. Cont.

Amino Acid	Class	Statistic	Natural (Nat.) Populations		Anthropogenic (Ant.) Populations	
			ZAB	ROS	SIL	SOP
			Non-proteogenic amino acids (µM)			
Ornithine (Orn)		$\bar{x} \pm SE$	94.95 ± 3.81	116.33 ± 2.55	67.20 ± 1.24	ND
		Q_1	78.61	109.38	64.80	ND
		Q_2 (IQR)	89.72 (37.50) ^{B}	114.95 (9.11) ^{A}	68.54 (6.44) ^{C}	ND
		Q_3	116.11	118.50	71.24	ND
Citrulline (Cit)		$\bar{x} \pm SE$	4.15 ± 0.49	ND	ND	ND
		Q_1	2.48	ND	ND	ND
		Q_2 (IQR)	4.45 (3.89)	ND	ND	ND
		Q_3	6.38	ND	ND	ND
Taurine (Tau)		$\bar{x} \pm SE$	16.18 ± 0.67	14.98 ± 0.51	10.32 ± 0.28	7.61 ± 0.18
		Q_1	13.92	12.59	9.38	6.60
		Q_2 (IQR)	15.69 (2.84) ^{A}	14.39 (3.90) ^{A}	10.18 (2.11) ^{B}	7.51 (1.90) ^{C}
		Q_3	16.76	16.50	11.50	8.50
α-aminobutyric acid (AABA)		$\bar{x} \pm SE$	11.67 ± 0.42 ^{a}	9.34 ± 0.36 ^{b}	7.04 ± 0.30 ^{c}	10.41 ± 0.34 ^{c}
		Q_1	9.95	8.40	5.50	8.91
		Q_2 (IQR)	11.50 (2.84)	8.57 (2.13)	7.45 (2.24)	10.02 (2.98)
		Q_3	12.8	10.53	7.74	11.89
		ZAB vs. ROS vs. Ant.	a	b	c	
		Nat. vs. SIL vs. SOP				
		Nat. vs. Ant.		a		differ significantly
β-aminobutyric acid (BABA)		$\bar{x} \pm SE$	14.01 ± 0.38	21.74 ± 0.64	8.55 ± 0.31	4.55 ± 0.23
		Q_1	12.51	18.76	7.52	3.58
		Q_2 (IQR)	13.68 (2.26) ^{B}	21.08 (4.52) ^{A}	8.54 (1.84) ^{D}	4.4 (1.92) ^{C}
		Q_3	14.77	23.28	9.36	5.50
		ZAB vs. ROS vs. Ant.	B	A	C	

Table 2. Cont.

Amino Acid	Class	Statistic	Natural (Nat.) Populations		Anthropogenic (Ant.) Populations	
			ZAB	ROS	SIL	SOP
γ-aminobutyric acid (GABA)		$\bar{x} \pm SE$	6.84 ± 0.55	7.29 ± 0.35	3.91 ± 0.42	ND
		Q_1	6.54	5.50	2.52	ND
		Q_2 (IQR)	7.89 (2.12) [A]	6.96 (3.01) [A]	3.72 (2.98) [B]	ND
		Q_3	8.66	8.50	5.5	ND
		ZAB vs. ROS vs. Ant.	A	A	B	B
		Nat. vs. SIL vs. SOP				
		Nat. vs. Ant.		differ significantly		
β-alanine (β-Ala)		$\bar{x} \pm SE$	22.81 ± 1.12	16.25 ± 0.53	ND	ND
		Q_1	18.46	14.52	ND	ND
		Q_2 (IQR)	21.53 (8.88) [A]	16.04 (2.96) [B]	ND	ND
		Q_3	27.34	17.48	ND	ND
Total amount of amino acids (mg/mL)						
		$\bar{x} \pm SE$	0.52 ± 0.003 [b]	0.60 ± 0.002 [a]	0.41 ± 0.002 [c]	0.39 ± 0.002 [d]
		Q_1	0.51	0.59	0.41	0.39
		Q_2 (IQR)	0.52 (0.01)	0.60 (0.01)	0.42 (0.01)	0.39 (0.01)
		Q_3	0.53	0.60	0.42	0.40

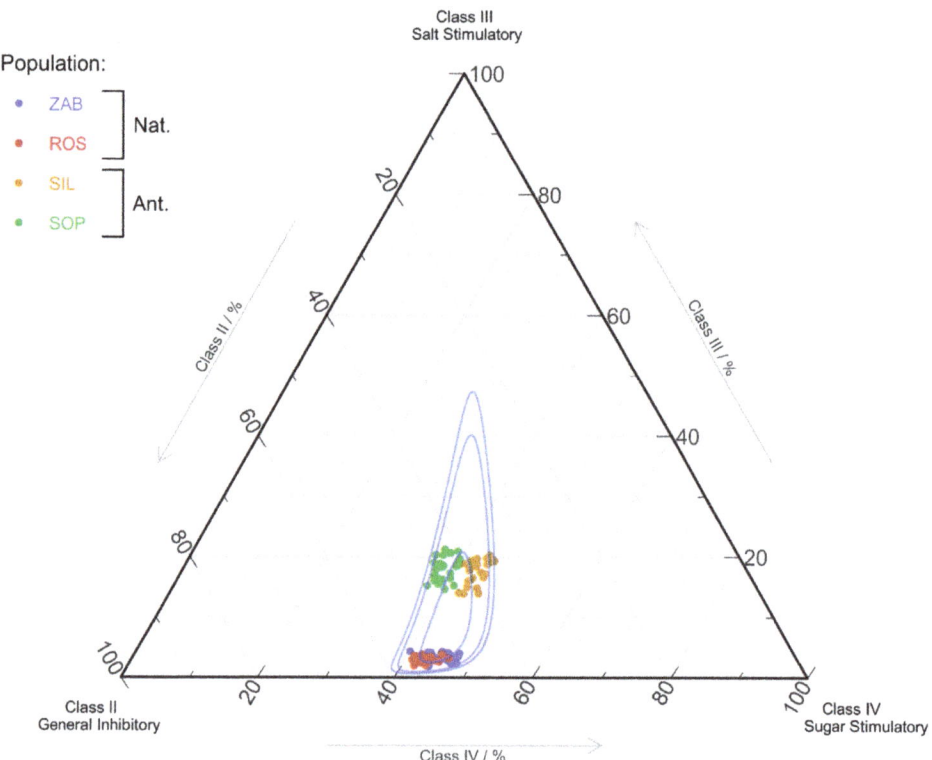

Figure 3. Ternary plot of amino acid classes for *Epipactis palustris* natural (Nat.) and anthropogenic (Ant.) populations: II (Asp, Glu, His, Arg, Lys), III (Pro), and IV (Val, Met, Trp, Phe, Ile, Leu). Blue lines show 50%, 90%, and 95% confidence intervals via the Mahalanobis Distance and use of the Log–Ratio Transformation. The first class of AAs (Asn, Gln, Ala, Cys, Gly, Ser, Thr, Tyr) does not affect the chemoreceptors of fly (data not shown). AAs' abbreviations and full names are present in Table 2.

Moreover, AAs were subjected to principal component analysis (PCA) and its preliminary tests. The *p*-value from Bartlett's test of sphericity was approximately equal to 0, while the calculated overall measure of sampling adequacy (MSA) from the Kaiser–Meyer–Olkin test was equal to 0.95 (Table S8). MSA for individual AAs ranged from 0.44 (Phe was the only AA with almost no interpopulation differences and very high data deviation, Figure S2) to 0.98 for Tyr, Trp, and Val (Table S8). Thus, according to Kaiser [77], the MSA value is high enough to perform PCA. According to Cattell's rule, one or two components should be selected (Figure S5) [78], while Kaiser's rule indicated that three components should be retained [79]. Finally, the first two components that explain about 75.5% of the variance were preserved. PCA grouped together anthropogenic populations (SIL and SOP), as they had much higher average Pro level and lower levels of other AAs (Figure 4). Differences between ZAB and ROS populations are also visible, e.g., much higher average Cit, His, β-Ala, and Ile levels, as well as lower Arg, Asn, Trp, β-aminobutyric acid (BABA), and Lys. On the basis of the first axis (Dim1), which accounts for 64.1% of the variation, there is a clear separation of the natural vs. anthropogenic populations, particularly due to differences in Pro. Based on the second axis (Dim2, 11.5%), the two natural (Nat.) populations are also separated, while the anthropogenic (Ant.) are not (Figure 4 and Figure S6).

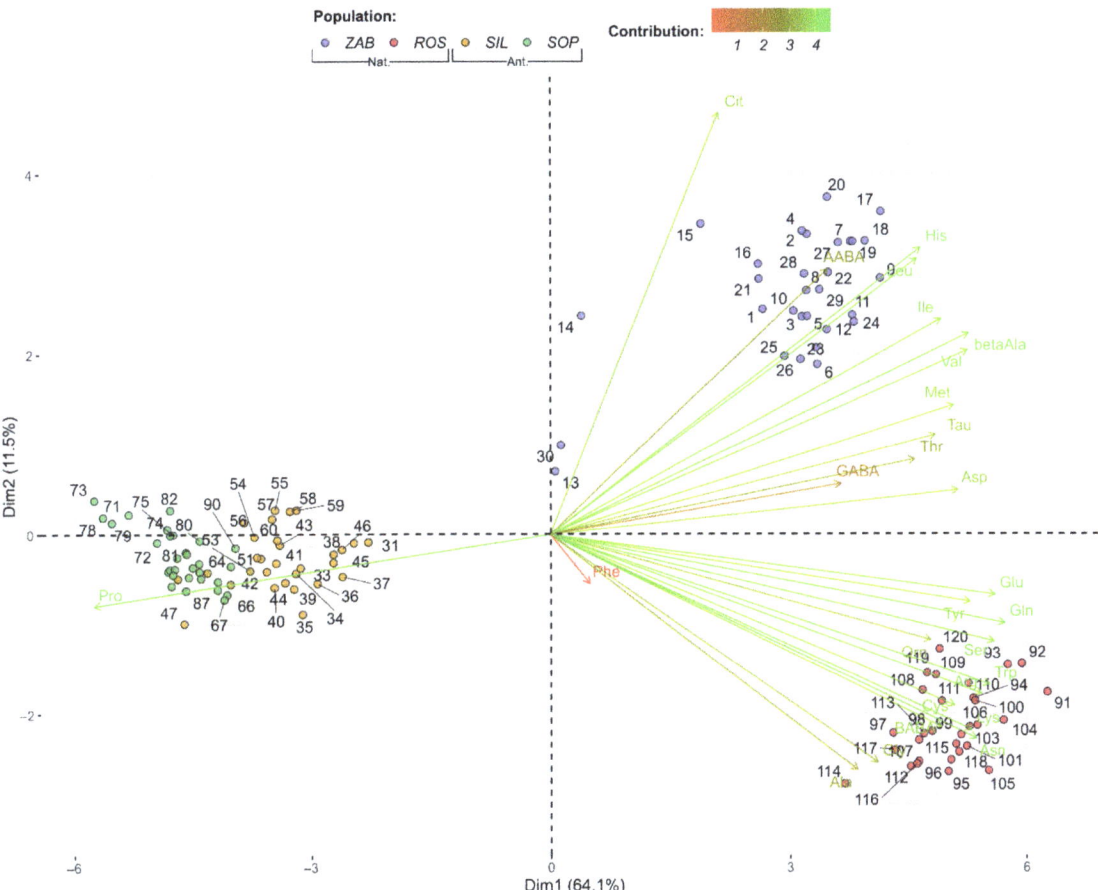

Figure 4. Biplot of amino acid profiles for *Epipactis palustris* natural (Nat.) and anthropogenic (Ant.) populations, showing the first two dimensions or factors (Dim1-2) of PCA that, together, explain 75.52% of the variance. Biplot vectors indicate the strength and direction of factor loading for the first two factors. Individuals (populations) are color-coded by population. Ellipses around the individuals show assumed 95% multivariate normal distribution.

2.3. Reproductive Success

Reproductive success in *E. palustris* populations was high (Table 3). Female reproductive success (FRS—the proportion of developed fruits to the number of flowers on the inflorescence) shaped from 81.47 ± 4.19% in ROS to 90.60 ± 2.49% in SOP (Table 3) and did not differ between populations (F = 0.862; p = 0.46). Pollinaria removal (PR) was also similar in all populations (F = 1.289; p = 0.28); although, activity of insects was about 10% higher in SOP (96.55 ± 1.61%) than in other populations (Table 3). The efficiency of pollination was high—PR to FRS ratio was equal to about 1 in all populations. Although average values of indexes of reproductive success are similar in the populations studied, we observed some differences in details of the pollination process at an individual level. In ROS, SIL, and SOP populations, about one third of individuals (11, 9, and 9, respectively) had higher PR than FRS, while in ZAB only five of them had PR larger than FRS. On the other hand, we noted higher FRS than PR in ROS and SIL (7 and 9 individuals, respectively). In ZAB, only two shoots had higher fruiting than PR, while in SOP, we did not observe such cases.

Table 3. Spatial and temporal variation of female reproductive success (FRS) and pollinaria removal (PR) in *Epipactis palustris* populations. Data (n = 30) show the mean ± standard error.

Parameter	Natural Populations		Anthropogenic Populations	
	ZAB	ROS	SIL	SOP
FRS (%)	94.40 ± 2.83	81.47 ± 4.19	87.99 ± 3.36	90.60 ± 2.49
PR (%)	97.03 ± 2.68	87.48 ± 3.46	85.66 ± 4.36	96.55 ± 1.61
PR/FRS	1.07 ± 0.06	1.32 ± 0.22	1.07 ± 0.10	1.08 ± 0.03

3. Discussion

In line with our expectations and the results of earlier studies [5,67,80,81], high levels of PR and FRS (above 80%) were found in both natural and anthropogenic *E. palustris* populations. Jacquemyn, et al. [67], on the basis of Claessens and Kleynen [5] data from 24 populations of this species, reported that the average fruit set shaped at 77.6% %, and Jacquemyn and Brys [82] noted 70% fruiting in Belgian populations. We also found the high RS and pollination efficiency (in all places PR to FRS ratio equaled about 1). This result contrasts with the founding of Jacquemyn and Brys [82], who noted that fruiting in *E. palustris* populations was higher than the level of pollinaria removal.

Although pollinator deficiency is considered the main factor restricting RS in orchids [11,13,57], a high level of RS and pollination efficiency in our studies suggest that pollinators of *E. palustris* are abundant in all populations. Nevertheless, assemblages of insects and the dominant pollinators may differ from one part of geographical range to another [5,6,83]. High number of pollinators of this orchid (142 species [5]) and a wide range of their sizes and requirements increase the probability of pollination. Through a diversity of potential pollinators, we can also explain the lack of differences in RS between natural and anthropogenic populations. Our results indicate that in each of them, pollinators assemblages are large and diverse enough to ensure RS at the observed level. This result contrasts with the results of other studies, where fruiting was lower in anthropogenic than in natural populations. Exceptionally low levels of fruiting were observed by Jermakowicz and Brzosko [59] in anthropogenic populations of *Malaxis monophyllos*. On the other hand, Pellegrino and Bellusci [84] noted an almost seven times lower fruit set in anthropogenic than in natural populations of *Serapias cordigera* in Italy. In a population of *Oncidium ascendens* from rainforest from Mexico, fruit production was almost two times higher than in populations from synanthropic habitats [85]. The authors of these studies recognized that pollinator deficiency in altered habitats was the main factor, which decreased RS in these species. In our opinion, differences between species characters of orchids studied by Pellegrino and Bellusci [84] and Parra-Tabla, et al. [85] and *E. palustris* could also cause distinct answers for habitat types. *S. cordigera* is deceptive species and relies on relatively restricted groups of pollinators in comparison with *E. palustris*; additionally, *O. ascendens* is a self-incompatible species, whose sexual reproduction depends on cross-pollination by the native bee *Trigona nigra*. It could be suggested that the properties of *E. palustris* (its generalist character, presence of nectar, and spontaneous autogamy) and pollinators behavior (penetration of many flowers on inflorescence) are advantages, which ensure effective pollination regardless of habitat. Anthropogenic habitats are generally recognized as those with poorer assemblages of pollinators, which negatively influences plant RS, but some of them seem to be suitable for plant–pollinator interactions. For example, Rewicz, et al. [86] reported higher RS in anthropogenic than in natural populations of *E. helleborine*, due to the larger diversity of pollinators in the first type of habitat. Although two *E. palustris* populations exist within the city border, in changed places, within populations area, and in neighboring communities, other flowering species grew, which can attract many insects, including *E. palustris* pollinators. Moreover, allotments are placed in the vicinity of SIL population, which may increase pollinator numbers.

In the light of our results, it seems that generalists are less sensitive for pollinator deficiency, even in anthropogenic habitats. This is in accordance with findings of other authors, who state that a decrease of fruit set as a result of the reduction in insect movements is particularly strong for specialists that show a high degree of dependence on their pollinators for fruit production [11,84,87]. Higher specialization levels and anthropogenic declines in pollinator populations can also intensify selection on floral traits [51]. The generalist character of *E. palustris* and pollinator efficiency could explain the very weak selection on flower and floral display traits. Only ten correlations between them and parameters of reproductive success were observed (among 152 tested cases), and only four of these concerned flower structures. Five statistically significant correlations were found in the anthropogenic SOP population and all of them were negative. In this population, lower individuals with shorter inflorescences and a lower number of flowers were favored. This can reflect the behavior of pollinators in this locality. First, they may operate at the lower part of vegetation, and secondly, penetrating a given inflorescence, they are able (or need) to acquire nectar from a restricted number of flowers. On the other hand, in ZAB, fruiting was higher on longer inflorescences. In this place, *E. palustris* grows in tall sedges, and probably shoots should be higher than neighboring plants to be recognized by pollinators. Stronger pollinator-mediated selection on inflorescence was noted in taller than in shorter vegetation [46,88,89]. The case of ZAB population is also in accordance with the common expectation that more fruits often develop on larger inflorescences because they attract more pollinators, which visit more flowers on larger inflorescences [48,90–92]. On the other hand, the SOP population may illustrate situations, that smaller inflorescences are favored by natural selection when larger inflorescences suffer factors decreasing fitness, such as the higher probability of geitonogamy or intense herbivore activity [92–94].

One of the most important evolutionary mechanisms, crucial for successful pollination, is the mechanical fit between plants and their pollinators [2,7,8,41,95,96]. Such a match is generally stronger in specialized systems [51,97], which confirms, for example, the results of studies on long-spurred orchids [41,45,47,48]. These findings confirm the results of our studies because only four distinct flower traits influenced RS in three among four populations. These traits seem unimportant for pollinators, which may suggest that observed correlations are random or their functions are difficult to explain. The lack of strong selection pressures on these traits maintains variation in flower traits [98].

Nevertheless, it should be noted that the isthmus area was significantly larger in the anthropogenic population than in the natural population, suggesting that pollinators with distinct mouth apparatus operate in two population groups. The important point, which enables understanding the evolution of plant–pollinator interactions, is knowledge about the importance of floral rewards, including nectar, for RS [99,100]. Nectar properties shape the growth, survival, reproduction, and behavior of nectar-feeding animals [30,31,35,101,102]. Our results suggest stronger dependence of RS in *E. palustris* populations on nectar properties than flower structure. Nectar characteristics influenced mainly PR (16 statistically significant correlations observed among 21). This is in contrast to our study on another generalist orchid, *Neottia ovata*, where nectar properties shaped mainly FRS [19]. Similarly to Percival [21], we found three main sugars in *E. palustris* nectar (sucrose, fructose, and glucose) with their amount shown to be larger in natural than anthropogenic populations. Sugar components influenced RS only in the natural ROS population, where PR was positively influenced by hexose (i.e., fructose and glucose) amounts and FRS was positively influenced by fructose amount. This is interesting because in both natural populations, sucrose percentage is significantly higher than in anthropogenic ones. Positive selection on hexoses in ROS may suggest that in this population insects, which prefer nectar rich in monosaccharides are important pollinators, and the amount of hexoses in this population is not enough to provide for their needs. Similar insects could be abundant in anthropogenic populations, where hexoses were more abundant than in natural populations. Preferences for hexoses, taken up more easily than sucrose, show nonspecialized insects, i.e., syrphids, flies, and beetles [25]. Insects from these groups were noted as *E. palustris* pollinators [6,67,83,99]. Nonspecialized insects choose hexose-rich nectar (especially fructose) because it is easier absorbed due to lower viscosity [32]. Moreover, some ants (often observed by us on *E. palustris* shoots and noted by Jakubska-Busse and Kadej [103]) even prefer sucrose-free nectar because they are not able to assimilate this sugar due to lack of invertase [29]. The lack of selection on nectar sugars in three populations may suggest that these nectar components are not aimed at any of the pollinator group and sugar composition met the requirements of pollinated insects. Similar results were obtained for another generalist orchid, *N. ovata* [19]. Different sucrose to (fructose and glucose) ratios (~1 in natural and ~0.5 in anthropogenic populations) suggest distinct pollinator assemblages in these two population groups. Larger sucrose content in natural populations could indicate that such insects as honey bees and bumblebees, which prefer this sugar, are main pollinators in these places. These insects were recognized as main pollinators in some Polish *E. palustris* populations by Jakubska-Busse and Kadej [103]. It should be noted that different sugar ratios in natural and anthropogenic *E. palustris* populations only partially confirms the statement that nectar secreted in open flowers is dominated by glucose and fructose [36,40].

Nectar in particular populations also differed according to AAs composition. Their total amount was higher in natural populations, but proteogenic AAs have larger participation in anthropogenic ones. At the species level, we noted a high number of different AAs (27, including 20 proteogenic and 7 non-proteogenic), similarly to another generalist orchid *N. ovata* (28 AAs; [19]). Pais, et al. [40] found only 17 AAs in *E. atropurpurea*. Fewer AAs than those in our study were observed in the nectar of specialist spurred orchids [15,17]. Additionally, we found domination of different AAs in the two population groups—Glu, Tyr, Arg, and β-Ala were more abundant in natural populations and Pro, Ala, and Phe were more abundant in anthropogenic populations. The most common AAs in *E. palustris* nectar are Gln, Glu, and serine (Ser), which are always above 10%. These AAs were also found to be among the most abundant in the nectar of generalist *N. ovata* [19]. Gln and Glu are needed for energetically exhaustive flights, while Glu and Ser influence pollinator behavior [17,18,28].

Available studies document preferences of pollinators both to total AAs amount in nectar and to particular AAs. Although the importance of AAs for pollinator life is poorly studied, the role some of them are known. First of all, they play a nutritional function and attract or discourage pollinators. One of the most common AAs in plants [24,104]—important for many pollinators, especially Hymenoptera—is Pro, production of which is more expensive than other nectar components [105]. Its participation constitutes the greatest difference between natural and anthropogenic populations among measured nectar components. In anthropogenic populations, it was one of the three most abundant AAs, and its amount was about five times higher than in natural populations. Like the majority of amino acids, proline can be used in energy production [104]. This AA rewards pollinators, propels the lift phase of the flight [105,106], and stimulates insects' salt receptors, which initiate feeding [26,35,107]. Carter, et al. [105] found that Pro accumulation is a plants' answer to stress factors. Through the last function could be explained a few times, larger amounts of this AA in were found anthropogenic populations than in natural population, since changed habitats are stressful for plants. It can indicate that Pro plays an important role in the metabolism of *E. palustris* pollinators in anthropogenic populations. Bertazzini, et al. [108] documented a preference of honey bees for proline-enriched artificial nectar. The other AA, whose amount was larger in the anthropogenic than in the natural population, is Phe. It has a strong phagostimulatory effect on bees and its concentration is highly variable [102,109]. Petanidou, et al. [102] attributed the dominance of Phe in the Mediterranean to the high number of bees, especially long-tongued bees. The authors suggest that in the Mediterranean, such bees act as crucial selective factors for Phe-rich nectars. The last abundant AA in anthropogenic populations is Ala, which influences insect growth [17]. Aspartic acid (Asp), like Glu and Ser, influences pollinator behavior [17,28] and disgusts pollinators [102]. The second importance of Asp may explain its negative influence on PR in ROS. In the same way, a negative correlation between the percentage of Ser and FRS in SIL population could be explained. A negative response of honey bees to Ser was reported [108]. On the other hand, Kim and Smith [110] showed that Gly elicited a feeding response in honeybees.

Variation in amount and participation of nectar components, and the differentiation of selection on distinct constituents in particular populations suggest, again, that different pollinators with different nutritional needs operate in distinct populations. This supposition could be strengthened by the results showing preferences of pollinators to different taste classes. The most sensitive for nectar taste were pollinators in ROS, where we found positive correlations between PR and percentage of AAs from taste class IV (stimulation the sugar receptor cell), while negative between PR and taste class II (inhibition of the three types of chemosensory cell: salt, sugar, and water). Among the remaining three populations, only in SIL AAs from taste class I positively shaped PR. This confirms the results of earlier studies [28,111], in which it was recognized that nectar amino acids might be detectable by pollinators and may contribute to the overall taste of nectar. We noted the importance of nectar taste for pollinators, and consequently for RS, in other orchids, both in generalist (with open nectaries) and in specialist (with the accumulation of nectar in the spur) [15,19].

The nectar composition can be modified by habitat properties, especially soil nutrients [18,104]. These authors documented changes in the total concentration of amino acids, as well as changes in the amount of some of them after fertilization. Moreover, they stated that such changes may have implications for plant–insect interactions, as local populations of pollinators may benefit from the increased amino acid content of the nectar and preferentially visit plants growing in high nutrient conditions. The influence of soil parameters, especially carbon and the carbon to nitrogen ratio, influenced flower structure and nectar chemistry in generalist *N. ovata* [19]. We did not analyze soil chemistry because differences between natural and anthropogenic habitats are so evident (see Section 4.2), which, with high probability, may be recognized as one of the causes of differentiation of nectar traits between them. Nevertheless, soil in natural populations seems to be richer in elements required to produce more sugars and AAs.

Results of our studies confirm the generalist character of *E. palustris*. High levels of RS in all populations indicate that both flower traits and nectar chemistry, and variation of these properties meet the needs of wide, differentiated pollinator groups. Simultaneously, the lack of selection on flower traits and stronger selection on nectar components suggest that pollinators are more sensitive to nectar properties, including taste. Moreover, selection on distinct nectar characters in particular populations may indicate that different pollinator assemblages operate within them.

The most important finding of our research is documentation (to our knowledge, it is the first such report) of significant differences in nectar properties between natural and anthropogenic orchid populations. We suggest that they are caused by the differentiation of pollinators in these two habitat types or are stronger depending on soil characters (or both). However, to precisely point out the most important factor, more detailed studies should be conducted. The results of our studies importantly enrich the knowledge needed to explain mechanisms, which underlie plant–pollinator interactions.

4. Materials and Methods

4.1. Study Species

E. palustris is widely distributed throughout most of Europe but is absent in the Southern Mediterranean regions and extreme north [112]. This species usually occupies calcareous, nutrient-poor, and moist–wet substrates, mainly in full sunlight. It exists in wet dune slacks, calcareous fens, and on peat, but may also occur on sandy substrates overlying heavy clay or loamy soils [67].

E. palustris is rhizomatous species. Each shoot has about six leaves in the lower part and a few more scattered ones below the inflorescence. The inflorescence produces about twelve flowers (sometimes more than 20). The flowers are usually more or less tinged with a rose, red, or brown coloration [67,81]. A wide spectrum of insects visits *E. palustris* flowers, but the main pollinators are Diptera (i.e., *Empis* sp., and *Episyrphus* sp.) and Hymenoptera (i.e., *Apis mellifera*, *Bombus lapidarius*, and *Bombus lucorum*) [6,83]. Claessens and Kleynen [5] found 142 pollinators of this orchid in the literature. Insects are attracted by strong scent (with eugenol and vanillin, as the crucial components attracting Diptera) and nectar with attractants such as nonanal (pelargonaldehyde), decanal, eicosanol, and its derivatives [83]. The shallow nectary is located on the labellum, and nectar is secreted on the whole surface of lip callus and abaxial side of isthmus in hypochile [113]. E. palustris has the potential for spontaneous autogamy [67,80]. Fruiting takes place August–September, depending on location and weather conditions [81].

4.2. Study Area

This study was performed in July and August 2021 in four populations of *E. palustris* in northeast Poland, two of which were localized in well-preserved natural peat bogs. ZAB population is localized in the Biebrza National Park—one of the biggest areas of peatlands in Europe. ROS population is localized in Rospuda valley. It is a vast, moss-free, low, and transitional peat bog, with rich and unique flora (four plant species listed in Annex 2 of the EU Habitats Directive, fifteen plant species listed in *The Polish Red Book of Plants*). The bog is also the only refuge in Poland of a rare orchid species in Europe—*Herminium monorchis*. Natural populations are distanced c.a. 70–100 km from Bialystok, the largest city in NE Poland, and at least a few kilometers from the nearest villages. Two other populations (SIL and SOP) exist in anthropogenic habitats at the border of Bialystok city. The SIL population exists in an abandoned gravel pit (c.a. 3 ha). The SOP population exists within the highly damaged soligenic peat bog (0.5 ha), which is under advanced secondary succession, with a lowered level of groundwater and the presence of alien species. Both anthropogenic populations are surrounded by human-changed habitats, typical for urban areas.

4.3. Fieldwork and Floral Trait Measurements

In populations studies, 30–32 flowering individuals were chosen and marked. In the field, the floral display traits, i.e., the height of shoots, length of inflorescence, and the number of flowers, were quantified. The five lowest flowers from each inflorescence were collected and used for the evaluation of nectar composition and measurement of the morphological variables of flowers (full names of abbreviations are present in Table 4). The area of isthmus (AI), as the product of LI and WI, was amounted. The isthmus size was considered as a measure of nectar quantity. Flower traits of one individual are given as average from five measurements.

Table 4. Measured flower structure properties.

Abbreviation	Full Name
AI	isthmus area
FH	length of flowers
FW	width of flowers
HL	length of hypochile
HW	width of hypochile
LDS	length of dorsal sepal
LE	length of epichile
LI	length of isthmus
LL	length of labellum
LLS	length of lateral sepal
LP	length of petal
WDS	width of dorsal sepal
WE	width of epichile
WI	width of isthmus
WLS	width of lateral sepal
WP	width of petal

Samples from all populations were collected 7–12 July during the peak of flowering under sunny and hot weather (the temperature of each day was about 30 °C).

The morphological measures were taken using an opto-digital microscope DSX110 (Olympus Life Science, Waltham, MA, USA) in the Laboratory of Insect Evolutionary Biology and Ecology, Faculty of Biology, University of Bialystok.

To assess the level of reproductive success (RS), the shoots were marked and the number of flowers per inflorescence in full blooming were counted. During the maturation of capsules, FRS and PR were quantified. FRS was evaluated as the proportion of developed fruits to the number of flowers on the inflorescence and was given in percentages. PR was determined in percentages (PR to the total number of pollinaria for each inflorescence). The efficiencies of pollination were also evaluated, found as the ratio of PR to FRS—the higher the index, the lower the pollination efficiency within a population.

4.4. Nectar Analysis

4.4.1. Nectar Isolation

Flower nectar isolation was performed using a water washing method [114]. Five flowers per sample were placed into a 2 mL Eppendorf tube, containing 1 mL of distilled water and shaken in a laboratory thermomixer (120 rpm, 21 °C, 45 min; Eppendorf Corporate, Hamburg, Germany) for the nectar efflux. Then, the flowers were removed from the tubes, and the mixture of water with nectar was evaporated to dryness by centrifugal vacuum concentrator (45 °C, Eppendorf Concentrator Plus, Eppendorf Corporate, Hamburg, Germany). The obtained pellet was dissolved in 20 μL of distilled water, then transferred into the centrifuge tube with a filter and centrifuged to remove impurities ($9000 \times g$, 5 min; MPW-55 Med. Instruments, Gliwice, Poland). The purged extract was collected in a glass vial with a 250 μL insert with polymer feet.

4.4.2. Sugar and Amino Acid Determination

Determination and quantification of sugars and AAs were performed using the high-performance liquid chromatography (HPLC) method. An Agilent 1260 Infinity Series HPLC apparat (Agilent Technologies, Inc., Santa Clara, CA, USA) with quaternary pump with an in-line vacuum degasser, a thermostatted column, and a refrigerated autosampler with an autoinjector sample loop was used.

For sugar analysis, a ZORBAX Carbohydrate Analysis Column (4.6 mm × 250 mm, 5 μm) (Agilent Technologies, Inc., Santa Clara, CA, USA), at a temperature of 30 °C and a refractive index detector, was applied. The mobile phase was a solution of acetonitrile and water (70:30, v/v) at a flow rate of 1.4 mL/min. The injection volume was 10 μL. The total time of analysis was 15 min [15].

Meanwhile, for AA detection (Table 5), an automatic program of derivatization was set. Thus, the o-phthalaldehyde and 9-fluorenylmethyl chloroformate reagents were used for the derivatization of primary and secondary AAs [15]. The Agilent Zorbax Eclipse Plus C_{18} (4.6 × 150 mm, 5 μm) column (Agilent Technologies, Inc., Santa Clara, CA, USA), at a temperature of 40 °C, was used to separate individual AAs. Detection of primary AAs was performed by a photodiode array detector at 388 nm, while detection of secondary AAs was performed by a fluorescence detector with an excitation wavelength of 266 nm and an emission wavelength of 305 nm. The injection volume was 5 μL. The flow rate was 1 mL/min. Eluent A of the mobile phase was 40 mM NaH_2PO_4 (pH 7.8, adjusted by 10 M NaOH solution), while eluent B was a mixture including acetonitrile, methanol, and water (45:45:10, $v/v/v$). The gradient was the following: 0–5 min, 100–90% A; 5–25 min, 90–59.5% A; 25–30 min, 59.5–37% A; 30–35 min, 37–18% A; 35–37 min, 18–0% A; 37–40 min, 0% A; and 40–43 min, 100% A.

Table 5. Amino acids evaluated during HPLC analysis.

Abbreviation	Full Name
AABA	α-aminobutyric acid
Ala	alanine
Arg	arginine
Asn	asparagine
Asp	aspartic acid
BABA	β-aminobutyric acid
Cit	citrulline
Cys	cysteine
GABA	γ-aminobutyric acid
Gln	glutamine
Glu	glutamic acid
Gly	glycine
His	histidine
Ile	isoleucine
Leu	leucine
Lys	lysine
Met	methionine
Orn	ornithine
Phe	phenylalanine
Pro	proline
Ser	serine
Tau	taurine
Thr	threonine
Trp	tryptophan
Tyr	tyrosine
Val	valine
β-Ala	β-alanine

The analytical data were integrated using the Agilent OpenLab CDS ChemStation software (Agilent Technologies, Inc., Santa Clara, CA, USA) for liquid chromatography systems. Identification of sugars and AAs was performed by comparing retention times of individual sugars and AAs in the reference vs. test solution. The concentration of these compounds was assayed based on comparisons of peak areas obtained for the samples, investigated with those of the reference solutions.

4.5. Statistical Analysis

The R programming language or statistical environment was used to perform all statistical computations and analyses, as well as to prepare graphics and transform data for tabular representation [115,116]. The dataset of sugars was subjected to two-way ANOVA followed by Tukey's post-hoc test, while AAs, floral display, and flower structure datasets were supplied to either (a) two-way ANOVA followed by Tukey's post-hoc test or (b) the Kruskal–Wallis test followed by a pairwise Wilcoxon Rank Sum test with Benjamini–Hochberg adjustment, which compared the median values of different parameters between populations, depending on ANOVA pre-conditions (verified using Shapiro–Wilk test and Bartlett's test) (Table S1, Table S5, Table S6, Figure S1, Figure S4) [116–120]. Furthermore, a set of descriptive statistics (mean, standard error, quartiles, and interquartile range) was calculated for AAs, sugars, floral display, and flower structure. For all tests, the significance level was $\alpha = 0.05$. In order to check if a monotonic relationship exists between floral display and flower structure parameters, Spearman's rank correlations were calculated (Table S2) using the 'rcorr' function from the 'Hmisc' package. Spearman's correlations were also calculated between AAs (Table S7). Correlations were considered significant for $p < 0.05$.

To analyze the effect of AAs on insect chemoreceptors, all identified and determined AAs were grouped into four classes [24] (full names of abbreviations are present in Table 5): I. Asn, Gln, Ala, Cys, Gly, Ser, Thr, and Tyr (no effect on the chemoreceptors of fly);

II. Arg, Asp, Glu, His, and Lys (inhibition of fly chemoreceptors); III. Pro (stimulation of the salt cell); and IV. Ile, Leu, Met, Phe, Trp, and Val (ability to stimulate the sugar cell) and presented as a ternary plot [121]. Principal component analysis (PCA) was used to simplify the exploration of AAs. To build the PCA model, the FactoMineR package was used [122]. Two tests that indicate the suitability of the AA dataset for structure detection and reduction were performed—Bartlett's test of sphericity [123] and the Kaiser–Meyer–Olkin test of factorial adequacy (psych package [124]). Unit variance scaling of the data was applied; thus, PCA was performed on a correlation matrix, rather than on a covariance matrix. Number of principal components to retain was selected with the help of Cattell's and Kaiser's rules [78,79]. All biplots were created using the factoextra package [125]. Moreover a PCA was also applies to flower structure dataset using an approach identical to that used for AAs dataset.

Supplementary Materials: The following are available online at https://www.mdpi.com/article/10.3390/ijms222212164/s1.

Author Contributions: Conceptualization, E.B.; methodology, E.B., A.B. and M.C.; software, A.B.; validation, E.B., J.B. and M.C.; formal analysis, E.B. and A.B.; investigation, J.B. and M.C.; resources, J.B. and M.C.; data curation, E.B. and A.B.; writing—original draft preparation, E.B.; writing—review and editing, E.B. and A.B.; visualization, A.B.; supervision, E.B.; project administration, E.B.; funding acquisition, E.B. and A.B. All authors have read and agreed to the published version of the manuscript.

Funding: This work was funded by the Ministry of Education and Science as part of subsidies for maintaining research potential awarded to the Faculty of Biology of the University of Bialystok.

Informed Consent Statement: Not applicable.

Data Availability Statement: Data is contained within the current article and supplementary material.

Acknowledgments: We thank the Rector of University in Bialystok and Dean of Department of Biology of University in Bialystok for financial support. A.B. thanks Adam Bajguz for statistical and programming consultations.

Conflicts of Interest: The authors declare no conflict of interest.

References

1. Joffard, N.; Massol, F.; Grenié, M.; Montgelard, C.; Schatz, B. Effect of pollination strategy, phylogeny and distribution on pollination niches of Euro-Mediterranean orchids. *J. Ecol.* **2019**, *107*, 478–490. [CrossRef]
2. Ollerton, J.; Winfree, R.; Tarrant, S. How many flowering plants are pollinated by animals? *Oikos* **2011**, *120*, 321–326. [CrossRef]
3. Phillips, R.D.; Reiter, N.; Peakall, R. Orchid conservation: From theory to practice. *Ann. Bot.* **2020**, *126*, 345–362. [CrossRef] [PubMed]
4. Tremblay, R.L. Trends in the pollination ecology of the Orchidaceae: Evolution and systematics. *Can. J. Bot.* **1992**, *70*, 642–650. [CrossRef]
5. Claessens, J.; Kleynen, J. *The Flower of the European Orchid. Form and Function*; Claessens & Kleynen: Guelle, The Netherlands, 2011; p. 440.
6. Nilsson, L.A. Pollination ecology of *Epipactis palustris* (L.) Crantz (Orchidaceae). *Bot. Not.* **1978**, *131*, 355–368.
7. Ackerman, J.D. Mechanisms and evolution of food-deceptive pollination systems in orchids. *Lindleyana* **1986**, *1*, 108–113.
8. Cozzolino, S.; Widmer, A. Orchid diversity: An evolutionary consequence of deception? *Trends Ecol. Evol.* **2005**, *20*, 487–494. [CrossRef] [PubMed]
9. Jersáková, J.; Johnson, S.D.; Kindlmann, P. Mechanisms and evolution of deceptive pollination in orchids. *Biol. Rev.* **2006**, *81*, 219–235. [CrossRef]
10. Renner, S.S. Rewardless flowers in the angiosperms and the role of insect cognition in their evolution. In *Plant-Pollinator Interactions: From Specialization to generalization*; Waser, N.M., Ollerton, J., Eds.; The University of Chicago Press: Chicago, IL, USA, 2006; pp. 123–144.
11. Tremblay, R.L.; Ackerman, J.D.; Zimmerman, J.K.; Calvo, R.N. Variation in sexual reproduction in orchids and its evolutionary consequences: A spasmodic journey to diversification. *Biol. J. Linn. Soc.* **2005**, *84*, 1–54. [CrossRef]
12. Dressler, R. *The Orchids: Natural History and Classification*; Harvard University Press: Cambridge, MA, USA, 1981; p. 356. [CrossRef]
13. Neiland, M.R.M.; Wilcock, C.C. Fruit set, nectar reward, and rarity in the Orchidaceae. *Am. J. Bot.* **1998**, *85*, 1657–1671. [CrossRef] [PubMed]

14. Jersáková, J.; Johnson, S.D.; Kindlmann, P.; Pupin, A.C. Effect of nectar supplementation on male and female components of pollination success in the deceptive orchid *Dactylorhiza sambucina*. *Acta Oecol.* **2008**, *33*, 300–306. [CrossRef]
15. Brzosko, E.; Bajguz, A. Nectar composition in moth-pollinated *Platanthera bifolia* and *P. chlorantha* and its importance for reproductive success. *Planta* **2019**, *250*, 263–279. [CrossRef] [PubMed]
16. Gijbels, P.; Ceulemans, T.; Van den Ende, W.; Honnay, O. Experimental fertilization increases amino acid content in floral nectar, fruit set and degree of selfing in the orchid *Gymnadenia conopsea*. *Oecologia* **2015**, *179*, 785–795. [CrossRef]
17. Gijbels, P.; Van den Ende, W.; Honnay, O. Landscape scale variation in nectar amino acid and sugar composition in a Lepidoptera pollinated orchid species and its relation with fruit set. *J. Ecol.* **2014**, *102*, 136–144. [CrossRef]
18. Gijbels, P.; Van den Ende, W.; Honnay, O. Phenotypic selection on nectar amino acid composition in the Lepidoptera pollinated orchid species *Gymnadenia conopsea*. *Oikos* **2015**, *124*, 421–427. [CrossRef]
19. Brzosko, E.; Bajguz, A.; Chmur, M.; Burzyńska, J.; Jermakowicz, E.; Mirski, P.; Zieliński, P. How are the flower structure and nectar composition of the generalistic orchid *Neottia ovata* adapted to a wide range of pollinators? *Int. J. Mol. Sci.* **2021**, *22*, 2214. [CrossRef] [PubMed]
20. Jeffrey, D.C.; Arditti, J.; Koopowitz, H. Sugar content in floral and extrafloral exudates of orchids: Pollination, myrmecology and chemotaxonomy implication. *New Phytol.* **1970**, *69*, 187–195. [CrossRef]
21. Percival, M.S. Types of nectar in angiosperms. *New Phytol.* **1961**, *60*, 235–281. [CrossRef]
22. Baker, H.G.; Baker, I. Floral nectar sugar constituents in relation to pollinator type. In *Handbook of Experimental Pollination Biology*; Jones, C.E., Little, R.J., Eds.; Van Nostrand Reinhold Company Inc.: New York, NY, USA, 1983; pp. 131–141.
23. Baker, H.G.; Baker, I. The predictive value of nectar chemistry to the recognition of pollinator types. *Isr. J. Bot.* **1990**, *39*, 157–166. [CrossRef]
24. Nicolson, S.W.; Thornburg, R.W. Nectar chemistry. In *Nectaries and Nectar*; Nicolson, S.W., Nepi, M., Pacini, E., Eds.; Springer: Dordrecht, The Netherlands, 2007; pp. 215–264. [CrossRef]
25. Petanidou, T. Sugars in Mediterranean floral nectars: An ecological and evolutionary approach. *J. Chem. Ecol.* **2005**, *31*, 1065–1088. [CrossRef]
26. Willmer, P. Pollination by butterflies and moths. In *Pollination and Floral Ecology*; Willmer, P., Ed.; Princeton University Press: Princeton, NJ, USA, 2011; pp. 322–336. [CrossRef]
27. Adler, L.S. The ecological significance of toxic nectar. *Oikos* **2000**, *91*, 409–420. [CrossRef]
28. Gardener, M.C.; Gillman, M.P. The taste of nectar—A neglected area of pollination ecology. *Oikos* **2002**, *98*, 552–557. [CrossRef]
29. Heil, M. Postsecretory hydrolysis of nectar sucrose and specialization in ant/plant mutualism. *Science* **2005**, *308*, 560–563. [CrossRef] [PubMed]
30. Levin, E.; McCue, M.D.; Davidowitz, G. More than just sugar: Allocation of nectar amino acids and fatty acids in a Lepidopteran. *Proc. Biol. Sci.* **2017**, *284*, 20162126. [CrossRef] [PubMed]
31. Mevi-Schütz, J.; Erhardt, A. Amino acids in nectar enhance butterfly fecundity: A long-awaited link. *Am. Nat.* **2005**, *165*, 411–419. [CrossRef] [PubMed]
32. Heyneman, A.J. Optimal sugar concentrations of floral nectars—Dependence on sugar intake efficiency and foraging costs. *Oecologia* **1983**, *60*, 198–213. [CrossRef] [PubMed]
33. Pyke, G.H.; Waser, N.M. The production of dilute nectars by hummingbird and honeyeater flowers. *Biotropica* **1981**, *13*, 260–270. [CrossRef]
34. Brzosko, E.; Mirski, P. Floral nectar chemistry in orchids: A short review and meta-analysis. *Plants* **2021**, *10*, 2315. [CrossRef]
35. Heil, M. Nectar: Generation, regulation and ecological functions. *Trends Plant Sci.* **2011**, *16*, 191–200. [CrossRef]
36. Gottsberger, G.; Schrauwen, J.; Linskens, H.F. Amino acids and sugars in nectar, and their putative evolutionary significance. *Plant Syst. Evol.* **1984**, *145*, 55–77. [CrossRef]
37. Pacini, E.; Nepi, M.; Vesprini, J.L. Nectar biodiversity: A short review. *Plant Syst. Evol.* **2003**, *238*, 7–21. [CrossRef]
38. Pais, M.S.; Figueiredo, A.C.S. Floral nectaries from *Limodorum abortivum* (L.) Sw and *Epipactis atropurpurea* Rafin (Orchidaceae)—Ultrastructural changes in plastids during the secretory process. *Apidologie* **1994**, *25*, 615–626. [CrossRef]
39. Vandelook, F.; Janssens, S.B.; Gijbels, P.; Fischer, E.; Van den Ende, W.; Honnay, O.; Abrahamczyk, S. Nectar traits differ between pollination syndromes in Balsaminaceae. *Ann. Bot.* **2019**, *124*, 269–279. [CrossRef] [PubMed]
40. Pais, M.; Neves, H.; Maria, P.; Vasconcelos, A. Amino acid and sugar content of the nectar exudate from *Limodorum abortivum* (Orchidaceae). Comparison with *Epipactis atropurpurea* nectar composition. *Apidologie* **1986**, *17*, 125–136. [CrossRef]
41. Moré, M.; Amorim, F.W.; Benitez-Vieyra, S.; Medina, A.M.; Sazima, M.; Cocucci, A.A. Armament imbalances: Match and mismatch in plant-pollinator traits of highly specialized long-spurred orchids. *PLoS ONE* **2012**, *7*, e41878. [CrossRef]
42. Trunschke, J.; Sletvold, N.; Ågren, J. Manipulation of trait expression and pollination regime reveals the adaptive significance of spur length. *Evolution* **2020**, *74*, 597–609. [CrossRef]
43. Alexandersson, R.; Johnson, S.D. Pollinator-mediated selection on flower-tube length in a hawkmoth-pollinated *Gladiolus* (Iridaceae). *Proc. Biol. Sci.* **2002**, *269*, 631–636. [CrossRef]
44. Anderson, B.; Alexandersson, R.; Johnson, S.D. Evolution and coexistence of pollination ecotypes in an African *Gladiolus* (Iridaceae). *Evolution* **2010**, *64*, 960–972. [CrossRef] [PubMed]
45. Boberg, E.; Ågren, J. Despite their apparent integration, spur length but not perianth size affects reproductive success in the moth-pollinated orchid *Platanthera bifolia*. *Funct. Ecol.* **2009**, *23*, 1022–1028. [CrossRef]

46. Boberg, E.; Alexandersson, R.; Jonsson, M.; Maad, J.; Ägren, J.; Nilsson, L.A. Pollinator shifts and the evolution of spur length in the moth-pollinated orchid *Platanthera bifolia*. *Ann. Bot.* **2014**, *113*, 267–275. [CrossRef]
47. Little, K.J.; Dieringer, G.; Romano, M. Pollination ecology, genetic diversity and selection on nectar spur length in *Platanthera lacera* (Orchidaceae). *Plant Spec. Biol.* **2005**, *20*, 183–190. [CrossRef]
48. Maad, J. Phenotypic selection in hawkmoth-pollinated *Platanthera bifolia*: Targets and fitness surfaces. *Evolution* **2000**, *54*, 112–123. [CrossRef]
49. Maad, J.; Alexandersson, R. Variable selection in *Platanthera bifolia* (Orchidaceae): Phenotypic selection differed between sex functions in a drought year. *J. Evol. Biol.* **2004**, *17*, 642–650. [CrossRef] [PubMed]
50. Whittall, J.B.; Hodges, S.A. Pollinator shifts drive increasingly long nectar spurs in columbine flowers. *Nature* **2007**, *447*, 706–709. [CrossRef]
51. Caruso, C.M.; Eisen, K.E.; Martin, R.A.; Sletvold, N. A meta-analysis of the agents of selection on floral traits. *Evolution* **2018**, *73*, 4–14. [CrossRef] [PubMed]
52. De Jager, M.L.; Peakall, R. Experimental examination of pollinator-mediated selection in a sexually deceptive orchid. *Ann. Bot.* **2019**, *123*, 347–354. [CrossRef]
53. Paulus, H.F. Deceived males-pollination biology of the Mediterranean orchid genus *Ophrys* (Orchidaceae). *J. Eur. Orchid.* **2006**, *38*, 303–353.
54. Scopece, G.; Juillet, N.; Lexer, C.; Cozzolino, S. Fluctuating selection across years and phenotypic variation in food-deceptive orchids. *PeerJ* **2017**, *5*, e3704. [CrossRef] [PubMed]
55. Sletvold, N.; Ågren, J. Nonadditive effects of floral display and spur length on reproductive success in a deceptive orchid. *Ecology* **2011**, *92*, 2167–2174. [CrossRef] [PubMed]
56. Sletvold, N.; Trunschke, J.; Smit, M.; Verbeek, J.; Agren, J. Strong pollinator-mediated selection for increased flower brightness and contrast in a deceptive orchid. *Evolution* **2016**, *70*, 716–724. [CrossRef]
57. Bernhardt, P.; Edens-Meier, R. What we think we know vs. what we need to know about orchid pollination and conservation: *Cypripedium* L. as a model lineage. *Bot. Rev.* **2010**, *76*, 204–219. [CrossRef]
58. Potts, S.G.; Biesmeijer, J.C.; Kremen, C.; Neumann, P.; Schweiger, O.; Kunin, W.E. Global pollinator declines: Trends, impacts and drivers. *Trends Ecol. Evol.* **2010**, *25*, 345–353. [CrossRef] [PubMed]
59. Jermakowicz, E.; Brzosko, E. Demographic responses of boreal-montane orchid *Malaxis monophyllos* (L.) Sw. populations to contrasting environmental conditions. *Acta Soc. Bot. Pol.* **2016**, *85*, 1. [CrossRef]
60. Mitchell, R.J.; Flanagan, R.J.; Brown, B.J.; Waser, N.M.; Karron, J.D. New frontiers in competition for pollination. *Ann. Bot.* **2009**, *103*, 1403–1413. [CrossRef]
61. Vamosi, J.C.; Knight, T.M.; Steets, J.A.; Mazer, S.J.; Burd, M.; Ashman, T.-L. Pollination decays in biodiversity hotspots. *Proc. Natl. Acad. Sci. USA* **2006**, *103*, 956–961. [CrossRef]
62. Brzosko, E. The dynamics of *Listera ovata* populations on mineral islands in the Biebrza National Park. *Acta Soc. Bot. Pol.* **2002**, *71*, 243–251. [CrossRef]
63. Brzosko, E. Dynamics of island populations of *Cypripedium calceolus* in the Biebrza river valley (north-east Poland). *Bot. J. Linn. Soc.* **2002**, *139*, 67–77. [CrossRef]
64. Brzosko, E. The dynamics of island populations of *Platanthera bifolia* in the Biebrza National Park (NE Poland). *Ann. Bot. Fenn.* **2003**, *40*, 243–253.
65. Ackerman, J.D. Invasive orchids: Weeds we hate to love? *Lankesteriana* **2015**, *7*, 19–21. [CrossRef]
66. Adamowski, W. Expansion of native orchids in anthropogenous habitats. *Pol. Bot. Stud.* **2006**, *22*, 35–44.
67. Jacquemyn, H.; Brys, R.; Hutchings, M.J. Biological flora of the British Isles: *Epipactis palustris*. *J. Ecol.* **2014**, *102*, 1341–1355. [CrossRef]
68. Rewicz, A.; Bomanowska, A.; Shevera, M.; Kurowski, J.; Krasoń, K.; Zielińska, K. Cities and disturbed areas as man-made shelters for orchid communities. *Not. Bot. Horti Agrobot. Cluj-Napoca* **2017**, *45*, 126–139. [CrossRef]
69. The IUCN Red List of Threatened Species, Version 2021-2. Available online: https://www.iucnredlist.org/ (accessed on 27 October 2021).
70. Fay, M.F. Orchid conservation: How can we meet the challenges in the twenty-first century? *Bot. Stud.* **2018**, *59*, 16. [CrossRef]
71. Kull, T.; Selgis, U.; Peciña, M.V.; Metsare, M.; Ilves, A.; Tali, K.; Sepp, K.; Kull, K.; Shefferson, R.P. Factors influencing IUCN threat levels to orchids across Europe on the basis of national red lists. *Ecol. Evol.* **2016**, *6*, 6245–6265. [CrossRef]
72. Zizka, A.; Silvestro, D.; Vitt, P.; Knight, T.M. Automated conservation assessment of the orchid family with deep learning. *Conserv. Biol.* **2021**, *35*, 897–908. [CrossRef]
73. Swarts, N.D.; Dixon, K.W. Terrestrial orchid conservation in the age of extinction. *Ann. Bot.* **2009**, *104*, 543–556. [CrossRef] [PubMed]
74. Kull, T.; Hutchings, M.J. A comparative analysis of decline in the distribution ranges of orchid species in Estonia and the United Kingdom. *Biol. Conserv.* **2006**, *129*, 31–39. [CrossRef]
75. Jacquemyn, H.; Brys, R.; Hermy, M.; Willems, J. Does nectar reward affect rarity and extinction probabilities of orchid species? An assessment using historical records from Belgium and the Netherlands. *Biol. Conserv.* **2005**, *121*, 257–263. [CrossRef]
76. Goulson, D.; Lye, G.C.; Darvill, B. Decline and conservation of bumble bees. *Annu. Rev. Entomol.* **2008**, *53*, 191–208. [CrossRef] [PubMed]

77. Kaiser, H.F. An index of factorial simplicity. *Psychometrika* **1974**, *39*, 31–36. [CrossRef]
78. Cattell, R.B. The scree test for the number of factors. *Multivar. Behav. Res.* **1966**, *1*, 245–276. [CrossRef] [PubMed]
79. Kaiser, H.F. The application of electronic computers to factor analysis. *Educ. Psychol. Meas.* **1960**, *20*, 141–151. [CrossRef]
80. Tałałaj, I.; Brzosko, E. Selfing potential in *Epipactis palustris*, *E. helleborine* and *E. atrorubens* (Orchidaceae). *Plant Syst. Evol.* **2008**, *276*, 21–29. [CrossRef]
81. Vakhrameeva, M.G.; Tatarenko, I.V.; Varlygina, T.I.; Torosyan, G.K.; Zagulski, M.N. *Orchids of Russia and Adjacent Countries (within the Borders of the Former USSR)*; A.R.G. Gantner Verlag K.G.: Ruggell, Liechtenstein, 2008; p. 690.
82. Jacquemyn, H.; Brys, R. Pollen limitation and the contribution of autonomous selfing to fruit and seed set in a rewarding orchid. *Am. J. Bot.* **2015**, *102*, 67–72. [CrossRef] [PubMed]
83. Jakubska-Busse, A.; Kadej, M. The pollination of *Epipactis* Zinn, 1757 (Orchidaceae) species in Central Europe—The significance of chemical attractants, floral morphology and concomitant insects. *Acta Soc. Bot. Pol.* **2011**, *80*, 49–57. [CrossRef]
84. Pellegrino, G.; Bellusci, F. Effects of human disturbance on reproductive success and population viability of *Serapias cordigera* (Orchidaceae). *Bot. J. Linn. Soc.* **2014**, *176*, 408–420. [CrossRef]
85. Parra-Tabla, V.; Vargas, C.F.; Magaña-Rueda, S.; Navarro, J. Female and male pollination success of *Oncidium ascendens* Lindey (Orchidaceae) in two contrasting habitat patches. *Biol. Conserv.* **2000**, *94*, 335–340. [CrossRef]
86. Rewicz, A.; Jaskuła, R.; Rewicz, T.; Tończyk, G. Pollinator diversity and reproductive success of *Epipactis helleborine* (L.) Crantz (Orchidaceae) in anthropogenic and natural habitats. *PeerJ* **2017**, *5*, e3159. [CrossRef]
87. Ollerton, J.; Killick, A.; Lamborn, E.; Watts, S.; Whiston, M. Multiple meanings and modes: On the many ways to be a generalist flower. *Taxon* **2007**, *56*, 717–728. [CrossRef]
88. Ågren, J.; Fortunel, C.; Ehrlén, J. Selection on floral display in insect-pollinated *Primula farinosa*: Effects of vegetation height and litter accumulation. *Oecologia* **2006**, *150*, 225–232. [CrossRef]
89. Sletvold, N.; Grindeland, J.M.; Ågren, J. Vegetation context influences the strength and targets of pollinator-mediated selection in a deceptive orchid. *Ecology* **2013**, *94*, 1236–1242. [CrossRef] [PubMed]
90. Grindeland, J.M.; Sletvold, N.; Ims, R.A. Effects of floral display size and plant density on pollinator visitation rate in a natural population of *Digitalis purpurea*. *Funct. Ecol.* **2005**, *19*, 383–390. [CrossRef]
91. Kindlmann, P.; Jersáková, J. Effect of floral display on reproductive success in terrestrial orchids. *Folia Geobot.* **2006**, *41*, 47–60. [CrossRef]
92. Vallius, E.; Arminen, S.; Salonen, V. Are There Fitness Advantages Associated with a Large Inflorescence in *Gymnadenia conopsea* ssp. *conopsea*? 2006. Available online: http://www.r-b-o.eu/rbo_public?Vallius_et_al_2006.html (accessed on 29 November 2020).
93. Calvo, R.N. Inflorescence size and fruit distribution among individuals in three orchid species. *Am. J. Bot.* **1990**, *77*, 1378–1381. [CrossRef]
94. Pellegrino, G.; Bellusci, F.; Musacchio, A. The effects of inflorescence size and flower position on female reproductive success in three deceptive orchids. *Bot. Stud.* **2010**, *51*, 351–356.
95. Van der Niet, T.; Peakall, R.; Johnson, S.D. Pollinator-driven ecological speciation in plants: New evidence and future perspectives. *Ann. Bot.* **2014**, *113*, 199–211. [CrossRef]
96. Vereecken, N.J.; Cozzolino, S.; Schiestl, F.P. Hybrid floral scent novelty drives pollinator shift in sexually deceptive orchids. *BMC Evol. Biol.* **2010**, *10*, 103. [CrossRef] [PubMed]
97. Peralta, G.; Vázquez, D.P.; Chacoff, N.P.; Lomáscolo, S.B.; Perry, G.L.W.; Tylianakis, J.M. Trait matching and phenological overlap increase the spatio-temporal stability and functionality of plant-pollinator interactions. *Ecol. Lett.* **2020**, *23*, 1107–1116. [CrossRef] [PubMed]
98. Jacquemyn, H.; Brys, R. Lack of strong selection pressures maintains wide variation in floral traits in a food-deceptive orchid. *Ann. Bot.* **2020**, *126*, 445–453. [CrossRef] [PubMed]
99. Abrahamczyk, S.; Kessler, M.; Hanley, D.; Karger, D.N.; Müller, M.P.J.; Knauer, A.C.; Keller, F.; Schwerdtfeger, M.; Humphreys, A.M. Pollinator adaptation and the evolution of floral nectar sugar composition. *J. Evol. Biol.* **2016**, *30*, 112–127. [CrossRef]
100. Parachnowitsch, A.L.; Manson, J.S.; Sletvold, N. Evolutionary ecology of nectar. *Ann. Bot.* **2019**, *123*, 247–261. [CrossRef]
101. Fowler, R.E.; Rotheray, E.L.; Goulson, D. Floral abundance and resource quality influence pollinator choice. *Insect Conserv. Divers.* **2016**, *9*, 481–494. [CrossRef]
102. Petanidou, T.; Van Laere, A.; Ellis, W.N.; Smets, E. What shapes amino acid and sugar composition in Mediterranean floral nectars? *Oikos* **2006**, *115*, 155–169. [CrossRef]
103. Jakubska-Busse, A.; Kadej, M. Pollination ecology of marsh helleborine *Epipactis palustris* (L.) Crantz on the Polish side of the Orlickie Mts. (Central Sudety Mts.). In *Environmental Changes and Biological Assessments IV*; Kočárek, P., Plášek, V., Malachová, K., Eds.; Scripra Facultatis Rerum Naturalium Universitatis Ostraviensis: Cambridge, UK, 2008; Volume 186, pp. 247–253.
104. Gardener, M.C.; Gillman, M.P. The effects of soil fertilizer on amino acids in the floral nectar of corncockle, *Agrostemma githago* (Caryophyllaceae). *Oikos* **2001**, *92*, 101–106. [CrossRef]
105. Carter, C.; Shafir, S.; Yehonatan, L.; Palmer, R.G.; Thornburg, R. A novel role for proline in plant floral nectars. *Naturwissenschaften* **2006**, *93*, 72–79. [CrossRef]
106. Nepi, M.; Soligo, C.; Nocentini, D.; Abate, M.; Guarnieri, M.; Cai, G.; Bini, L.; Puglia, M.; Bianchi, L.; Pacini, E. Amino acids and protein profile in floral nectar: Much more than a simple reward. *Flora* **2012**, *207*, 475–481. [CrossRef]

107. Nocentini, D.; Pacini, E.; Guarnieri, M.; Martelli, D.; Nepi, M. Intrapopulation heterogeneity in floral nectar attributes and foraging insects of an ecotonal Mediterranean species. *Plant Ecol.* **2013**, *214*, 799–809. [CrossRef]
108. Bertazzini, M.; Medrzycki, P.; Bortolotti, L.; Maistrello, L.; Forlani, G. Amino acid content and nectar choice by forager honeybees (*Apis mellifera* L.). *Amino Acids* **2010**, *39*, 315–318. [CrossRef]
109. Tiedge, K.; Lohaus, G. Nectar sugars and amino acids in day- and night-flowering *Nicotiana* species are more strongly shaped by pollinators' preferences than organic acids and inorganic ions. *PLoS ONE* **2017**, *12*, e0176865. [CrossRef]
110. Kim, Y.S.; Smith, B.H. Effect of an amino acid on feeding preferences and learning behavior in the honey bee, *Apis mellifera*. *J. Insect Physiol.* **2000**, *46*, 793–801. [CrossRef]
111. Baker, H.G.; Baker, I. Intraspecific constancy of floral nectar amino acid complements. *Bot. Gaz.* **1977**, *138*, 183–191. [CrossRef]
112. Hultén, E.; Fries, M. *Atlas of North European Vascular Plants: North of the Tropic of Cancer. Volumes I-III*; Koeltz Scientific Books: Königstein, Germany, 1986; p. 1172.
113. Kowalkowska, A.K.; Kostelecka, J.; Bohdanowicz, J.; Kapusta, M.; Rojek, J. Studies on floral nectary, tepals' structure, and gynostemium morphology of *Epipactis palustris* (L.) Crantz (Orchidaceae). *Protoplasma* **2015**, *252*, 321–333. [CrossRef] [PubMed]
114. Morrant, D.S.; Schumann, R.; Petit, S. Field methods for sampling and storing nectar from flowers with low nectar volumes. *Ann. Bot.* **2009**, *103*, 533–542. [CrossRef]
115. Wickham, H.; Averick, M.; Bryan, J.; Chang, W.; McGowan, L.; François, R.; Grolemund, G.; Hayes, A.; Henry, L.; Hester, J.; et al. Welcome to the Tidyverse. *J. Open Source Softw.* **2019**, *4*, 1686. [CrossRef]
116. R Core Team. *R: A Language and Environment for Statistical Computing, R Version 4.1.1, Kick Things*; R Foundation for Statistical Computing: Vienna, Austria, 2021; Available online: https://www.R-project.org/ (accessed on 5 October 2021).
117. Fox, J.; Weisberg, S. *An R Companion to Applied Regression*, 3rd ed.; SAGE Publications, Inc.: Thousand Oaks, CA, USA, 2019.
118. Burda, M. *Paircompviz: Multiple Comparison Test Visualization (R Package Version 1.28.0)*; 2020; Available online: https://www.bioconductor.org/packages/release/bioc/html/paircompviz.html (accessed on 5 October 2021).
119. Graves, S.; Piepho, H.-P.; Selzer, L.; Dorai-Raj, S. *multcompView: Visualizations of Paired Comparisons, R Package Version 0.1-8*; 2019; Available online: https://cran.r-project.org/web/packages/multcompView/index.html (accessed on 5 October 2021).
120. Mangiafico, S. *Rcompanion: Functions to Support Extension Education Program Evaluation (R Package Version 2.3.26)*; 2020; Available online: https://rdrr.io/cran/rcompanion/ (accessed on 5 October 2021).
121. Hamilton, N.E.; Ferry, M. ggtern: Ternary diagrams using ggplot2. *J. Stat. Softw.* **2018**, *87*, 1–17. [CrossRef]
122. Lê, S.; Josse, J.; Husson, F. FactoMineR: An R package for multivariate analysis. *J. Stat. Softw.* **2008**, *25*, 1–18. [CrossRef]
123. Bartlett, M.S. Tests of significance in factor analysis. *Br. J. Stat. Psychol.* **1950**, *3*, 77–85. [CrossRef]
124. Revelle, W. *psych: Procedures for Personality and Psychological Research, R Package Version 1.8.10*; Northwestern University: Evanston, IL, USA, 2018.
125. Kassambara, A.; Mundt, F. *Factoextra: Extract and Visualize the Results of Multivariate Data Analyses, R Package Version 1.0.6*; 2019; Available online: https://cran.r-project.org/web/packages/factoextra/index.html (accessed on 5 October 2021).

Article

Extending the Toolkit for Beauty: Differential Co-Expression of *DROOPING LEAF*-like and Class B MADS-box Genes during *Phalaenopsis* Flower Development

Francesca Lucibelli [1], Maria Carmen Valoroso [1], Günter Theißen [2], Susanne Nolden [2], Mariana Mondragon-Palomino [3,*,†] and Serena Aceto [1,*]

1 Department of Biology, University of Naples Federico II, 80126 Napoli, Italy; francesca.lucibelli@unina.it (F.L.); mariacarmen.valoroso@unina.it (M.C.V.)
2 Matthias Schleiden Institute of Genetics, Friedrich Schiller University Jena, 07743 Jena, Germany; guenter.theissen@uni-jena.de (G.T.); nolden_susanne@yahoo.de (S.N.)
3 Department of Cell Biology and Plant Biochemistry, University of Regensburg, 93040 Regensburg, Germany
* Correspondence: mariana.mondragon@biomax.com (M.M.-P.); serena.aceto@unina.it (S.A.)
† Current address: Biomax Informatics AG, 82152 Planegg, Germany.

Citation: Lucibelli, F.; Valoroso, M.C.; Theißen, G.; Nolden, S.; Mondragon-Palomino, M.; Aceto, S. Extending the Toolkit for Beauty: Differential Co-Expression of *DROOPING LEAF*-like and Class B MADS-box Genes during *Phalaenopsis* Flower Development. *Int. J. Mol. Sci.* 2021, 22, 7025. https://doi.org/10.3390/ijms22137025

Academic Editor: Pedro Martínez-Gómez

Received: 3 June 2021
Accepted: 27 June 2021
Published: 29 June 2021

Publisher's Note: MDPI stays neutral with regard to jurisdictional claims in published maps and institutional affiliations.

Copyright: © 2021 by the authors. Licensee MDPI, Basel, Switzerland. This article is an open access article distributed under the terms and conditions of the Creative Commons Attribution (CC BY) license (https://creativecommons.org/licenses/by/4.0/).

Abstract: The molecular basis of orchid flower development is accomplished through a specific regulatory program in which the class B MADS-box *AP3/DEF* genes play a central role. In particular, the differential expression of four class B *AP3/DEF* genes is responsible for specification of organ identities in the orchid perianth. Other MADS-box genes (*AGL6* and *SEP*-like) enrich the molecular program underpinning the orchid perianth development, resulting in the expansion of the original "orchid code" in an even more complex gene regulatory network. To identify candidates that could interact with the *AP3/DEF* genes in orchids, we conducted an in silico differential expression analysis in wild-type and peloric *Phalaenopsis*. The results suggest that a YABBY *DL*-like gene could be involved in the molecular program leading to the development of the orchid perianth, particularly the labellum. Two YABBY *DL/CRC* homologs are present in the genome of *Phalaenopsis equestris*, *PeDL1* and *PeDL2*, and both express two alternative isoforms. Quantitative real-time PCR analyses revealed that both genes are expressed in column and ovary. In addition, *PeDL2* is more strongly expressed the labellum than in the other tepals of wild-type flowers. This pattern is similar to that of the *AP3/DEF* genes *PeMADS3/4* and opposite to that of *PeMADS2/5*. In peloric mutant *Phalaenopsis*, where labellum-like structures substitute the lateral inner tepals, *PeDL2* is expressed at similar levels of the *PeMADS2-5* genes, suggesting the involvement of *PeDL2* in the development of the labellum, together with the *PeMADS2-PeMADS5* genes. Although the yeast two-hybrid analysis did not reveal the ability of PeDL2 to bind the PeMADS2-PeMADS5 proteins directly, the existence of regulatory interactions is suggested by the presence of CArG-boxes and other MADS-box transcription factor binding sites within the putative promoter of the orchid *DL2* gene.

Keywords: *DROOPING LEAF*; flower development; gene expression; Orchidaceae; YABBY transcription factors

1. Introduction

The Orchidaceae is one of the widely distributed and most diversified families of angiosperms. Their evolutionary success is possibly due to sundry causes such as epiphytism, extraordinary adaptive capacities to different habitats, highly specialized pollination strategies, and diversified flower morphology [1–3]. Despite the diversity of flower colors, sizes, shapes, and appendages, the floral organs of orchids share a common organization (Figure 1). There are three outer tepals in the first floral whorl; in the second whorl, the three tepals are distinguished into two lateral inner tepals and a median inner tepal called lip or labellum. This organ often has a peculiar morphology and bears distinct color patterns (Figure 1a,b). Female and male reproductive organs are fused to form the gynostemium or column, at whose apex are located the pollinia. The ovary is placed at the base of the gynostemium, and its development is activated by pollination [4].

Figure 1. Wild-type and peloric mutants of *Phalaenopsis*. (**a**) Wild-type *P. aphrodite*; (**b**) floral buds at stages B1–B5 and floral organs at the OF stage of the wild-type *P. aphrodite*; (**c**) flower of the wild-type *Phalaenopsis* hyb. "Athens"; (**d**) flower of the peloric mutant *Phalaenopsis* hyb. "Athens"; (**e**) flower of the peloric mutant *Phalaenopsis* hyb. "Joy Fairy Tale". The arrow in (**a**) indicates the point of rotation of the pedicel during resupination. Size of the developmental stages: B1 (0.5–1 cm), B2 (1–1.5 cm), B3 (1.5–2 cm), B4 (2–2.5 cm), B5 (2.5–3 cm), OF (open flower). 1, outer tepals; 2, lateral inner tepals; 3, labellum; 4, column; 5, ovary; 2/3, labellum-like organs.

The labellum is a central organ in orchid pollination because of its strikingly distinct morphology and its direct opposition to the gynostemium. Therefore, its showy color patterns and structures are visual attractants and it act as a landing platform that guides pollinators towards the gynostemium. Because the labellum is the uppermost perianth organ, its role in pollination depends on becoming the lowermost through resupination, a 180° developmental rotation of the flower pedicel or ovary (Figure 1a) [5].

Bilateral symmetry or zygomorphy in orchids is a syndrome defined by the association of several characters (e.g., labellum and the developmental suppression of adaxial stamens). This association took place early in Orchidaceae evolution and became the basis for the progressive addition of further innovations like pollinaria, a spur, or showy markings on the labellum [4]. The concurrence of these floral features is considered a key morphological innovation in the two most derived and diverse orchid subfamilies Epidendroideae and Orchidoideae [4]. Together they mediate the specialized relationships of this family with pollinators, facilitating the processes of prezygotic reproductive isolation [6,7].

Because of the central role of the labellum in orchid reproduction, its developmental origin is a subject of intense study [4,8–10]. In the last decade, several gene regulatory models inspired by the more general angiosperm ABC model [11,12] helped to explain the developmental specification of the distinct orchid perianth organs [13–18]. Specifically, the "orchid code" argues that the diversification of the organs of the orchid perianth is due to the combined differential expression of class B MADS-box genes belonging to the *AP3/DEF* group [13–15]. The "homeotic orchid tepals" (HOT model) proposes a combinatorial action of homeotic MADS-box proteins consistent with the "orchid code" [16]. The more recent "P-code" model hypothesizes a pivotal role of the class B and *AGL6* MADS-box genes in forming the orchid perianth [19].

In order to understand the more extensive regulatory network behind orchid flower development, we and others have found that, like *AP3/DEF*s, also candidate *SEP*-, *FUL*-, *AG*-, and *STK*-like MADS-box genes have been duplicated in the Orchidaceae. However, only some of them are differentially expressed in association with the distinct flower organs. For instance, in developing *Phalaenopsis* flowers, we observed that *SEP3*-like and *DEF*-like genes have common expression domains. This shared domain of expression suggests that both candidates are associated with labellum specification, and that similar positional cues determine their expression domains [20]. Elucidating the nature of the positional cues behind the development of specific orchid flower organs is a central question to understand the developmental program of this family.

Top candidates for providing the positional information for differentially expressed MADS-box genes are CYCLOIDEA-like (CYC-like) transcription factors [13], which are well known for their role in flower bilateral symmetry specification in core eudicots [21–25]. Comparative studies of *CYC*-like genes identified several major, well-supported monocot-specific clades and reported the first *CYC*-like genes in orchid species [26–28]. Additional studies also showed that the DDR regulatory module composed of the MYB factors DIVARICATA, DRIF, and RADIALIS, responsible in *Antirrhinum majus* for bilateral flower symmetry [29,30], seems to be conserved in orchids [31–33]. However, the critical CYC-like transcription factor that activates the transcription of *RADIALIS* in *A. majus* [21,34] is not conserved in orchids. Moreover, the current literature reports contrasting results [26–28,35], possibly because the functional equivalent of *CYC*, if it exists, has not yet been identified in orchids.

Our interest in identifying additional components of the regulatory network determining orchid flower organ identity prompted us to conduct a preliminary in silico differential expression study using RNA-seq data of *Phalaenopsis*. This preliminary work suggests a scenario where MADS-box genes and members of the plant-specific family of transcription factors termed YABBY contribute to labellum development.

During the course of angiosperm evolution, the YABBY *DROOPING LEAF/CRABS CLAW* (*DL/CRC*) genes came to regulate the development of different structures like the carpel, nectaries, or the leaf mid-rib [36]. In addition, *DL/CRC* and other members of the YABBY gene family like *FILAMENTOUS FLOWER* (*FIL*) [37,38] respectively determine flower meristem and organ identity in *Arabidopsis* and rice [37,38]. In the rice flower meristem, the expression domain of *DL* is delimited by the class B MADS-box gene *SUPERWOMAN1* [37,39], thus suggesting a regulatory relationship between them. Additional evidence of regulatory interaction between *DL/CRC* and MADS-box genes comes from maize. In this species, the co-orthologs *drl1* and *drl2* have a potential antagonistic relationship with *silky1*, the ortholog of the class B *APETALA3/DEFICIENS* gene, during floral patterning and establishment of floral bilateral symmetry [40].

The existence of a regulatory relationship between *DL/CRC* and MADS-box genes in model dicot and monocot species inspired us to explore the role of *DL*-like genes in orchids. In this family, gene duplication and differential expression of *DEF*-like class B MADS-box genes play a pivotal role in modularizing the perianth [13–19,41,42]. However, it is not yet clear as to which positional cues determine their expression domains, resulting in flower bilateral symmetry. The present study tests the hypothesis that *DL*-like orchid genes are associated with the development of distinct orchid flower organs. To this purpose, we compared their patterns of expression with those of *DEFICIENS*-like MADS-box genes *PeMADS2, PeMADS3, PeMAD4,* and *PeMADS5* (*PeMADS2-PeMADS5*) in wild-type and peloric *Phalaenopsis* flowers. These mutants have labellum-like structures that substitute the lateral inner tepals, thus lacking the bilateral symmetry of the perianth, and are especially useful to study genes possibly involved in orchid perianth formation. Next, we tested whether the co-expression of *DL*-like and *DEF*-like genes also involves direct protein–

protein interactions via yeast two-hybrid assays. Finally, we scanned the putative promoters of the *DL*-like genes of *Phalaenopsis* and *Dendrobium* to identify conserved motifs with possible regulatory functions.

2. Results

2.1. Identification of Transcription Factors Differentially Expressed in the Labellum

Our initial RNA-seq screening of the *Phalaenopsis* hyb. "Athens" (Figure 1c,d) inner-perianth transcriptome showed over 78% of the read pairs mapped to the *Phalaenopsis equestris* genome v 1.0 [43]. About 68% of the transcripts annotated (21,200 genes) are expressed in the flower organs analyzed with at least 1 TPM (transcripts per kilobase million). Labellum-like lateral inner tepals of peloric flowers and wild-type labella share 98% of all expressed genes. This indicates that these organs express almost the same genes, strongly suggesting that they have the same organ identity (Supplementary Figure S1a).

Analyses of differential gene expression yielded an interesting group of transcripts significantly up- or downregulated in wild-type lateral inner tepals compared to the labellum (Supplementary Figure S1b and Data S1). Among them, we identified transcripts that are possibly associated with labellum development, encoding DROOPING LEAF-like proteins (DL-like) and the class B MADS-domain protein PeMADS2 (Supplementary Figure S1c). In our analysis, two DL-like transcripts are downregulated in wild-type lateral inner tepals, in comparison to their wild-type and peloric labella levels. Transcripts of class B MADS-box gene PeMADS2 are upregulated in wild-type lateral inner tepals, just as documented by qPCR in the "orchid code" [13–15]. Furthermore, *CYC-TB1*-like genes are expressed in lateral inner tepals and labellum at levels under 1 TPM. This extremely low level of expression of *CYC-TB1*-like genes during orchid development has also been observed in previous studies [26].

We then conducted an in silico differential expression analysis using publicly available reads of the perianth organs of wild-type and peloric mutant *Phalaenopsis* hyb. "Brother Spring Dancer" KHM190 [44]. We mapped and quantified the reads against the transcriptome of *Phalaenopsis* hyb. "Brother Spring Dancer" assembled from the Illumina raw reads. In this case we also found transcripts encoding class B MADS-box proteins differentially expressed among the organs of the wild-type plant, and detected differential expression for a transcript encoding a DL-like protein (Supplementary Data S2). In particular, in the wild-type *Phalaenopsis* this *DL*-like transcript showed a 3 to 4 $\log_2$ FC expression in labellum than in lateral inner tepals. No significant difference was observed between the transcripts of this gene in the labellum and labellum-like lateral inner tepals of the peloric mutant *Phalaenopsis* hyb. "Brother Spring Dancer" (Supplementary Data S2).

The differential pattern of expression of the *DL*-like transcript is analogous to those observed in MADS-box *DEF*-like genes *PeMADS3* and *PeMADS4*, which are highly expressed in the wild-type labellum and labellum-like structures of peloric mutants [14,45]. This similarity suggests an association between the activity of *DL*-like and *DEF*-like homeotic genes and the development of the labellum.

Further in silico analyses of the reference transcriptome of *Phalaenopsis* hyb. "Brother Spring Dancer" identified two *DL*-like transcripts, *PeDL1* and *PeDL2*, each with two different isoforms.

We confirmed the presence of these transcripts by the PCR amplification of cDNA from perianth tissues of *Phalaenopsis* hyb. "Athens" followed by cloning and sequencing, and deposited the sequences in GenBank with the accession numbers MW574592, MW574593 (*PeDL1_1* and *PeDL1_2*), MW574594, and MW574595 (*PeDL2_1* and *PeDL2_2*). The longest isoforms of both transcripts (*PeDL1_1* and *PeDL2_1*) encode proteins containing a C2C2 zinc-finger domain at the N-terminus and a YABBY domain, whereas both the alternative isoforms encode proteins missing the C2C2 zinc-finger domain completely (*PeDL1_2*) or partially (*PeDL2_2*) (Supplementary Figure S2). The PeDL1_1 (189 aa) and PeDL2_1 (196 aa) proteins are 64.3% similar, with highly conserved YABBY domains and more variable C2C2 zinc-finger domains. In comparison, the region spanning from the C2C2 to the YABBY domain and the C-terminal region are the less-conserved parts of these proteins (Supplementary Figure S2).

2.2. Genomic Organization of the PeDL1 and PeDL2 Genes

Reconstruction of the genomic organization of the *PeDL1* and *PeDL2* genes based on BLAST analyses of the longest *PeDL* transcripts against the assembled genome of *Phalaenopsis equestris* [43] showed the *PeDL* genes have seven exons and six introns (Figure 2). The large intron 4 is particularly rich in repetitive sequences. This feature has affected the correct assembly of the *PeDL1* and *PeDL2* genes, which were both split in two different genomic scaffolds (Scaffold000404_23 and Scaffold000404_21 for *PeDL1*; Scaffold000061_46 and Scaffold000061_45 for *PeDL2*).

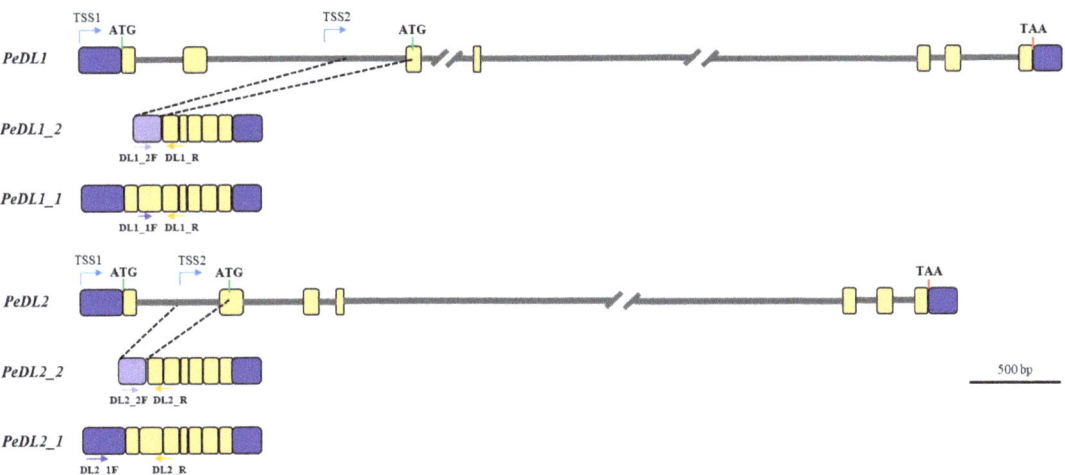

Figure 2. Genomic organization of the *PeDL1* and *PeDL2* transcribed regions of *P. equestris* and diagram of the corresponding alternative transcripts. The blue boxes represent the 5'- and 3'-UTRs; the yellow boxes represent the coding regions, the gray lines represent the introns. Introns of unknown size are shown as interrupted gray lines. The green and red bars indicate the position of the translation start (ATG) and stop (TAA) codons, respectively. TSS1 and 2 are the putative alternative transcription start sites of the different isoforms. The blue and yellow arrows indicate the position of the isoform-specific primer pairs.

The alignment of the short transcripts *PeDL1_2* and *PeDL2_2* with the corresponding genomic region revealed the presence of a putative alternative transcription start site within intron 1 of *PeDL2* and intron 2 of *PeDL1*, resulting in transcripts whose ATG start codon is located within exon 2 and exon 3, respectively (Figure 2).

2.3. Differential Expression of the PeDL1 and PeDL2 Genes

To analyze the expression pattern of *PeDL1* and *PeDL2* in the floral organs of *Phalaenopsis*, we performed quantitative real-time PCR on cDNA from different organs of the wild-type *Phalaenopsis* hyb. "Athens" dissected from floral buds of ~1 cm (B2 stage, Figure 1c). Both genes are highly expressed in the column and ovary. However, the *PeDL2* isoforms are also highly expressed in the labellum relative to outer and the other inner tepals. These results confirm the initial in silico differential expression analysis (Figure 3).

Then, to verify the conservation of these expression patterns and follow them along with flower development, we examined the expression profile of *PeDL1* and *PeDL2* in the perianth tissues of *P. aphrodite* at different developmental stages (Figure 1a,b). As shown in Figure 4, all but *PeDL2_1* have low expression levels in all the perianth organs (outer tepals, inner tepals, and labellum) from the earliest stage B1 to OF (open flower). Interestingly, the isoform *PeDL2_1* is expressed at high levels in the labellum during the first developmental stages. Its expression decreases over time, with a statistically significant negative correlation between expression level and stage (Spearman correlation $r = -1$, $p = 0.0028$).

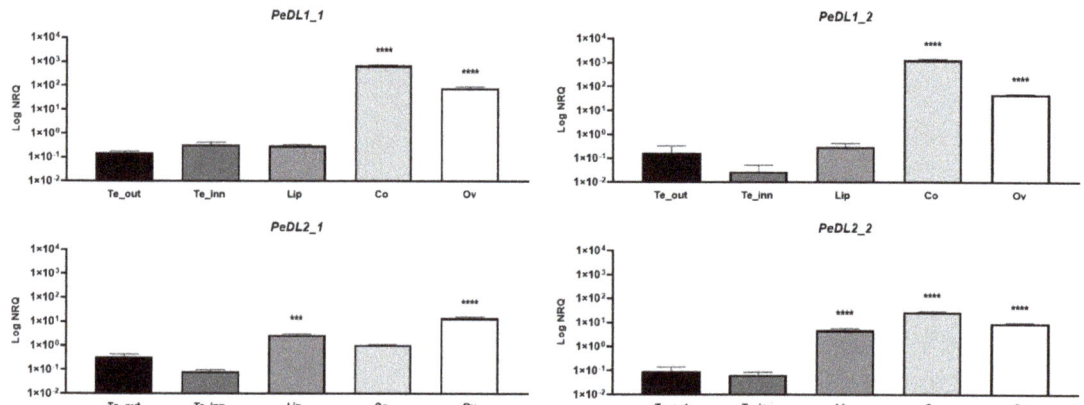

Figure 3. Relative expression of the different isoforms of the *PeDL1* and *PeDL2* genes in the floral tissues of the wild-type *Phalaenopsis* hyb. "Athens" at the B2 developmental stage (bud size 1–1.5 cm). The expression is reported as logarithm of the normalized relative quantity (Log NRQ). The bars represent the SEM of the biological and technical replicates. The asterisks indicate the statistically significant difference of the expression compared to outer tepals. *p*-Values *** <0.001, **** <0.0001. Te_out, outer tepals; Te_inn, lateral inner tepals; Co, column; Ov, ovary.

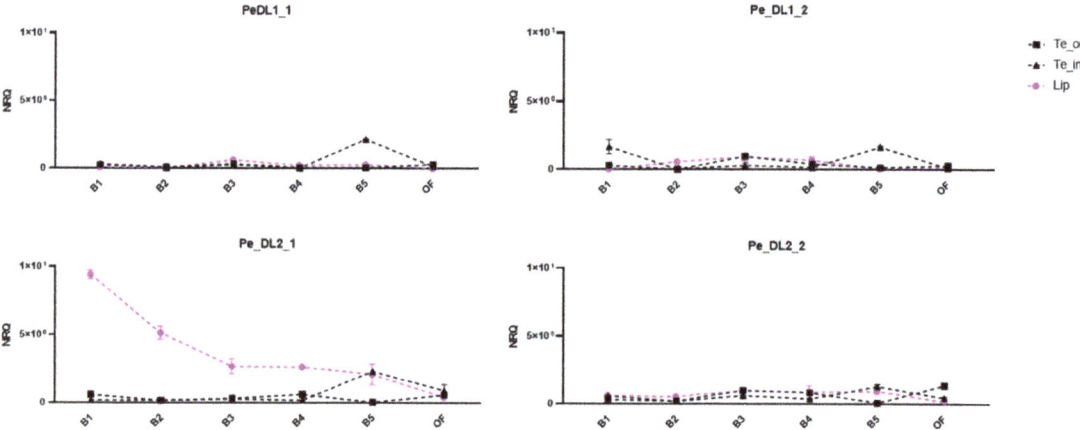

Figure 4. Relative expression of the different isoforms of the *PeDL1* and *PeDL2* genes in the perianth of the wild-type *P. aphrodite* at different developmental stages. The expression is reported as normalized relative quantity (NRQ). The bars represent the SEM of the biological and technical replicates. Bud size of the developmental stages: B1 (0.5–1 cm), B2 (1–1.5 cm), B3 (1.5–2 cm), B4 (2–2.5 cm), B5 (2.5–3 cm), OF (open flower). Te_out, outer tepals; Te_inn, lateral inner tepals.

To test the hypothesis that *PeDL2* is associated with the development of distinct perianth organs, we analyzed the expression pattern of the isoforms *PeDL2_1* and *PeDL2_2* in two *Phalaenopsis* peloric mutants bearing labellum-like structures in place of lateral inner tepals. The peloric *Phalaenopsis* hyb. "Athens" shows an increased expression of both *PeDL2* isoforms in the labellum-like structures compared to the lateral inner tepals of the wild-type (Figure 5). In particular, the mean difference of the expression between lateral inner tepals and labellum decreases from −2.71 (wild-type) to −1.93 (peloric) for *PeDL2_1* and from −4.82 (wild-type) to −0.83 (peloric) for *PeDL2_2*. In the peloric *Phalaenopsis* hyb. "Joy Fairy Tale" there are no significant differences found in the expression levels of *PeDL2* in the inner and outer perianth organs (Figure 6). Additionally, no significant differences were detected in the expression of *PeDL1_1* and *PeDL1_2* in the perianth of wild-type and both peloric *Phalaenopsis* mutants (Supplementary Figure S3).

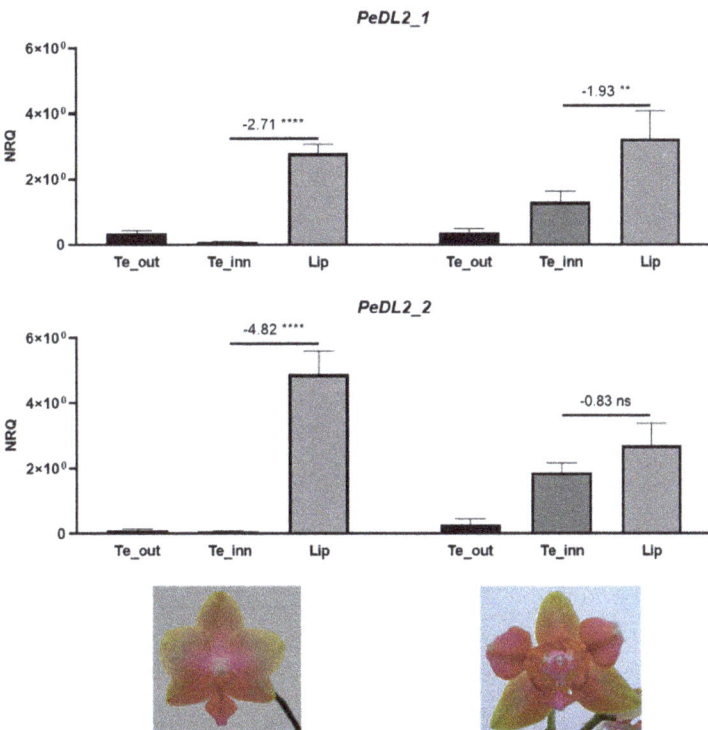

Figure 5. Relative expression of the isoforms *PeDL2_1* and *PeDL2_2* in the perianth of the wild-type (left) and peloric (right) *Phalaenopsis* hyb. "Athens" at the B2 developmental stage (bud size 1–1.5 cm). The expression is reported as normalized relative quantity (NRQ). The vertical bars represent the SEM of the biological and technical replicates. The numbers above the horizontal lines are the mean differences of the expression between lateral inner tepals and labellum (Te_inn - Lip). *p*-Values ** <0.01, **** <0.0001; ns, not significant. Te_out, outer tepals; Te_inn, lateral inner tepals or labellum-like structures that substitute the lateral inner tepals in the peloric mutant.

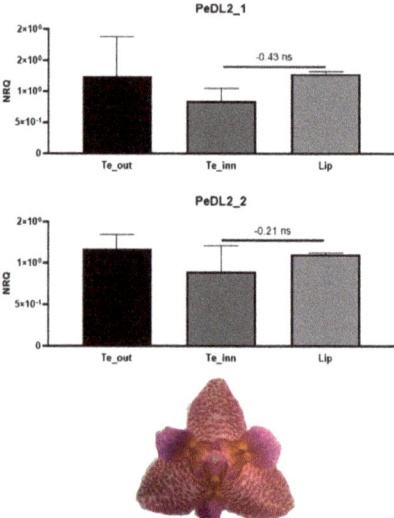

Figure 6. Relative expression of the isoforms *PeDL2_1* and *PeDL2_2* in the perianth of the peloric *Phalaenopsis* hyb. "Joy Fairy Tale" at the B2 developmental stage (bud size 1–1.5 cm). The expression is reported as normalized relative quantity (NRQ). The vertical bars represent the SEMs of the biological and technical replicates. The numbers above the horizontal lines are the mean differences of the expression between labellum-like structures and labellum (Te_inn-Lip). ns, not significant. Te_out, outer tepals; Te_inn, labellum-like structures that substitute the lateral inner tepals in the peloric mutant.

2.4. Differential Expression of the PeMADS2-PeMADS5 Genes

To compare the expression profile of the *DEF*-like genes in the perianth organs of wild-type and peloric *Phalaenopsis*, we performed real-time PCR experiments on cDNA of wild-type and peloric *Phalaenopsis* hyb. "Athens" (Figure 7) and of the peloric *Phalaenopsis* hyb. "Joy Fairy Tale" (Figure 8). As expected, *PeMADS2* and *PeMADS5* are less expressed in labellum than in outer and inner tepals in the wild-type *Phalaenopsis*. Genes *PeMADS3* and *PeMADS4* show an opposite behavior, being more expressed in the labellum than in other organs of the wild-type perianth. In the peloric *Phalaenopsis* "Athens", the mean difference between the expression levels of the *PeMADS2-PeMADS5* genes in labellum-like structures and labellum decreases due to the reduced (for *PeMADS2* and *PeMADS5*) or the increased (for *PeMADS3* and *PeMADS4*) expression in the labellum-like structures (Figure 7).

In the peloric *Phalaenopsis* hyb. "Joy Fairy Tale", the differences in expression level between the labellum-like structures and lip are not significant, except for *PeMADS4*, which shows a higher expression in the labellum-like structures than in labellum (Figure 8).

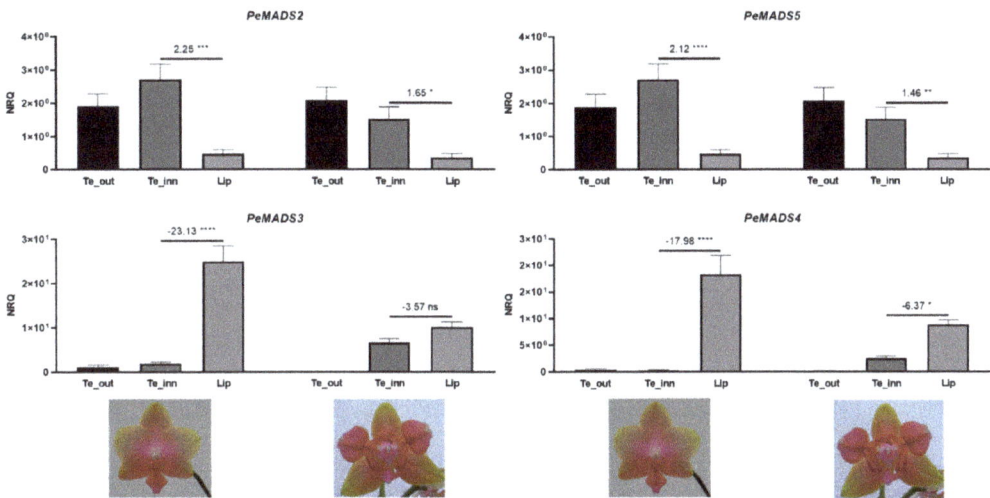

Figure 7. Relative expression of the class B MADS-box genes *PeMADS2-PeMADS5* in the perianth of *Phalaenopsis* hyb. "Athens" wild-type (left) and peloric mutant (right) at the B2 developmental stage (bud size 1–1.5 cm). The expression is reported as normalized relative quantity (NRQ). The vertical bars represent the SEM of the biological and technical replicates. The numbers above the horizontal lines are the mean differences of the expression between lateral inner tepals and labellum (Te_inn-Lip). *p*-Values * <0.05, ** <0.01, *** <0.001, **** <0.0001; ns, not significant. Te_out, outer tepals; Te_inn, lateral inner tepals or labellum-like structures that substitute the lateral inner tepals in the peloric mutant.

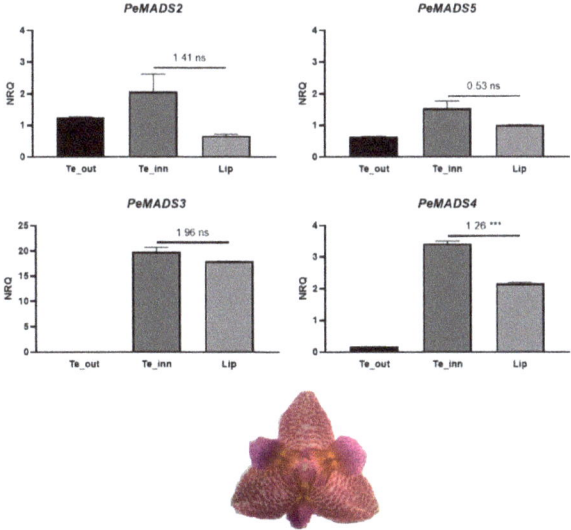

Figure 8. Relative expression of the class B MADS-box genes *PeMADS2–5* in the perianth of *Phalaenopsis* hyb. "Joy Fairy Tale" at the B2 developmental stage (bud size 1–1.5 cm). The expression is reported as normalized relative quantity (NRQ). The vertical bars represent the SEM of the biological and technical replicates. The numbers above the horizontal lines are the mean differences of the expression between lateral inner tepals and labellum-like structures (Te_inn-Lip). *p*-Values *** <0.001; ns, not significant. Te_out, outer tepals; Te_inn, labellum-like structures that substitute the lateral inner tepals in the peloric mutant. Note the different scale for *PeMADS3*.

2.5. Protein Interaction: Y2H Analysis

We used the yeast two-hybrid (Y2H) assay to determine if the proteins PeMADS2-PeMADS5 and PeDL2_1 can interact (Supplementary Figure S4). Our results show that the DEF-like proteins of *Phalaenopsis* do not directly interact with PeDL2_1. We also checked the ability of PeDL2_1 to bind the GLO protein PeMADS6, equally expressed in all the perianth organs [14], also revealing the absence of direct interaction (Supplementary Figure S4). In addition, we verified the ability of both the isoforms of PeDL1 and PeDL2 to interact with each other, showing the absence of direct interaction in all the possible combinations (Supplementary Figure S4). As a positive control of the Y2H experiments, we tested the ability of PeMADS2-PeMADS5 to interact with PeMADS6. The results confirm that PeMADS6 can interact with each of the DEF-like proteins of *Phalaenopsis*, although with different strengths, as previously reported (Supplementary Figure S5) [46].

2.6. Conserved Regulatory Motifs

To search for conserved motifs within the promoters of the *PeDL* genes, we analyzed the 3000 bp upstream of the translation start site of the *DL2* genes of *Phalaenopsis equestris* (*PeDL2*) and *Dendrobium catenatum* (*DcDL2*). The MEME analysis revealed motifs shared by the putative promoters of *PeDL2* and *DcDL2* (Figure 9). Two motifs (Motifs 1 and 3) have a relatively well-conserved position within the ~300 bp upstream of the translation start site. These motifs were not found when the analysis was repeated using the shuffled sequences of the putative promoters (Supplementary Figure S6) and are not present within the putative promoter of the *DL1* gene (Supplementary Table S1).

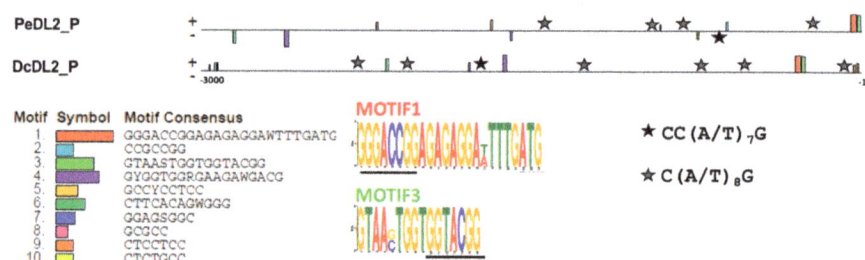

Figure 9. Conserved motifs within the putative promoters of the *DL2* genes of *P. equestris* and *D. catenatum*. PeDL2_P and DcDL2_P are the nucleotide sequences spanning 3000 bp upstream of the ATG translation start site, numbered from −1 to −3000. In the sequence logo of Motifs 1 and 3, the predicted binding site of the TCP factor (JASPAR IDs MA1096.1 and MA1035.1) and of the SBP-type zinc finger (JASPAR ID MA0955.1) are underlined. The black and gray stars indicate the CArG-box variants $CC(A/T)_7G$ and $C(A/T)_8G$, respectively.

The TOMTOM analysis of Motif 1 against the JASPAR Core Plants database shows that it contains a putative binding site for a TCP protein. The same analysis conducted on Motif 3 revealed that it contains a putative binding site for an SBP-type zinc-finger protein (Figure 9).

The search of known transcription factor binding sites (TFBSs) within the putative promoters of the *DL2* genes of *Phalaenopsis* and *Dendrobium* through PLANTPAN 3.0 [47] identified putative conserved elements belonging to different transcription factor families. For example, in addition to the TCP and SBP binding sites, AP2/ERF, MYB/SANT, and MADS-box binding sites (CArG-boxes) were identified.

The specific search of CArG-boxes gave positive results for the variants $CC(A/T)_7G$ and $C(A/T)_8G$. In particular, one $CC(A/T)_7G$ site is present in both the *PeDL2* and *DcDL2* putative promoters. In addition, four and six $C(A/T)_8G$ sites are located within the *PeDL2* and *DcDL2* promoters, respectively (Figure 9). One variant $CC(A/T)_7G$ and four $C(A/T)_8G$ CArG-boxes are also present within the putative promoter of *DcDL1*.

3. Discussion

Flower formation is the outcome of a complex developmental program in which environmental and genetic factors cooperate. The genetic pathway that drives the correct formation of the floral organs and the establishment of floral symmetry has been studied in detail in model species, where some transcription factor families play a relevant role, mainly MADS-box [11,12], TCP [21], MYB [29], and YABBY [38]. In orchids, the morphology of the flower organs and the establishment of bilateral floral symmetry have been widely studied, resulting in orchid-specific regulatory models where the coordinated action of MADS-box genes explains the formation of the orchid outer, lateral inner tepals, and labellum [13–19]. In the perspective of a broader, integrated view of these models, recent studies have suggested a possible involvement of TCP [26–28,35,48] and MYB [31,33] transcription factors in the developmental program leading to the formation of the orchid perianth, in particular of the labellum. In contrast, the possible involvement of the YABBY transcription factors in this developmental process is still unexplored. Based on these premises and the existence of a regulatory interaction between the YABBY transcription factor DL/CRC and the class B MADS domain transcription factors in rice (OsMADS16) [37,39] and maize (silky) [40], we tested the hypothesis of a similar regulatory relationship in orchids during the formation of the perianth organs, in particular of the labellum.

3.1. Paralogous DL-Like Genes in Orchidaceae

Our results support the identification of two *DL*-like genes in the genome of *P. equestris*: *PeDL1* and *PeDL2* [49]. These genes belong to the CRABS CLAW/DROOPING LEAF clade. Each of them is part of one of the sister clades resulting from an Orchidaceae-specific duplication early after the divergence of subfamilies Apostasioideae and Vanilloideae (Supplementary Figure S7). Our results agree with the finding that *PeDL1* and *PeDL2* are expressed in the column and ovary of *Phalaenopsis* (Figure 3) [49]. This expression profile suggests that like in *Oryza*, *Zea*, *Triticum*, *Sorghum*, and *Arabidopsis*, *PeDL1* and *PeDL2* are also involved in carpel development [49–51].

3.2. Different Transcripts of DL-Like Genes

We found two differentially spliced transcripts of the *PeDL1* and *PeDL2* genes of *Phalaenopsis*, differing at the 5′ terminus (Figure 2) and encoding proteins completely (PeDL1_2) or partially (PeDL2_2) missing the C2C2 zinc-finger domain (Supplementary Figure S2). Although we scanned the transcriptomes of various orchid species present in the orchid-specific database Orchidstra 2.0 [52] and OrchidBase 2.0 [53], we did not find similar alternative short transcripts of the *DL* homolog genes. Our initial in silico identification of the *PeDL1_2* and *PeDL2_2* isoforms was verified by PCR, sequencing, and real-time PCR experiments using isoform-specific primers. Our results confirmed the existence of differentially spliced isoforms for both *PeDL* genes. The failure to find alternative transcripts of *DLs* in other orchids might be due to the kind of transcriptomes deposited in the orchid-specific database. This data generally represents transcripts of the whole inflorescence, with possible under-representation of isoforms expressed specifically in few types of cells or organs. Outside orchids, we found the annotation of two isoforms of both the DL genes of *Zea mays drl1* (https://maizegdb.org/gene_center/gene/GRMZM2G088309, access date 18 January 2021) and *drl2* (https://www.maizegdb.org/gene_center/gene/GRMZM2G102218, access date 18 January 2021). In particular, the predicted alternative isoform of the *drl2* gene encodes a short protein missing the C2C2 zinc-finger domain, as in *Phalaenopsis*. Unfortunately, functional or expression data for the *drl* isoforms are not available, and their role is still unknown. Further analyses are needed to assess the function of the truncated isoforms that might work as competitive inhibitors and thus have a regulatory function.

In *Arabidopsis*, YABBY proteins form homo and heterodimers [54]. In particular, the CRC protein forms homodimers and can interact with the YABBY protein INO [55]. In contrast to CRC, our results indicate that PeDL1, PeDL2, and their short isoforms, form neither homo- nor heterodimers (Supplementary Figure S4), showing that the ability of CRC/DL proteins to homo- and heterodimerize is not conserved among plants. This unexpected result is in agreement with that reported in a recent study on the *DL*-like genes of *Phalaenopsis* [56] and might be due to sequence divergence after duplication, resulting in the loss of the ability to form homo- and heterodimers.

3.3. Divergent Patterns of Expression of PeDL1 and PeDL2 during Flower Development

Interestingly, the expression of the two *PeDL* genes in the perianth organs of wild-type *Phalaenopsis* is not overlapping. In contrast to very low expression levels of *PeDL1* in all perianth organs from early to late floral buds, *PeDL2* has a higher level of expression in the labellum than in outer and lateral inner tepals. This trend decreases steadily towards anthesis (Figure 3). Considering that the expression of *DL* in *O. sativa* is restricted to the flower meristem and developing carpels, the expression of *PeDL2* in the perianth is unusual for a *DL*-like gene, and is the first evidence of a possible novel regulatory function acquired by these genes after duplication early in orchid evolution. Our hypothesis of the recruitment of *PeDL2* in orchid perianth development is supported by the expression pattern of the gene in orchid peloric mutants where the inner tepals are substituted by labellum-like structures. In the peloric *Phalaenopsis* "Athens" (Figure 1d), early expression of both *PeDL2* isoforms increases in labellum-like inner lateral tepals compared to the wild type (Figure 5). In addition, the peloric *Phalaenopsis* "Joy Fairy Tale" (Figure 1e) shows similar expression of *PeDL2* in labellum-like structures and labellum (Figure 6). These results support the relationship between the combinatory expression of *PeDL2*, *PeMADS3*, and *PeMADS4* transcripts and labellum development.

3.4. The "Orchid Code" beyond MADS

The idea of an "orchid code" enriched by the function of *PeDL2* during the labellum development fully fits with the regulatory profile of the other well-known components of this model: the *DEF*-like MADS-box genes *PeMADS2-PeMADS5*. In wild-type *Phalaenopsis*, the expression in the perianth of *PeDL2* has a similar pattern in the labellum and lateral outer tepals as *PeMADS3* and *PeMADS4* and is opposite to that of *PeMADS2* and *PeMADS5*.

The transcription patterns of *PeDL2* and *PeMADS2-PeMADS5* in wild-type and peloric *Phalaenopsis* allow us to suggest that during the formation of the labellum there could be regulatory interactions between PeMADS2-PeMADS5 and PeDL2, based on different possible molecular mechanisms: either PeMADS proteins bind to regulatory DNA of the *PeDL2* gene (i.e., protein–DNA interactions), or PeMADS and PeDL2 proteins interact (protein–protein interactions) (Figure 10). Although our results from the Y2H analysis do not reveal the ability of PeDL2 to bind any of the PeMADS2-PeMADS5 proteins, a direct protein–protein interaction cannot be definitely excluded, as it could require the formation of a multimeric protein complex.

Alternatively, the regulatory interaction between PeDL2 and PeMADS2-PeMADS5 might be carried out at the transcriptional level, with PeMADS2/5 functioning as transcriptional repressors or PeMADS3/4 as transcriptional activators of the *PeDL2* gene. The MADS-box proteins are transcription factors that bind conserved sites on DNA with the consensus sequence $CC(A/T)_6GG$, the canonical CArG-box, or its variants [57]. The presence of the variant CArG-boxes $CC(A/T)_7G$ and $C(A/T)_8G$ (Figure 9), known transcription factor binding sites of MADS-domain proteins, in multiple sites of the putative promoter of *DL2* of *Phalaenopsis* and *Dendrobium* suggests that orchid DEF-like proteins or other MIKC-MADS domain genes could regulate the transcription of *DL2*. The presence of multiple CArG-boxes even strongly suggests that tetrameric complexes of MIKC-type proteins ("floral quartets") are constituted [19,58]. In addition, the presence of shared transcription factor binding sites of TCP and SBP proteins, conserved in sequence and spatial organization in the putative promoters of *DL2*, suggests that other transcription factors could also modulate the expression of this gene to the expression domains here documented.

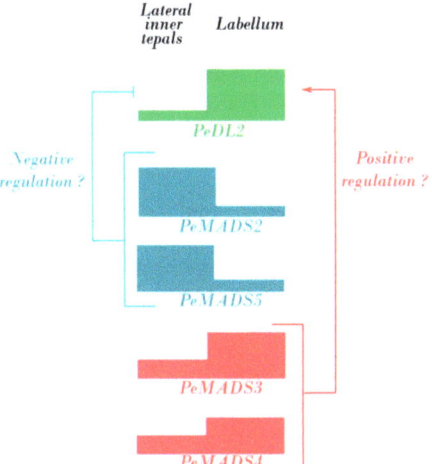

Figure 10. Possible regulatory interaction between PeDL2 and PeMADS2-PeMADS5 during the formation of the labellum of wild-type *Phalaenopsis*.

3.5. Conclusions

The molecular basis of orchid flower development is only partially understood. The main components of the orchid "toolkit for beauty" are MADS-box transcription factors; however, other transcription factor families (TCP and MYB) contribute to the differentiation of the organs of the orchid perianth. Our study proposes further expanding this complex developmental program, including the YABBY *PeDL2* of *Phalaenopsis* among the genes responsible for the labellum differentiation. Future studies should be focused on understanding the way of interaction among the different players of this fascinating developmental program to shed light on their regulatory connections.

4. Materials and Methods
4.1. Plant Material

The orchids used in this study were the wild-type *Phalaenopsis* hyb. "Athens" and *P. aphrodite* and the peloric mutants *Phalaenopsis* hyb. "Athens" and *Phalaenopsis* hyb. "Joy Fairy Tale". All the plants were grown under natural light and temperature in the greenhouse of the Department of Biology (University of Naples Federico II, Napoli, Italy) or of the Department of Cell Biology and Plant Biochemistry (University of Regensburg, Regensburg, Germany).

The wild-type *Phalaenopsis* has the second floral whorl clearly distinguished into two lateral inner tepals and one median inner tepal (labellum or lip) (Figure 1a–c). Both peloric mutants have two labellum-like organs in substitution of the lateral inner tepals (Figure 1d,e).

Single flowers from three different plants of the wild-type *P. aphrodite* were collected before anthesis at different developmental stages: B1 (bud length 0.5–1 cm), B2 (1–1.5 cm), B3 (1.5–2 cm), B4 (2–2.5 cm), and B5 (2.5–3 cm) (Figure 1a,b). Open flowers (OFs) were collected soon after anthesis (Figure 1a,b). Single flowers of six wild-type *Phalaenopsis* hyb. "Athens" and of the peloric mutants were collected at developmental stage B2.

The perianth tissues (outer tepals, lateral inner tepals, and labellum) of all the collected flowers at the different developmental stages as well as the column and ovary were dissected and immediately frozen in liquid nitrogen or immersed in RNA*later* (Ambion, Austin, TX, USA) and stored at $-80\,°C$ until RNA extraction.

4.2. In Silico Identification of the PeDL1 and PeDL2 Genes

Total RNA was extracted from inner lateral tepals and labellum of wild-type and labellum-like lateral inner tepals from peloric *Phalaenopsis* hyb. "Athens" collected from 3 individual plants at the B2 developmental stage using Trizol (Ambion) followed by DNase treatment. After extraction, RNA was analyzed with the 2100 Bioanalyzer system (Agilent Technologies, Santa Clara, CA, USA) for sizing, quantitation, and quality control. Samples between 1 and 1.5 µg with an RIN (RNA integrity number) between 8.5 and 9.0 were sequenced by Macrogen (Seoul, Korea). Illumina TruSeq RNA (Oligo dT) mate paired-end libraries were generated and individually sequenced in a lane with a coverage >150 million 100 bp pair-end reads. For each sample 220 million paired-end reads were obtained. Analysis with FastQC showed that 94% of them had a quality score over 30. Trimming and mapping to the *Phalaenopsis equestris* genome v 1.0 (ASM126359v1) were carried out with the CLC Genomics Workbench (v11.01).

The Illumina raw reads of wild-type and peloric mutant *Phalaenopsis* hyb. "Brother Spring Dancer" KHM190 [59] were downloaded from the Sequence Read Archive. Paired-end reads from wild-type and peloric mutant outer tepal (accession numbers SRR1055198 and SRR1055947), inner tepal (SRR1055945 and SRR1055948), and labellum (SRR1055946 and SRR1055949) were assembled using the Trinity v2.3.0 software [60]. The Annocript v2.0.1 software [61] was used to obtain the functional annotation of the transcripts, and differential gene expression analysis between wild-type and peloric mutant tissues was performed with the edgeR v3.13 software [62].

The genomic organization of the *PeDL1* and *PeDL2* genes was reconstructed through BLAST analyses against the genome of *P. equestris* (assembly ASM126359v1), using as query the nucleotide sequence of the *DL*-like transcripts present in the transcriptome of *Phalaenopsis* hyb. "Brother Spring Dancer".

4.3. Quantitative Expression Analysis

Total RNA was extracted from the tissues collected at the different developmental stages using Trizol (Ambion) followed by DNase treatment. After RNA extraction and quantification, equal amounts of total RNA were pooled, producing two pools for each tissue, each made of three different RNAs. Then, 500 ng of total RNA from each pool were reverse-transcribed using the Advantage RT-PCR kit (Clontech, Mountain View, CA, USA) and a mix of oligo dT and random hexamer primers.

The nucleotide sequences of the *PeDL1* and *PeDL2* transcripts and of their alternatively spliced isoforms identified by in silico analysis were verified through PCR amplification of the cDNA of *Phalaenopsis* hyb. "Athens" using gene- and isoform-specific primer pairs (Supplementary Table S2). The amplification products were cloned into pSC-A-amp/kan vector (Agilent Technologies, Santa Clara, CA, USA) and sequenced using the T3 and T7 primers (Eurofins Genomics, Ebersberg, Germany). The nucleotide sequences were deposited in GenBank with the following accession numbers: MW574592 (*PeDL1_1*), MW574593 (*PeDL1_2*), MW574594 (*PeDL2_1*), MW574595 (*PeDL2_2*).

Relative expression of *PeDL1* (two isoforms: *PeDL1_1* and *PeDL1_2*), *PeDL2* (two isoforms: *PeDL2_1* and *PeDL2_2*), *PeMADS2*, *PeMADS5*, *PeMADS3*, and *PeMADS4* was evaluated in all the collected tissues by qPCR experiments, using 18S, *Actin*, and *Elongation Factor 1α* as reference genes, as previously described [20]. The gene- and isoform-specific primer pairs used are listed in the Supplementary Table S2. At least one primer for each pair was constructed spanning two adjacent exons (Figure 2). The reactions were conducted in technical triplicates. Normalized relative quantity (NRQ) ± SEM was calculated for each replicate to the geometric average expression of three internal control genes [20].

ANOVA analysis followed by Holm–Sidak post-hoc test was performed to assess the statistical significance of the differences of NRQ among the different tissues.

4.4. Yeast Two-Hybrid Analysis

The GAL4-based yeast two-hybrid (Y2H) system (Matchmaker two-hybrid system; Clontech) was used to analyze protein–protein interactions between PeDL2_1 and PeMADS2-PeMADS6, and between the different isoforms of PeDL1 and PeDL2. As positive control, Y2H analysis was used to check the ability of PeMADS6 to form heterodimers with PeMADS2-PeMADS5.

The full-length coding regions of *PeDL1_1* (MW574592), PeDL1_2 (MW574593), PeDL2_1 (MW574594), PeDL2_2 (MW574595), *PeMADS2* (AY378149), *PeMADS3* (AY378150), *PeMADS4* (AY378147), *PeMADS5* (AY378148), and *PeMADS6* (AY678299) were amplified by PCR using the primer pairs listed in Supplementary Table S2 and sub-cloned into the yeast expression vectors pGADT7 (prey) and pGBKT7 (bait) from the MATCHMAKER two-hybrid system 3 (Clontech), in frame with the sequence of either the transcription-activating (AD) or DNA-binding domains (BD) of the transcription factor GAL4. *Saccharomyces cerevisiae* strain AH109 was transformed with all the prey and bait recombinant vector combinations [63], conducting each experiment in triplicate.

Plasmid presence after double yeast transformations was verified by growing cells in Synthetic Defined (SD) medium lacking tryptophan and leucine. Protein–protein interaction was tested in SD medium lacking tryptophan, leucine, and histidine. Possible transcriptional activation activity of PeDLs and PeMADS2–6 proteins fused to the binding domain of GAL4 (pGBKT7 vector) was checked by monitoring the growth of yeast transformed cells in SD medium without histidine, in the presence of 20 mM 3-aminotriazole (3AT). Empty vectors pGBKT7 or pGADT7 were transformed in combination with the recombinant vectors as negative controls.

4.5. Identification of Conserved Motifs

Nucleotide sequences (3000 bp) upstream of the *PeDL2* gene of *P. equestris* were downloaded, as were the 3000 bp upstream of the *DcDL2* and *DcDL1* of *D. catenatum*. Unfortunately, it was impossible to download the sequence upstream of the *PeDL1* gene because the genomic scaffold starts with the coding sequence of this gene.

Putative promoter sequences were scanned for the presence of conserved motifs using the online tool MEME v5.3.3 [64] with the following parameters: motif width between 5 and 25, one occurrence of motif per sequence, and the maximum number of motifs 10. The search was repeated with the same parameters on the shuffled sequences as negative control. The identified conserved motifs were then checked against the JASPAR2020 Core Plants database (http://jaspar.genereg.net/, access date 18 January 2021) through TOMTOM v5.3.3 [65].

The search of known transcription factor binding sites within the putative promoters was conducted in PLANTPAN 3.0 [47]. In addition, using the FUZZNUC software (http://emboss.bioinformatics.nl/cgi-bin/emboss/fuzznuc), the putative promoters were scanned for the presence of perfect CArG-boxes CC(A/T)6GG and for the variants CC(A/T)7G and C(A/T)8G.

Supplementary Materials: Supplementary materials can be found at https://www.mdpi.com/article/10.3390/ijms22137025/s1. Figure S1. Transcripts expressed in perianth organs of wild-type (WT) and peloric *Phalaenopsis* hyb. "Athens" with at least 1 TPM. Figure S2. Amino acid alignment of the different isoforms of the PeDL1 and PeDL2 proteins of *P. equestris*. Figure S3. Relative expression of the isoforms *PeDL1_1* and *PeDL1_2* in the perianth of the wild-type (a) and peloric (b) *Phalaenopsis* hyb. "Athens" and of the peloric *Phalaenopsis* hyb. "Joy Fairy Tale" (c) at the B2 developmental stage (bud size 1–1.5 cm). Figure S4. Interactions of the PeDL2_1 and PeMADS2–6 (left) and the different isoforms of PeDL1/2 (right) of *Phalaenopsis* in yeast two-hybrid analysis. Figure S5. Interactions of PeMADS2-PeMADS5 and PeMAD6 of *Phalaenopsis* in yeast two-hybrid analysis. Figure S6. Conserved motifs within the shuffled sequences of the putative promoters of the *DL2* genes of *P. equestris* and *D. catenatum*. Figure S7. Neighbor joining tree of the CRC/DL proteins. Table S1. Conserved motifs of the putative promoters of the *PeDL2* and *DcDL2* genes found within the putative promoter of the *DcDL1* gene. Table S2. List of the primer sequences used.

Data S1. Differentially expressed transcripts (FDR < 0.05) between lateral inner tepals and labellum of wild-type *Phalaenopsis* hyb. Data S2. Selected differentially expressed transcripts (FDR < 0.05) between labellum and inner tepals of wild-type and peloric *Phalaenopsis* hyb.

Author Contributions: Conceptualization, M.M.-P. and S.A.; methodology, F.L., M.C.V. and S.N.; validation, F.L. and M.C.V.; formal analysis, M.M.-P., S.A., G.T., S.N., F.L. and M.C.V.; writing—original draft preparation, M.M.-P. and S.A.; supervision, M.M.-P. and S.A.; funding acquisition, S.A. All authors have read and agreed to the published version of the manuscript.

Funding: This research was funded by Università degli Studi di Napoli Federico II, grant number 000020-2019.

Institutional Review Board Statement: Not applicable.

Informed Consent Statement: Not applicable.

Data Availability Statement: The data presented in this study are openly available in NCBI (https://www.ncbi.nlm.nih.gov/) with the following accession numbers: PeDL1_1 (MW574592), PeDL1_2 (MW574593), PeDL2_1 (MW574594), PeDL2_2 (MW574595), PeMADS2 (AY378149), PeMADS3 (AY378150), PeMADS4 (AY378147), PeMADS5 (AY378148), PeMADS6 (AY678299), Phalaenopsis equestris genome v 1.0 (ASM126359v1), reads from wild-type and peloric mutant outer tepal (accession numbers SRR1055198 and SRR1055947), Phalaenopsis hyb. "Brother Spring Dancer" KHM190 Illumina reads of inner tepal (SRR1055945, wild-type, and SRR1055948, peloric), and labellum (SRR1055946, wild-type, and SRR1055949, peloric).

Conflicts of Interest: The authors declare no conflict of interest.

References

1. Aceto, S.; Gaudio, L. The MADS and the Beauty: Genes Involved in the Development of Orchid Flowers. *Curr. Genom.* **2011**, *12*, 342–356. [CrossRef]
2. Cozzolino, S.; Widmer, A. Orchid diversity: An evolutionary consequence of deception? *Trends Ecol. Evol.* **2005**, *20*, 487–494. [CrossRef]
3. Tremblay, R.L.; Ackerman, J.D.; Zimmerman, J.K.; Calvo, R.N. Variation in sexual reproduction in orchids and its evolutionary consequences: A spasmodic journey to diversification. *Biol. J. Linn. Soc.* **2005**, *84*, 1–54. [CrossRef]
4. Rudall, P.J.; Bateman, R.M. Roles of synorganisation, zygomorphy and heterotopy in floral evolution: The gynostemium and labellum of orchids and other lilioid monocots. *Biol. Rev.* **2002**, *77*, 403–441. [CrossRef] [PubMed]
5. Bateman, R.M.; Rudall, P.J. The good, the bad and the ugly: Using naturally occurring terata to distinguish the possible from the impossible in orchid floral evolution. In *Monocots: Comparative Biology and Evolution. Excluding Poales*; Columbus, J.T., Friar, E.A., Porter, J.M., Prince, L.M., Simpson, M.G., Eds.; Rancho Santa Ana Botanical Garden: Claremont, CA, USA, 2006; Volume I, pp. 481–496.
6. Kocyan, A.; Endress, P.K. Floral structure and development and systematic aspects of some 'lower' Asparagales. *Plant Syst. Evol.* **2001**, *229*, 187–216. [CrossRef]
7. Burnsbalogh, P.; Bernhardt, P. Evolutionary Trends in the Androecium of the Orchidaceae. *Plant Syst. Evol.* **1985**, *149*, 119–134. [CrossRef]
8. Darwin, C. *On the Various Contrivances by which British and Foreign Orchids Are Fertilised by Insects*; Murray: London, UK, 1862.
9. Wordsell, W.C. *Principles of Plant Teratology*; The Ray Society: London, UK, 1916.
10. Endress, P.K. *Diversity and Evolutionary Biology of Tropical Flowers*; Cambridge Univeristy Press: Cambridge, UK, 1994.
11. Bowman, J.L.; Smyth, D.R.; Meyerowitz, E.M. Genetic Interactions among Floral Homeotic Genes of Arabidopsis. *Development* **1991**, *112*, 1–20. [CrossRef]
12. Coen, E.S.; Meyerowitz, E.M. The War of the Whorls-Genetic Interactions Controlling Flower Development. *Nature* **1991**, *353*, 31–37. [CrossRef]
13. Mondragon-Palomino, M.; Theissen, G. Why are orchid flowers so diverse? Reduction of evolutionary constraints by paralogues of class B floral homeotic genes. *Ann. Bot.-Lond.* **2009**, *104*, 583–594. [CrossRef]
14. Mondragon-Palomino, M.; Theissen, G. Conserved differential expression of paralogous DEFICIENS- and GLOBOSA-like MADS-box genes in the flowers of Orchidaceae: Refining the 'orchid code'. *Plant J.* **2011**, *66*, 1008–1019. [CrossRef]
15. Mondragon-Palomino, M.; Theissen, G. MADS about the evolution of orchid flowers. *Trends Plant Sci.* **2008**, *13*, 51–59. [CrossRef]
16. Pan, Z.J.; Cheng, C.C.; Tsai, W.C.; Chung, M.C.; Chen, W.H.; Hu, J.M.; Chen, H.H. The duplicated B-class MADS-box genes display dualistic characters in orchid floral organ identity and growth. *Plant Cell Physiol.* **2011**, *52*, 1515–1531. [CrossRef] [PubMed]
17. Chang, Y.Y.; Kao, N.H.; Li, J.Y.; Hsu, W.H.; Liang, Y.L.; Wu, J.W.; Yang, C.H. Characterization of the possible roles for B class MADS box genes in regulation of perianth formation in orchid. *Plant Physiol.* **2010**, *152*, 837–853. [CrossRef] [PubMed]

18. Su, C.L.; Chen, W.C.; Lee, A.Y.; Chen, C.Y.; Chang, Y.C.; Chao, Y.T.; Shih, M.C. A modified ABCDE model of flowering in orchids based on gene expression profiling studies of the moth orchid Phalaenopsis aphrodite. *PLoS ONE* **2013**, *8*, e80462. [CrossRef] [PubMed]
19. Hsu, H.F.; Hsu, W.H.; Lee, Y.I.; Mao, W.T.; Yang, J.Y.; Li, J.Y.; Yang, C.H. Model for perianth formation in orchids. *Nat. Plants* **2015**, *1*. [CrossRef]
20. Acri-Nunes-Miranda, R.; Mondragón Palomino, M. Expression of paralogous *SEP*-, *FUL*-, *AG*- and *STK*-like MADS-box genes in wild-type and peloric *Phalaenopsis* flowers. *Front. Plant Sci.* **2014**, 1–17. [CrossRef]
21. Luo, D.; Carpenter, R.; Copsey, L.; Vincent, C.; Clark, J.; Coen, E. Control of organ asymmetry in flowers of Antirrhinum. *Cell* **1999**, *99*, 367–376. [CrossRef]
22. Feng, X.Z.; Zhao, Z.; Tian, Z.X.; Xu, S.L.; Luo, Y.H.; Cai, Z.G.; Wang, Y.M.; Yang, J.; Wang, Z.; Weng, L.; et al. Control of petal shape and floral zygomorphy in Lotus japonicus. *Proc. Natl. Acad. Sci. USA* **2006**, *103*, 4970–4975. [CrossRef]
23. Busch, A.; Zachgo, S. Control of corolla monosymmetry in the Brassicaceae Iberis amara. *Proc. Natl. Acad. Sci. USA* **2007**, *104*, 16714–16719. [CrossRef]
24. Broholm, S.K.; Tahtiharju, S.; Laitinen, R.A.E.; Albert, V.A.; Teeri, T.H.; Elomaa, P. A TCP domain transcription factor controls flower type specification along the radial axis of the Gerbera (Asteraceae) inflorescence. *Proc. Natl. Acad. Sci. USA* **2008**, *105*, 9117–9122. [CrossRef]
25. Zhang, W.H.; Kramer, E.M.; Davis, C.C. Floral symmetry genes and the origin and maintenance of zygomorphy in a plant-pollinator mutualism. *Proc. Natl. Acad. Sci. USA* **2010**, *107*, 6388–6393. [CrossRef] [PubMed]
26. De Paolo, S.; Gaudio, L.; Aceto, S. Analysis of the TCP genes expressed in the inflorescence of the orchid Orchis italica. *Sci. Rep.* **2015**, *5*, 16265. [CrossRef] [PubMed]
27. Lin, Y.F.; Chen, Y.Y.; Hsiao, Y.Y.; Shen, C.Y.; Hsu, J.L.; Yeh, C.M.; Mitsuda, N.; Ohme-Takagi, M.; Liu, Z.J.; Tsai, W.C. Genome-wide identification and characterization of TCP genes involved in ovule development of Phalaenopsis equestris. *J. Exp. Bot.* **2016**, *67*, 5051–5066. [CrossRef] [PubMed]
28. Madrigal, Y.; Alzate, J.F.; Pabon-Mora, N. Evolution and Expression Patterns of TCP Genes in Asparagales. *Front. Plant Sci.* **2017**, *8*. [CrossRef]
29. Raimundo, J.; Sobral, R.; Bailey, P.; Azevedo, H.; Galego, L.; Almeida, J.; Coen, E.; Costa, M.M.R. A subcellular tug of war involving three MYB-like proteins underlies a molecular antagonism in Antirrhinum flower asymmetry. *Plant J.* **2013**, *75*, 527–538. [CrossRef]
30. Raimundo, J.; Sobral, R.; Laranjeira, S.; Costa, M.M.R. Successive Domain Rearrangements Underlie the Evolution of a Regulatory Module Controlled by a Small Interfering Peptide. *Mol. Biol. Evol.* **2018**, *35*, 2873–2885. [CrossRef]
31. Valoroso, M.C.; De Paolo, S.; Iazzetti, G.; Aceto, S. Transcriptome-Wide Identification and Expression Analysis of DIVARICATA- and RADIALIS-Like Genes of the Mediterranean Orchid Orchis italica. *Genome Biol. Evol.* **2017**, *9*. [CrossRef]
32. Lucibelli, F.; Valoroso, M.C.; Aceto, S. Radial or Bilateral? The Molecular Basis of Floral Symmetry. *Genes-Basel* **2020**, *11*, 395. [CrossRef]
33. Valoroso, M.C.; Sobral, R.; Saccone, G.; Salvemini, M.; Costa, M.M.R.; Aceto, S. Evolutionary Conservation of the Orchid MYB Transcription Factors DIV, RAD, and DRIF. *Front. Plant Sci.* **2019**, *10*. [CrossRef]
34. Costa, M.M.R.; Fox, S.; Hanna, A.I.; Baxter, C.; Coen, E. Evolution of regulatory interactions controlling floral asymmetry. *Development* **2005**, *132*, 5093–5101. [CrossRef] [PubMed]
35. Mondragon-Palomino, M.; Trontin, C. High time for a roll call: Gene duplication and phylogenetic relationships of TCP-like genes in monocots. *Ann. Bot.* **2011**, *107*, 1533–1544. [CrossRef] [PubMed]
36. Nakayama, H.; Yamaguchi, T.; Tsukaya, H. Expression patterns of AaDL, a CRABS CLAW ortholog in Asparagus asparagoides (Asparagaceae), demonstrate a stepwise evolution of CRC/DL subfamily of YABBY genes. *Am. J. Bot.* **2010**, *97*, 591–600. [CrossRef] [PubMed]
37. Yamaguchi, T.; Nagasawa, N.; Kawasaki, S.; Matsuoka, M.; Nagato, Y.; Hirano, H.Y. The YABBY gene DROOPING LEAF regulates carpel specification and midrib development in Oryza sativa. *Plant Cell* **2004**, *16*, 500–509. [CrossRef] [PubMed]
38. Sawa, S.; Ito, T.; Shimura, Y.; Okada, K. FILAMENTOUS FLOWER controls the formation and development of Arabidopsis inflorescences and floral meristems. *Plant Cell* **1999**, *11*, 69–86. [CrossRef] [PubMed]
39. Nagasawa, N.; Miyoshi, M.; Sano, Y.; Satoh, H.; Hirano, H.; Sakai, H.; Nagato, Y. SUPERWOMAN1 and DROOPING LEAF genes control floral organ identity in rice. *Development* **2003**, *130*, 705–718. [CrossRef] [PubMed]
40. Strable, J.; Vollbrecht, E. Maize YABBY genes drooping leaf1 and drooping leaf2 regulate floret development and floral meristem determinacy. *Development* **2019**, *146*. [CrossRef]
41. Cantone, C.; Gaudio, L.; Aceto, S. The PI/GLO-like locus in orchids: Duplication and purifying selection at synonymous sites within Orchidinae (Orchidaceae). *Gene* **2011**, *481*, 48–55. [CrossRef]
42. Cantone, C.; Sica, M.; Gaudio, L.; Aceto, S. The OrcPI locus: Genomic organization, expression pattern, and noncoding regions variability in Orchis italica (Orchidaceae) and related species. *Gene* **2009**, *434*, 9–15. [CrossRef]
43. Cai, J.; Liu, X.; Vanneste, K.; Proost, S.; Tsai, W.C.; Liu, K.W.; Chen, L.J.; He, Y.; Xu, Q.; Bian, C.; et al. The genome sequence of the orchid Phalaenopsis equestris. *Nat. Genet.* **2015**, *47*, 65–72. [CrossRef]
44. Huang, J.Z.; Lin, C.P.; Cheng, T.C.; Chang, B.C.H.; Cheng, S.Y.; Chen, Y.W.; Lee, C.Y.; Chin, S.W.; Chen, F.C. A De Novo Floral Transcriptome Rev. eals Clues into Phalaenopsis Orchid Flower Development. *PLoS ONE* **2015**, *10*. [CrossRef]

45. Tsai, W.C.; Kuoh, C.S.; Chuang, M.H.; Chen, W.H.; Chen, H.H. Four DEF-like MADS box genes displayed distinct floral morphogenetic roles in Phalaenopsis orchid. *Plant Cell Physiol.* **2004**, *45*, 831–844. [CrossRef]
46. Tsai, W.C.; Pan, Z.J.; Hsiao, Y.Y.; Jeng, M.F.; Wu, T.F.; Chen, W.H.; Chen, H.H. Interactions of B-class complex proteins involved in tepal development in Phalaenopsis orchid. *Plant Cell Physiol.* **2008**, *49*, 814–824. [CrossRef]
47. Chow, C.N.; Lee, T.Y.; Hung, Y.C.; Li, G.Z.; Tseng, K.C.; Liu, Y.H.; Kuo, P.L.; Zheng, H.Q.; Chang, W.C. PlantPAN3.0: A new and updated resource for reconstructing transcriptional regulatory networks from ChIP-seq experiments in plants. *Nucleic Acids Res.* **2019**, *47*, D1155–D1163. [CrossRef] [PubMed]
48. Chen, Y.H.; Tsai, Y.J.; Huang, J.Z.; Chen, F.C. Transcription analysis of peloric mutants of Phalaenopsis orchids derived from tissue culture. *Cell Res.* **2005**, *15*, 639–657. [CrossRef] [PubMed]
49. Chen, Y.Y.; Hsiao, Y.Y.; Chang, S.B.; Zhang, D.; Lan, S.R.; Liu, Z.J.; Tsai, W.C. Genome-Wide Identification of YABBY Genes in Orchidaceae and Their Expression Patterns in Phalaenopsis Orchid. *Genes* **2020**, *11*, 955. [CrossRef]
50. Ishikawa, M.; Ohmori, Y.; Tanaka, W.; Hirabayashi, C.; Murai, K.; Ogihara, Y.; Yamaguchi, T.; Hirano, H.Y. The spatial expression patterns of DROOPING LEAF orthologs suggest a conserved function in grasses. *Genes Genet. Syst.* **2009**, *84*, 137–146. [CrossRef]
51. Alvarez, J.; Smyth, D.R. CRABS CLAW and SPATULA, two Arabidopsis genes that control carpel development in parallel with AGAMOUS. *Development* **1999**, *126*, 2377–2386.
52. Chao, Y.T.; Yen, S.H.; Yeh, J.H.; Chen, W.C.; Shih, M.C. Orchidstra 2.0-A Transcriptomics Resource for the Orchid Family. *Plant Cell Physiol.* **2017**, *58*. [CrossRef]
53. Tsai, W.C.; Fu, C.H.; Hsiao, Y.Y.; Huang, Y.M.; Chen, L.J.; Wang, M.; Liu, Z.J.; Chen, H.H. OrchidBase 2.0: Comprehensive collection of Orchidaceae floral transcriptomes. *Plant Cell Physiol.* **2013**, *54*, e7. [CrossRef] [PubMed]
54. Stahle, M.I.; Kuehlich, J.; Staron, L.; von Arnim, A.G.; Golz, J.F. YABBYs and the transcriptional corepressors LEUNIG and LEUNIG_HOMOLOG maintain leaf polarity and meristem activity in Arabidopsis. *Plant Cell* **2009**, *21*, 3105–3118. [CrossRef]
55. Gross, T.; Broholm, S.; Becker, A. CRABS CLAW Acts as a Bifunctional Transcription Factor in Flower Development. *Front. Plant Sci.* **2018**, *9*, 835. [CrossRef] [PubMed]
56. Chen, Y.Y.; Hsiao, Y.Y.; Li, C.I.; Yeh, C.M.; Mitsuda, N.; Yang, H.X.; Chiu, C.C.; Chang, S.B.; Liu, Z.J.; Tsai, W.C. The ancestral duplicated DL/CRC orthologs, PeDL1 and PeDL2, function in orchid reproductive organ innovation. *J. Exp. Bot.* **2021**. [CrossRef]
57. Aerts, N.; de Bruijn, S.; van Mourik, H.; Angenent, G.C.; van Dijk, A.D.J. Comparative analysis of binding patterns of MADS-domain proteins in Arabidopsis thaliana. *BMC Plant Biol.* **2018**, *18*, 131. [CrossRef]
58. Theissen, G.; Melzer, R.; Rumpler, F. MADS-domain transcription factors and the floral quartet model of flower development: Linking plant development and evolution. *Development* **2016**, *143*, 3259–3271. [CrossRef] [PubMed]
59. Lee, C.Y.; Viswanath, K.K.; Huang, J.Z.; Lee, C.P.; Lin, C.P.; Cheng, T.C.; Chang, B.C.; Chin, S.W.; Chen, F.C. PhalDB:A comprehensive database for molecular mining of the Phalaenopsis genome, transcriptome and miRNome. *Genet. Mol. Res.* **2018**, *17*, gmr18051. [CrossRef]
60. Grabherr, M.G.; Haas, B.J.; Yassour, M.; Levin, J.Z.; Thompson, D.A.; Amit, I.; Adiconis, X.; Fan, L.; Raychowdhury, R.; Zeng, Q.; et al. Full-length transcriptome assembly from RNA-Seq data without a reference genome. *Nat. Biotechnol.* **2011**, *29*, 644–652. [CrossRef] [PubMed]
61. Musacchia, F.; Basu, S.; Petrosino, G.; Salvemini, M.; Sanges, R. Annocript: A flexible pipeline for the annotation of transcriptomes able to identify putative long noncoding RNAs. *Bioinformatics* **2015**, *31*, 2199–2201. [CrossRef]
62. Robinson, M.D.; McCarthy, D.J.; Smyth, G.K. edgeR: A Bioconductor package for differential expression analysis of digital gene expression data. *Bioinformatics* **2010**, *26*, 139–140. [CrossRef]
63. Gietz, R.D.; Schiestl, R.H.; Willems, A.R.; Woods, R.A. Studies on the Transformation of Intact Yeast-Cells by the Liac/S-DNA/Peg Procedure. *Yeast* **1995**, *11*, 355–360.
64. Bailey, T.L.; Boden, M.; Buske, F.A.; Frith, M.; Grant, C.E.; Clementi, L.; Ren, J.Y.; Li, W.W.; Noble, W.S. MEME SUITE: Tools for motif discovery and searching. *Nucleic Acids Res.* **2009**, *37*, W202–W208. [CrossRef]
65. Gupta, S.; Stamatoyannopoulos, J.A.; Bailey, T.L.; Noble, W.S. Quantifying similarity between motifs. *Genome Biol.* **2007**, *8*. [CrossRef] [PubMed]

Review

Evolution of Terpene Synthases in Orchidaceae

Li-Min Huang [1], Hsin Huang [1], Yu-Chen Chuang [1], Wen-Huei Chen [1,2], Chun-Neng Wang [3] and Hong-Hwa Chen [1,2,*]

[1] Department of Life Sciences, National Cheng Kung University, Tainan 701, Taiwan; limin925@gmail.com (L.-M.H.); n34545@gmail.com (H.H.); faseno@gmail.com (Y.-C.C.); wenhueic005@gmail.com (W.-H.C.)
[2] Orchid Research and Development Center, National Cheng Kung University, Tainan 701, Taiwan
[3] Department of Life Sciences, Institute of Ecology and Evolutionary Biology, National Taiwan University, Taipei 106, Taiwan; LEAFY@ntu.edu.tw
* Correspondence: hhchen@mail.ncku.edu.tw; Tel.: +886-6-275-7575 (ext. 58111); Fax: +886-6-235-6211

Abstract: Terpenoids are the largest class of plant secondary metabolites and are one of the major emitted volatile compounds released to the atmosphere. They have functions of attracting pollinators or defense function, insecticidal properties, and are even used as pharmaceutical agents. Because of the importance of terpenoids, an increasing number of plants are required to investigate the function and evolution of terpene synthases (*TPSs*) that are the key enzymes in terpenoids biosynthesis. Orchidacea, containing more than 800 genera and 28,000 species, is one of the largest and most diverse families of flowering plants, and is widely distributed. Here, the diversification of the *TPSs* evolution in Orchidaceae is revealed. A characterization and phylogeny of *TPSs* from four different species with whole genome sequences is available. Phylogenetic analysis of orchid *TPSs* indicates these genes are divided into *TPS-a*, *-b*, *-e/f*, and *g* subfamilies, and their duplicated copies are increased in derived orchid species compared to that in the early divergence orchid, *A. shenzhenica*. The large increase of both *TPS-a* and *TPS-b* copies can probably be attributed to the pro-duction of different volatile compounds for attracting pollinators or generating chemical defenses in derived orchid lineages; while the duplications of *TPS-g* and *TPS-e/f* copies occurred in a species-dependent manner.

Keywords: terpene synthase; Orchidaceae; evolution; phylogenetic tree

1. Introduction

Terpenoids are the largest group of natural metabolites in the plant kingdom, including more than 40,000 different compounds, and have multiple physiological and ecological roles. Terpene metabolites are not only essential for plant growth and development (e.g., gibberellin phytohormones), but also important intermediaries in the various interactions of plants with the environment [1]. For example, chlorophylls and carotenoids are photosynthetic pigments, while brassinosteroids, gibberellic acid, and abscisic acid are plant hormones [2,3]. Terpenoids can be classified based on the number of isoprene units, such as hemiterpene (C_5), monoterpene (C_{10}), sesquiterpene (C_{15}), diterpene (C_{20}), sesterterpene (25), triterpene (C_{30}), sesquarterpene (C_{35}), and tetraterpene (C_{40}) (Gershenzon and Dudareva, 2007). The increased number of cyclizations, possibly from a precursor with five additional carbon atoms, gives structural diversity. Terpenoid structures are extremely variable and most of them are low molecular weight like monoterpene (C_{10}), sesquiterpene (C_{15}), and diterpene (C_{20}) [4]. The approximate number of monoterpenes is 1000 and more than 7000 sesquiterpenes [5].

Terepene synthases (TPSs) are key enzymes in terpenoids biosynthesis. To date, TPSs have been studied in several typical plant genomes, such as *Arabidopsis thaliana* (Arabidopsis, 32 *TPSs*) [6], *Physcomitrella patens* (earthmoss, 1 *TPS*) [7], *Sorghum bicolor* (Sorghum, 24 *TPSs*) [8], *Vitis vinifera* (grape, 69 *TPSs*) [9], *Solanum lycopersicum* (tomato, 29 *TPSs*) [10],

Selaginella moellendorffii (spikemoss, 14 TPSs) [11], *Glycine max* (soybean, 23 TPSs) [12] *Populus trichocarpa* (poplar tree, 38 TPSs) [13], *Oryza sativa* (rice, 32 TPSs) [14], and *Dendrobium officinale* (Dendrobium orchid, 34 TPSs) [15]. According to the classification principle, TPSs can be generally classified into seven clades or subfamilies: *TPS-a*, *TPS-b*, *TPS-c*, *TPS-d*, *TPS-e/f*, *TPS-g*, and *TPS-h* [16]. *TPS-a*, *TPS-b*, and *TPS-g* are angiosperm-specific subfamilies, while the *TPS-e/f* subfamily is present in angiosperms and gymnosperms. *TPS-c* exists in land plants. *TPS-d* is a gymnosperm-specific subfamily, and the *TPS-h* subfamily only appears in *Selaginella moellendorffii* [16].

The full length of plant *TPS*s has three conserved motifs on C- and N-terminal regions. The conserved motif of N-terminal domain is R(R)X$_8$W (R, arginine, W, tryptophan and X, alternative amino acid) and the C-terminal domain contains two highly conserved aspartate-rich motifs. One of them is the DDXXD motif, which is involved in the coordination of divalent ion(s), water molecules, and the stabilization of the active site [17–19]. The second motif in the C-terminal domain is the NSE/DTE motif. These two motifs flank the entrance of the active site and function in binding a trinuclear magnesium cluster [20,21]. Most terpene synthases belong to monoterpene synthase (MTPSs) [22], sesquiterpene synthase (STPSs), and diterpene synthase (DTPSs) [23]. They all share three conserved domains in the active site, including 'DDXXD', 'DXDD', and 'EDXXD'. The 'R(R)X$_8$W' motif is also essential for monoterpene cyclization, while some MTPSs do not have it [16]. These circumstances can be seen in linalool synthase in rice (*Oryza sativa* L. cv. Nipponbare and Hinohikari) [24]; nerol synthase in soybean (*Glycine max* cv. 'Bagao'), which has a signal peptide and is believed to be functional in plastid [25]; and FaNES1, the cytosolic terpene synthase identified in strawberry, which is able to use cytosolic GDP and FDP to produce linalool and nerolidiol [26].

*TPS*s in the same subfamilies are similar in sequence and have similar functions. Based on the protein sequence, angiosperm STPSs and DTPSs belong to *TPS-a* subfamily and monoterpene synthases belong to *TPS-b* subfamily. Subfamilies in *TPS-c* and *e/f* have enzyme activities of DTPSs; Gymnosperm-specific *TPS-d* subfamily owns the enzyme activities for MTPSs, STPSs, and DTPSs. *TPS-g* encodes MTPSs, STPSs, and DTPSs that produce mainly acyclic terpenoids. *TPS-h* is *Selaginella moellendorffii*-specific subfamily and putative encodes DTPSs [16,27]. Recently, large amounts of TPSs have been identified by using BLAST and thus used for functional characterization assay to further confirm the activity of TPSs. The functions of TPSs can be mono- or multi-functional, and the enzymes can be highly identical to each other. For instance, the DTPs of levopimaradiene/abietadiene synthase and isopimaradiene synthase showed 91% identity in Norway spruce [28]. Moreover, the functional bifurcation of these two enzymes were proved to be caused by only four amino acid residues [28]. Some TPSs are responsible for producing compounds that are related to plant growth and development, such as gibberellin biosynthesis [29], others are responsible in secondary metabolism like monoterpenes and sesquiterpenes for pollination and defense [30,31]. Molecules catalyzed by *TPS* are usually further modified by cytochromes p450 (CYPs) to generate diverse structures [32].

Orchids show extraordinary morphological, structural, and physiological characteristics unique in the plant kingdom [33]. Containing more than 800 genera and 28,000 species, the Orchidaceae, classified in class Liliopsida, order Asparagales, is one of the largest and most diverse families of flowering plants [33]. They are widely distributed wherever sun shines except Antarctica, and with a variety of life forms from terrestrial to epiphytic [34]. According to molecular phylogenetic studies, Orchidaceae comprises five subfamilies, including Apostasioideae, Cypripedioideae, Vanilloideae, Orchidaideae, and Epidendroideae [35]. Orchids emit various volatile organic compounds (VOCs) to attract their pollinators, and/or the enemy of herbivores for olfactory capture. The emitted VOCs are plant secondary metabolites, and the major natural products include terpenoids, phenylpropenoids, benzenoids, and fatty acid derivatives. The floral scent composed of the VOCs plays an important role in plants, such as pollinator attraction, defense, and plant-to-plant communication, especially in insect-pollinated plants [30,36].

Floral VOCs are characterized into several orchids, including α- and β-pinene for *Cycnoches densiflorum* and *C. dianae* [37]; phenylpropanoids in *Bulbophyllum vinaceum* [38]; α-pinene and *e*-carvone oxide for *Catasetum integerrimum* [39]; *p*-dimethoxybenzene for *Cycnoches ventricosum* and *Mormodes lineata* [39]; β-bisabolene and 1,8-cineole for *Notylia barkeri* [39]; *e*-ocimene and linalool for *Gongora galeata* [39]; monoterpenes in *Orchis mascula* and *Orchis pauciflora* [40]; (Z)-11-eicosen-1-ol in *Dendrobium sinense* [41]; terpenoid of (E)-4,8-dimethylnona-1,3,7-triene (DMNT) in *Calanthe sylvatica* [42] and *Cyclopogon elatus* [43]; (E)-β-ocimene and (E)-epoxyocimene for *Catasetum cernuum* and *Gongora bufonia* [44]; and farnesol, methyl epi-jasmonate, nerolidol, and farnesene in *Cymbidium goeringii* [45].

Phalaenopsis spp. is very popular worldwide for its spectacular flower morphology and colors. Most *Phalaenopsis* orchids are scentless but some do emit scent VOCs [46]. The scented species have been extensively used as breeding parents for the production of scented cultivars, such as *P. amboinensis*, *P. bellina*, *P. javanica*, *P. lueddemanniana*, *P. schilleriana*, *P. stuartiana*, *P. venosa*, and *P. violace* [47]. *P. bellina* and *P. violacea* are two scented orchids that are very popular in breeding scented cultivars. *P. bellina* emits mainly monoterpenoids, including citronellol, geraniol, linalool, myrcene, nerol, and ocimene [47,48], while *P. violacea* emits monoterpenoids accompanied with a phenylpropanoid, cinnamyl alcohol [46]. The VOCs of *P. schilleriana* contain monoterpenoids as well, including citronellol, nerol, and neryl acetate [49]. Because of the importance of terpenoids in plants, an increasing number of plants are required to investigate the function and evolution of *TPSs*.

In the present review, we summarized the recent progress in the understanding of the biosynthesis and biological function of terpenoids, and the latest advances in research on the evolution and functional diversification of *TPSs* in Orchidaceae. TPSs from different orchid species are reported to explore the evolutionary history and the evolution diversification of Orchidaceae *TPSs*.

2. Terpenoids and Their Biosynthesis in Plants

There are two compartmentalized terpenoid biosynthesis pathways, the mevalonic acid (MVA) pathway that occurs in the cytosol, and the methylerythritol phosphate (MEP) pathway that occurs in plastids to produce isopentenyl diphosphate (IPP) and its allylic isomer-dimethylallyl diphosphate (DMAPP) converted by isopentenyl diphosphate isomerase (IDI) (Figure 1) [50–52]. There are four major steps involved in the biosynthesis of terpenoid, beginning with isoprene unit (IPP) formation, which has five carbons. Second, IPP combines to DMAPP by geranyl diphosphate synthase (GDPS), geranylgeranyl diphosphate synthases (GGDPS) or farnesyl diphosphate synthase (FDPS), and generates geranyl diphosphate (GDP), farnesyl diphosphate (FDP) or geranylgeranyl diphosphate (GGDP), respectively [1,27,53,54]. Third, the C_{10}-C_{20} diphosphates go through cyclization and rearrangement to produce the basic carbon skeletons for terpenoids catalyzed by *TPS* [53]. The *TPS* family consists of enzymes that use GDP to form cyclic and acylic monoterpenes (C_{10}), FDP for sesquiterpene (C_{15}), and GGDP for diterpene (C_{20}) [16]. Moreover, FDP and GGDP can be dimerized to form the precursors of C_{30} and C_{40}. The final step converts terpenes into different skeletons by oxidation, reduction, isomerization, conjugation, and other transformation [53]. *TPSs* are the key enzymes in terpenoid biosynthesis.

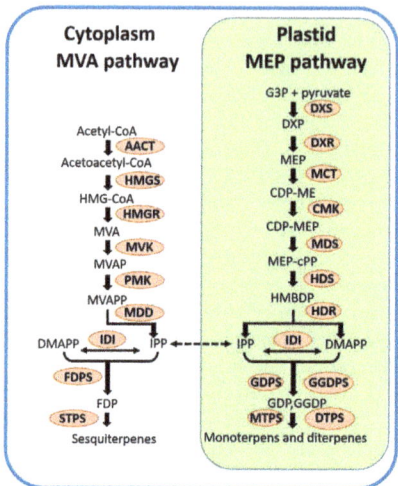

Figure 1. The MVA (left) and MEP (right) pathways responsible for IPP and DMAPP biosynthesis and monoterpene biosynthesis in plants. AACT, acetoacetyl-CoA thiolase; CMK, 4-(cytidine 5′-diphospho)-2-C-methyl-d-erythritol kinase; DMAPP, dimethylallyl diphosphate; DXR, 1-deoxy-d-xylulose 5-phosphate reductoisomerase; DXS, 1-deoxyd- xylulose 5-phosphate synthase; FDP, farnesyl diphosphate; FPPS, farnesyl diphosphate synthase; G3P, d-glyceraldehyde 3-phosphate; GDPS, geranyl diphosphate synthase; GDP, geranyl diphosphate; HDR, (E)-4-hydroxy-3-methylbut-2-enyl diphosphate reductase; HDS, (E)-4-hydroxy-3-methylbut-2-enyl diphosphate synthase; HMGR, 3-hydroxy-3-methylglutaryl-CoA reductase; HMGS, 3-hydroxy-3-methylglutaryl- CoA synthase; IDI, isopentenyl diphosphate isomerase; IPP, isopentenyl diphosphate; MCT, 2-C-methyl-d-erythritol 4-phosphate cytidylyltransferase; MDD, mevalonate diphosphate decarboxylase; MDS, 2-C-methyld-erythritol 2,4-cyclodiphosphate synthase; MVK, mevalonate kinase; MVAP, mevalonate 5-phosphate; MVAPP, mevalonate diphosphate; PMK, phosphomevalonate kinase; TPS, terpene synthase.

3. The Evolution of *TPS* Genes in Orchidaceae Species

We chose the whole genome sequences of four orchids, including *A. shenzhenica* [54] in Apostasioideae subfamily; *Vanilla planifolia* [55] in Vanilloideae subfamily; and *D. catenatum* [56] and *P. equestris* [57] in Epidendroideae subfamily. There were two justifications for this selection. First, these four orchids are distributed into three different subfamilies, and their whole genome sequences are available in NCBI (https://www.ncbi.nlm.nih.gov/ (accessed on 6 January 2021).) and OrchidBase database [58] (http://orchidbase.itps.ncku.edu.tw/est/home2012.aspx (accessed on 9 August 2020).). Second, *A. ashenzhenica* is the most original orchid, and *P. equestris* is the first whole genome sequenced orchid. *V. planifolia* produces vanillin and is important in the food industry, and *D. catenatum* is a medicinal orchid and produces important secondary metabolites for pharmaceutical purpose. We isolated the *TPS* genes of Orchidaceae through KAAS (http://www.genome.jp/tools/kaas/ (accessed on 21 February 2017).) annotation and BLASTp from the whole genome sequences of four orchids. Each full-length *TPS* is characterized by two conserved domains with Pfam [59] ID PF01397 (N-terminal) and PF03936 (C-terminal) [17]. A total of 9, 27, 35, and 15 *TPS* genes were identified from the whole genome sequences of *A. shenzhenica*, *V. planifolia*, *D. catenatum*, and *P. equestris*, respectively. In addition, *P. aphrodite* with white, scentless flowers and *P. bellina* scented flowers are native species. Their floral transcriptomes are available in Orchidstra and OrchidBase transcriptome database, respectively. 17 *TPS* genes in *P. aphrodite* and 11 *TPS* genes in *P. bellina* were identified from the transcriptome database. The *TPS* genes were denoted with numbers *Ash-*, *KAG-*, *Dca-*, *Peq-*, *PATC-*, and *PbTPS-* identified from *A. shenzhenica*, *V. planifolia*, *D. catenatum*, *P. equestris*, *P. aphrodite*, and *P. bellina*, respectively.

TPSs in *P. equestris* and *D. officinale* have been reported [15,60]. These TPSs are divided into four subfamilies (*TPS-a*, *TPS-b*, *TPS-c*, and *TPS-e/f*). So, we further investigated TPS evolution in Orchidaceae and provided insight into TPSs at the genome level. In this review, the encoded amino acid sequences of identified orchid TPS genes were aligned with those from *Arabidopsis* and *Abies grandis*, and those from *Selaginella moellendorffii* were used as outgroups (Appendix A Table A1). The phylogenetic tree was constructed using Neighbor-Joining method with Jones–Taylor–Thornton model and pairwise deletion with 1000 bootstrap replicates by using MEGA7 software. The orchid TPSs are grouped into *TPS-a*, *-b*, *-e/f*, and *g* subfamilies (Figure 2). Most of the orchid TPSs belong to *TPS-a* and *TPS-b* subfamilies (89/115, Table 1). In the *TPS-a* subfamily, copies from dicot and monocot species formed distinct subgroups, which is in accordance to previous studies [15,16]. However, compared to angiosperm dicot species, which have more TPSs in *TPS-a*, orchid (monocot) TPSs have more members in *TPS-b* subfamily than in *TPS-a* subfamily. Within *TPS-b* subfamily, these orchid TPSs form distinct clades separated from those of *Arabidopsis* (dicot) TPSs (Figure 2). Taken together, the persistence of dicot and monocot distinct clades within *TPS-a* and *TPS-b* implies that these TPSs have diverged since the ancestor of angiosperm. On the other hand, most of the duplicated orchid *TPS-a* and *TPS-b* copies were species-dependent (i.e., paralogs duplicated within each species). In particular, the number of duplicated orchid *TPS-a* and *TPS-b* copies increased in *V. planifolia* and *D. catenatum* (Figure 2). These data suggest that *TPS-a* and *TPS-b* copies evolved in a species-dependent manner and may have been positively selected to generate exceptionally more multiple copies. *TPS-a* and *TPS-b* are angiosperm-specific subfamilies that are responsible for sesquiterpene or diterpene and monoterpene synthases. These orchid volatile terpenes have critical roles in producing floral scents in order to be attractive to pollinators and to respond to environmental stresses [15]. It is therefore not surprising that *TPS-a* and *TPS-b* subfamilies have diverged greatly in orchid species.

Our phylogenetic analysis also reveals that the orchid *TPS-e/f* subfamily has increased copy numbers compared to that from *A. thaliana* (Table 1; Figure 2). Orchid *TPS-g* subfamily can only be found in *A. shenzhenica* and *V. planifolia* (Table 1; Figure 2), whereas those Epidendroideae *TPS-g* members have perhaps been lost during evolution. There are no orchid TPSs in *TPS-c* group that host copalyl diphosphate synthases (CPS) of angiosperm [61]. *TPS-d* and *TPS-h* are gymnosperm and *Selaginella moellendorffii* specific, respectively [16]. Our analysis showed that no orchid TPSs were grouped in these subfamilies, in accordance with previous conclusions by Chen et.al, and Trapp et.al. [16,62].

Motifs of identified orchid TPS proteins were predicted using MEME software (https://meme-suite.org/meme/tools/meme (accessed on 19 March 2021).) (Figure 3A), and five major functional conserved motifs of TPSs ($R(R)X_8W$, EDXXD, RXR, DDXXD, and NSE/DTE) were elucidated (Figure 3B). The *TPS-a* subfamily that encodes STPSs is mainly found in both dicot and monocot plants [9,11,16,63]. In this subfamily, STPSs contain the non-conserved secondary "R" (arginine) of motif $R(R)X_8W$ that functions in the initiation of the isomerization cyclization reaction [64], or in stabilizing the protein through electrostatic interactions [65]. Compared with *Arabidopsis*, most orchid TPSs contain motif $R(R)X_8W$, except PATC144727, Peq011664, Dca017107, and PATC155674 in *TPS-a* subfamily (Figure 4A). In contrast, the angiosperm-specific *TPS-b* subfamily that encodes MTPSs contains the highly conserved $R(R)X_8W$ motif. All TPSs in *Arabidopsis* TPS-b subfamily contain conserved $R(R)X_8W$ motif, except AtTPS02 (Figure 4B). However, several members of orchid TPS-b subfamily have lost the conserved $R(R)X_8W$ motif (Figure 4B). Motifs EDXXD, RXR, DDXXD, and NSE/DTE are highly conserved in *TPS-a* and *-b* subfamilies, while the conserved $R(R)X_8W$ motif of orchid TPSs is divergent in *TPS-b* subfamily.

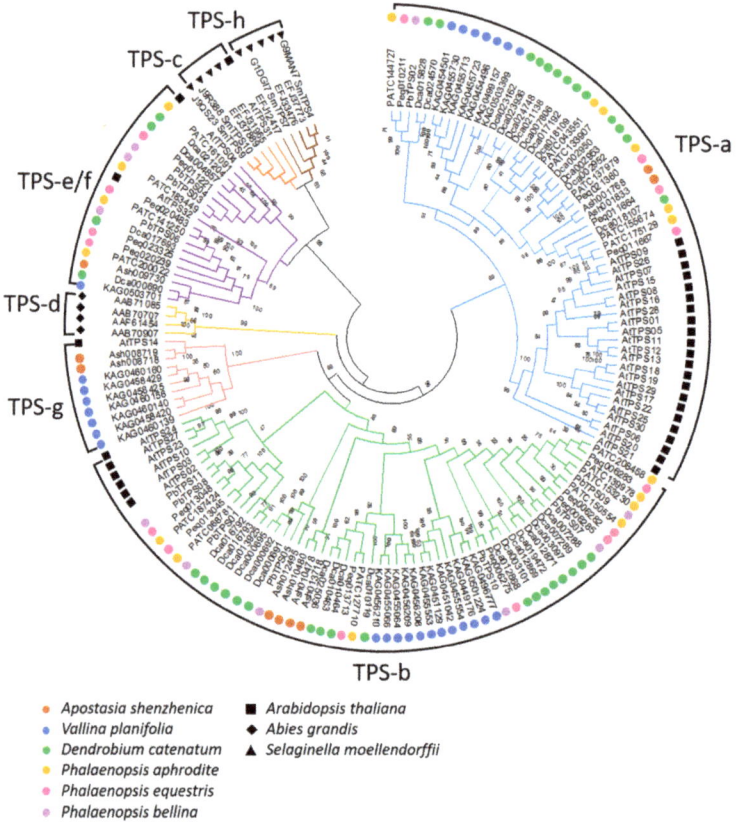

Figure 2. Phylogenetic analysis of terpene synthases. TPSs in Orchidaceae, including *A. shenzhenica*; *V. planifolia*; *D. catenatum*l *P. equestris* *Phalaenopsis aphrodite*; *P. bellina*, *Arabidopsis thaliana*, and *Abies grandis*; and *S. moellendorffii* were used. Sequence analysis was performed using MEGA 7.0 to create a tree using the nearest neighbor-joining method. The coding sequence was used for analysis. The numbers at each node represent the bootstrap values. Various colors mean distinct subfamilies and special symbols represent different plant species, with solid circles, tangle, diamond, and triangle illustrating Orchidaceae, *Arabidopsis thaliana*, *A. grandis*, and *S. moellendorffii*, respectively.

Table 1. The number of TPSs subfamilies in Orchidaceae and other plant species.

Species	TPS Subfamily							Total	Reference
	a	b	c	d	e/f	g	h		
Apostasia shenzhenica	2	4	0	0	1	2	0	9	This research
Vallina planifolia	7	12	0	0	1	7	0	27	This research
Dendrobium catenatum	13	18	0	0	4	0	0	35	This research
Phalaenopsis equestris	4	7	0	0	4	0	0	15	This research
Phalaenopsis aphrodite	6	7	0	0	4	0	0	17	This research
Phalaenopsis bellina	1	7	0	0	3	0	0	11	This research
Arabidopsis thaliana	22	6	1	0	2	1	0	32	Aubourg et al. (2002) [6]
Solanum lycopersicum	12	8	2	0	5	2	0	29	Falara et al. (2011) [10]
Oryza sativa	18	0	3	0	9	2	0	32	Chen et al. (2014) [14]
Sorghum bicolor	15	2	1	0	3	3	0	24	Paterson et al. (2009) [8]
Vitis vinifera	30	19	2	0	1	17	0	69	Martin et al. (2010) [9]
Populus trichocarpa	16	14	2	0	3	3	0	38	Irmisch et al., (2014) [13]
Selaginella moellendorffii	0	0	3	0	3	0	8	14	Li et al., (2012) [11]

Figure 3. The amino acid sequences of the predicted motifs in *TPS* proteins. (**A**) Twenty-five classical motifs in *TPS* proteins were analyzed using the MEME tool. The width of each motif ranges from 6 to 50 amino acids. The font size represents the strength of conservation. (**B**) The amino acid sequences of five highly conserved motifs in *TPS* proteins.

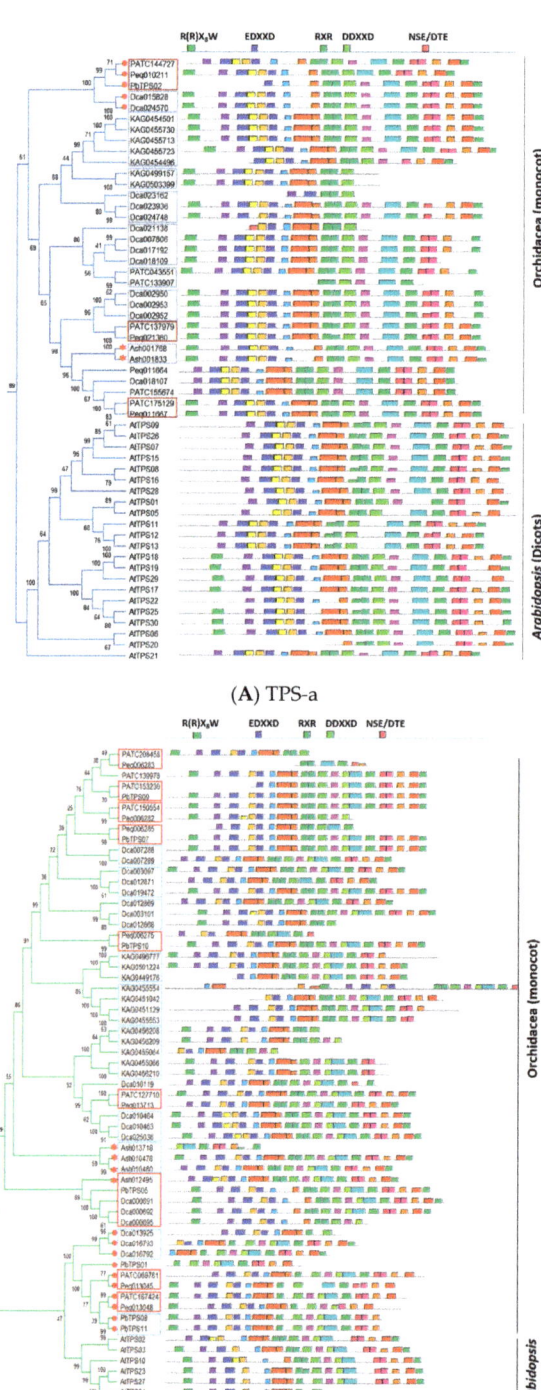

(A) TPS-a

(B) TPS-b

Figure 4. Cont.

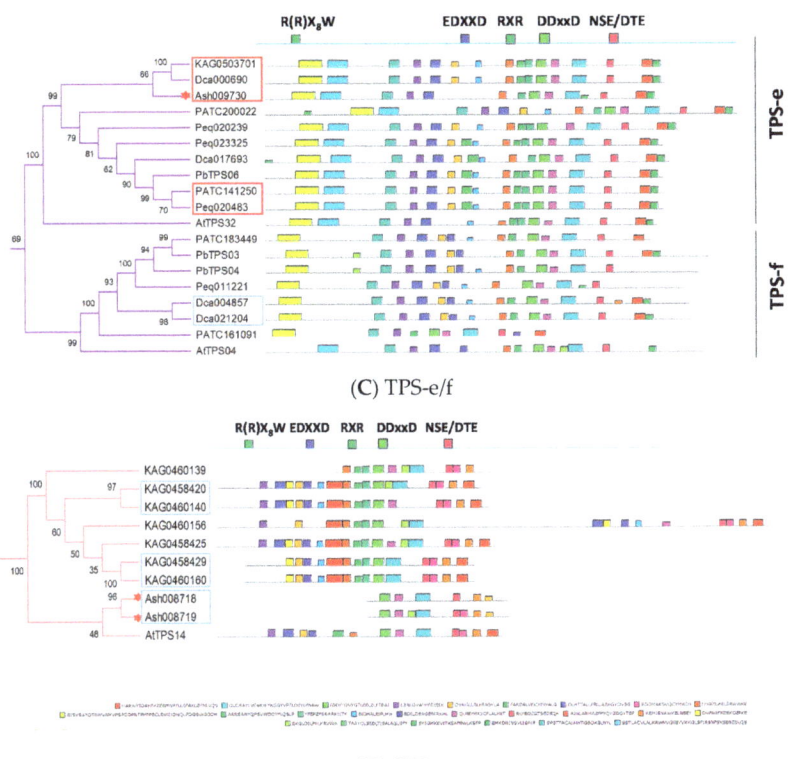

(C) TPS-e/f

(D) TPS-g

Figure 4. Motif structures of *TPS* proteins. (**A–D**) are *TPS-a, -b, -e/f,* and *-g* subfamilies, respectively. Twenty-five classical motifs in *TPS* proteins were analyzed by using the MEME tool. The width of each motif ranged from 6 to 50 amino acids. Different color blocks represent distinct motifs. Star indicates *TPS*s of *A. shenzhenica*, and the red solid circle indicates the out group of *Apostasia TPS*s. The red and blue rectangle squares reveal orthologous and paralogous gene pairs, respectively.

DTPSs are evolved from kaurene synthase (KS) and CPS. MTPSs and STPSs are evolved from ancestral DTPS through duplication and then sub- or neo-functionalization during evolution [66]. *A. shenzhenica* has clear evidence of whole-genome duplication that is shared by all orchids [54]. Yet, the copies of *TPS* in *A. shenzhenica* are among the fewest and are worthwhile for further investigation. For *Phalaenopsis* orchids, paralogs of *TPS* genes could be identified from each species, implying the duplications were attributed to their common ancestor, and some persisted or lost in current species (Figure 4). For example, *TPS-a* copies of *P. aphrodite, P. bellina,* and *P. equestris* species can be found (some lost) in three parallel clades of the phylogenetic tree (*PATC144727/Peq010211/PbTPS02, PATC137979/Peq021360,* and *PATC175129/Peq011667*) (red tangle, Figure 4A). Similarly, *TPS-b* copies of *P. aphrodite, P. bellina,* and *P. equestris* can be repeatedly identified (some lost) in eight parallel clades, indicating the *TPS-b* gene copy duplications could be traced back to the common ancestor of *Phalaenopsis* species (*PATC208458/Peq006283, PATC153230/PbTPS09, PATC150554/Peq006282, Peq006285/PbTPS07, Peq006275/PbTPS10, PATC127710/Peq013713, PATC068781/Peq013045* and *PATC187424/Peq013048*) (red tangle, Figure 4B).

Members of *TPS-e/f* subfamilies are mainly detected in angiosperm and conifers DTPSs of primary metabolism (i.e., gibberellin biosynthesis) [16,67]. Orchid *TPS-e/f* subfamilies comprise orthologous genes without $R(R)X_8W$ (Figure 4C), which are consistent with *Arabidopsis*. The *Ash009730* in *TPS-e/f* subfamily, predicted to be KS, was grouped with *KAG0503701* and *Dca000690* (red retangle with red star, Figure 4C). No *TPS*s were found in *A. shenzhenica* in *TPS-f* subclade. As copies of these orchid *TPS-e/f* subfamilies were duplicated within each species, the duplications seem to be species dependent.

TPS-g subfamily is closely related to the *TPS-b* but lacks the N-terminal "R(R)X$_8$W" motif and encodes MTPSs, STPSs, and DTPSs that produce mainly acyclic terpenoids [68,69]. A highly conserved arginine-rich RXR motif of sesquiterpene synthase reported that the motif is involved in producing a complex with the diphosphate group after the ionization of FPP in sesquiterpene biosynthesis [70]. *TPS-g* subfamily in *Arabidopsis* (AtTPS14) lacks both "R(R)X$_8$W" and "RXR" motifs. However, although *TPSs* of *V. planifolia* in *TPS-g* subfamily (those started with KAG in Figure 4D) lack the N-terminal "R(R)X$_8$W" motif, they still have the "RXR" motif (Figure 4D). This suggests that *TPS-g* subfamily of *V. planifolia* may have conserved enzyme activities that are capable of accepting a multi-substrate in terpene biosynthesis.

The pharmaceutical effective compounds in *D. catenatum*, a widely used Chinese herb, belong to terpenoid indole alkaloid (TIA) class [71], and many of them contain a terpene group. A sesquiterpene alkaloid-Dendrobine found in *Dendrobium* is believed to be responsible for its medical property [71]. Concomitantly, a significant increased number of *TPS-a TPSs* was detected in *D. catenatum*as as compared to that of other orchid species, which is responsible for sesquiterpene biosynthesis (Table 1). The increased number of *TPS-b* in *Dendrobium* may cause the floral fragrance in *D. catenatum* as well as the formation of TIA. *P. bellina* is a scented orchid with the main floral compounds of monoterpenes including linalool, geraniol, and their derivatives, which attract pollinators [48]. *PbTPSs* from the floral transcriptome database are majorly classified into the *TPS-b* subfamily (Table 1). Previously, the expression of both *PbTPS5* and *PbTPS10* were concomitant with the VOCs (monoterpene linalool and geraniol) emission in *P. bellina* [72]. This suggests that these genes may be involved in the biosynthesis of monoterpene in *P. bellina*. *TPS-e/f* enzymes have diverse functions, including linalool synthase, geranyllinalool synthase, and farnesene synthase in kiwifruit [73,74]. *TPSs* in the *TPS-e/f* subfamily are thought to be dicot-specific because so far no *TPS-e/f* activity has been reported in monocots. However, the number of TPS in *TPS-e/f* expands from 1 in *Apostasia* to 4 in *Phalaenopsis* (Table 1), suggesting that the duplication events of *TPS- b* and *TPS-e/f* have evolved in response to natural selection.

Together, our analyses suggest that orchid TPSs in each subfamily evolved from the early divergence orchid species, such as *A. shenzhenica* and/or *V. planifolia*. The large expansion of *TPS* copies in orchid groups such as *V. planifolia*, *D. catenatum*, and *Phalaenopsis* species might be due to high flexibility for adaptation and evolution through natural selection.

4. The Arrangement of *TPS*

The functional cluster phenomenon of *TPS* genes was detected in orchids. Orchid *TPS* gene clusters diverged with tandem or segmental duplications (Figure 5). Tandem duplication inferred that the duplication occurred in the same scaffold, such as *Ash012495* grouped with *Dca000691/Dca000692/Dca000697* cluster genes in *TPS-b* subfamily (Figures 4B and 5C). *TPS* genes duplicated on different scaffolds is thought to be segmental duplication, e.x.: *Ash008718/Ash008719* grouped with two cluster genes of *V. planifolia* (KAG0458420/KAG0458425/KAG0458429 and KAG0460140/KAG0460156/KAG0460160) in different scaffolds in the *TPS-g* subfamily (Figures 4D and 5A,B). We identified that 6, 24, 20, and 8 *TPSs* in *A. shenzhenica*, *V. planifolia*, *D. catenatum*, and *P. equestris*, respectively, form clusters in the same genome scaffold (Table 2, Figure 5A–D). In addition, these clusters were present with *TPSs* of the same subfamily and therefore the enhancement of functions was predicted. In *A. shenzhenica*, *V. planifolia*, *D. catenatum*, and *P. equestris*, *TPS* genes have three, nine, eight, and three clusters, respectively (Table 2, Figure 5). Each cluster contains two *TPS* genes in *A. shenzhenica*, while more genes are present in the clusters of *V. planifolia*, *D. catenatum*, and *P. equestris* (Figure 4). *TPS* genes in the same cluster usually belong to the same subfamily except that *V. planifolia* has one large scaffold containing *TPS* genes of *TPS-a*, *TPS-b*, and *TPS-e/f* subfamilies, yet with huge distance between each subfamily cluster (44 Mb and 5 Mb, respectively). The percentages of clustered *TPS* genes were 66.7%, 81.5%, 57.1%, and 53.3% for *A. shenzhenica*, *V. planifolia*, *D. catenatum*, and *P. equestris*, respectively, while that was 40.6% in *Arabidopsis thaliana* (Table 2). The cluster density of orchid TPSs could infer the event of *TPS* gene duplication occurred during evolution. The genome sizes of *A. shenzhenica*, *V. planifolia*, *D. catenatum*, and *P. equestris* are 349 Mb, 7449 Mb, 1104 Mb, and 1064 Mb, respectively (Table 3). The cluster densities of TPSs in orchids were 47.3%, 78.6%, 50.5%, and 38.9% for *A. shenzhenica*, *V. planifolia*, *D. catenatum*, and *P. equestris*, respectively (Table 3). Interestingly, orchids have more clusters and higher *TPS* gene density as compared to that of *Arabidopsis*, with that of *V. planifolia* having the highest cluster gene density of *TPS* among the four orchids analyzed. Even though *TPS*s copies of derived orchids (*D. catenatum* and *Phalaenopsis* spp.) were increased compared with those in *A. shenzhenica*, the total number was not linked to the increased genome size.

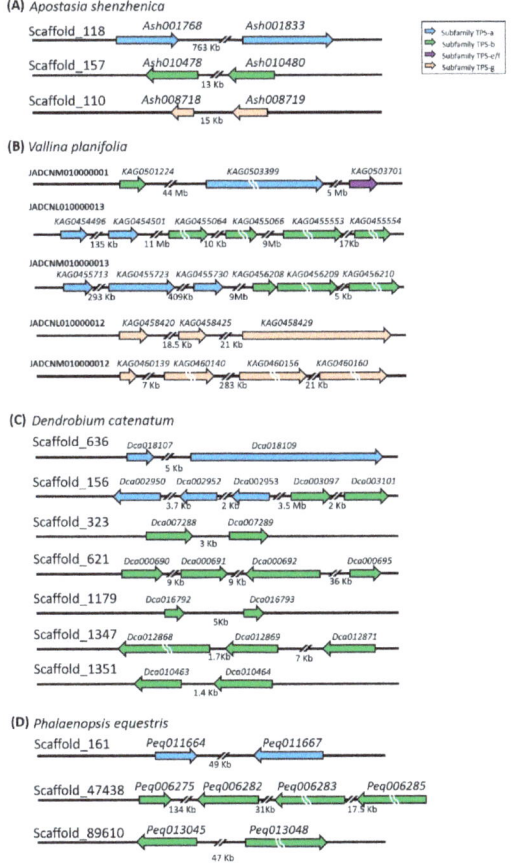

Figure 5. Gene clusters in Orchidaceae genome. Clustered genes in the genomic scaffolds of *A. shenzhenica* (**A**), *V. planifolia* (**B**), *D. catenatum* (**C**), and *P. equestris* (**D**), respectively. The *TPS* genes located on the scaffolds are identified from the assembled whole genome sequences of *A. shenzhenica*, *V. planifolia*, *D. catenatum*, and *P. equestris*. The direction of arrows illustrates the forward translation of genes in the scaffolds. Various colors indicate the distinct *TPS* subfamilies. Blue, green, purple, and bisque colors represent *TPS* genes in TPS-*a*, -*b*, -*e/f*, and -*g* subfamilies, respectively. Break lines indicate the shrink length of genes.

Table 2. The gene clusters of TPSs in the genome of Orchidaceae and *Arabidopsis thaliana*.

Species	Number of Clusters	Number of Scaffolds	Number of Clustered TPSs	Number of Total TPSs	Percentage of Clustered TPSs (%)
Apostasia shenzhenica	3	3	6	9	66.7
Vallina planifolia	7	5	22	27	81.5
Dendrobium catenatum	8	7	20	35	57.1
Phalaenopsis equestris	3	3	8	15	53.3
Arabidopsis thaliana [6]	5	5	13	32	40.6

Table 3. The gene density of TPSs in the genome of Orchidaceae and other plant species.

Species	Genome Size (Mb)	Cluster Length of TPSs (Kb)	Total Length of TPSs (Kb)	Cluster Density of TPSs (%)
Apostasia shenzhenica	349	26	56	47.3
Vallina planifolia	744	595	758	78.6
Dendrobium catenatum	1104	125	248	50.5
Phalaenopsis equestris	1064	62	158	38.9
Arabidopsis thaliana	120	43	109	39.9

In plants, gene clusters were often observed for metabolic pathways, such as gene clusters found in oat and *Arabidopsis* related to triterpene biosynthesis pathway [75]. Local duplication of *TPS* gene families in plants has been described and often results in tandem repeats, as an important driver for the expansion [16,76]. The genes related in terpene synthesis are usually lined together, forming functional clusters in plants [77]. The functional clusters of *TPS* genes have already been reported in several plant species, such as *Arabidopsis thaliana* [6], *Vitis vinifera* [9], *Solanum lycopersicum* [77], *Eucalpyus grandis* [78], and rice [79,80]. Genomic clusters of *TPS* genes in *E. grandis* are up to 20 genes [78]. In several *Solanum* species, the gene duplications and divergence give rise to *TPS* gene clusters for terpene biosynthesis [77]. A dense cluster of 45 *V. vinifera TPSs* are present on chromosome 18 [9]. *Arabidopsis TPS* genes are reported with the phenomenon of several gene clusters [6]. In addition, a gene cluster with three *TPS* members, including *Os08g07080*, *Os08g07100*, and *Os08g07120*, is observed in Asian rice *Oryza sativa* and also appears in various rice species including *O. glaberrima*, *O. rufipogon*, *O. nivara*, *O. barthii*, and *O. punctata*. [80]. Both conserved and species-specific expression patterns of the clustered rice *TPSs* indicate the functions in insect-damaged plants [80]. The expression of these rice *TPS* genes and their catalytic activities for emission patterns of volatile terpenes is induced by insect damage and is largely consistent [80]. Interestingly, the evolution of *TPSs* with other biosynthesis-related genes was also found to form unexpected connection with time passed. For instance, the evolution of *TPS/CYP* pairs is different in monocot and dicot [81]. *TPS/CYP* pairs duplicate with ancestral *TPS/CYP* pairs as templates to be evolved in dicots, but the evolutionary mechanism of monocot shows that the genome rearrangement of *TPS* and *CYP* occurred independently [81]. In *Solanum* spp., *TPS* forms functional clusters with *cis*-prenyl transferase [77]. Both tandem and segmental duplications significantly contribute toward family expansion and expression divergence and play important roles in the survival of these expanded genes. A functional gene cluster is a group of closely-related genes lined together in a genome, and the study of gene clusters is important for the understanding of evolution within species.

Together, the orchid *TPS* genes formed genomic clusters, and the clusters increased in *V. planifolia* and *D. catenatum*. Combining the results from phylogenetic analysis and functional gene clusters, orchid *TPSs* may be expanded by tandem or segmental duplications. Interestingly, the genome duplication events occurred all the way along the evolution from Apostasioideae to Vanilloideae and Epidendroideae; the *TPS* clusters and copy numbers increased in orchid lineages, such as the early divergence *A. shenzhenica*. The large expansion of orchid *TPS* copies in *V. planifolia*, and *D. catenatum* species might have high flexibility in secondary biosynthesis through natural selection.

5. Conclusions

The basic evolution of *TPS* is from duplication and loss of *TPS* genes. In Orchidaceae, we discover that the duplication event of *TPS* occurred among all *TPS* subfamilies. *TPs-a*, *TPS-b*, and *TPS-e/f* subfamilies went through gene duplication, while *TPS-g* duplicated from Apostaceae to Vaniloideae, and then lost from Vaniloideae to Epidendroideae. The driving force of *TPS* evolution in each subfamily may be different. For example, in *TPS-a* and *TPS-b*, the necessity of generating volatile compounds for the interaction of orchids with their pollinators, producing chemical defenses and being responsive to environmental stress, may be the major reason for their rapid evolution. On the other hand, the duplications of *TPS-g* and *TPS-e/f* copies were mainly species dependent and the reason remains to be uncovered.

Author Contributions: L.-M.H. performed the phylogenetic analysis and motif prediction of *TPS*s; H.H. performed the gene arrangement analysis; Y.-C.C. performed the identification of orchid *TPS*s; W.-H.C. provided the suggestions for plant materials.; C.-N.W. provided discussion and composed the *TPS*s evolution; H.-H.C. conceived research plans and composed the article with assistances of all the authors, completed the writing, and served as the corresponding author for communication. All authors have read and agreed to the published version of the manuscript.

Funding: This work was supported by a grant from Ministry of Science and Technology, Taiwan (MOST 107-2313-B-006-003-MY3) to H.-H.C.

Institutional Review Board Statement: Not applicable.

Informed Consent Statement: Not applicable.

Data Availability Statement: Not applicable.

Acknowledgments: We thank the people that finished the whole genome sequence of the four orchid species, which allowed us to undertake this detail analysis.

Conflicts of Interest: No conflict of interest declared.

Appendix A

Table A1. *TPS* genes used in phylogenetic analysis.

Species	Gene ID	Accession Number of *TPS* Gene
Apostasia shenzhenica [1]	Ash001768	Ash001768
	Ash001833	Ash001833
	Ash008718	Ash008718
	Ash008719	Ash008719
	Ash009730	Ash009730
	Ash010478	Ash010478
	Ash010480	Ash010480
	Ash012495	Ash012495
	Ash013718	Ash013718
Vallina planifolia [2]	KAG0449176	KAG0449176
	KAG0451042	KAG0451042
	KAG0451129	KAG0451129
	KAG0454496	KAG0454496
	KAG0454501	KAG0454501
	KAG0455064	KAG0455064
	KAG0455066	KAG0455066
	KAG0455553	KAG0455553
	KAG0455554	KAG0455554
	KAG0455713	KAG0455713

Table A1. Cont.

Species	Gene ID	Accession Number of TPS Gene
	KAG0455723	KAG0455723
	KAG0455730	KAG0455730
	KAG0456208	KAG0456208
	KAG0456209	KAG0456209
	KAG0456210	KAG0456210
	KAG0458420	KAG0458420
	KAG0458425	KAG0458425
	KAG0458429	KAG0458429
	KAG0460139	KAG0460139
	KAG0460140	KAG0460140
	KAG0460156	KAG0460156
	KAG0460160	KAG0460160
	KAG0496777	KAG0496777
	KAG0499157	KAG0499157
	KAG0501224	KAG0501224
	KAG0503399	KAG0503399
	KAG0503701	KAG0503701
Dendrobium catenatum [1]	Dca000690	Dca000690
	Dca000691	Dca000691
	Dca000692	Dca000692
	Dca000695	Dca000695
	Dca002950	Dca002950
	Dca002952	Dca002952
	Dca002953	Dca002953
	Dca003097	Dca003097
	Dca003101	Dca003101
	Dca004857	Dca004857
	Dca007288	Dca007288
	Dca007289	Dca007289
	Dca007806	Dca007806
	Dca010119	Dca010119
	Dca010463	Dca010463
	Dca010464	Dca010464
	Dca012868	Dca012868
	Dca012869	Dca012869
	Dca012871	Dca012871
	Dca013925	Dca013925
	Dca015828	Dca015828
	Dca016792	Dca016792
	Dca016793	Dca016793
	Dca017192	Dca017192
	Dca017693	Dca017693
	Dca018107	Dca018107
	Dca018109	Dca018109
	Dca019472	Dca019472
	Dca021138	Dca021138
	Dca021204	Dca021204
	Dca023162	Dca023162
	Dca023936	Dca023936
	Dca024570	Dca024570
	Dca024748	Dca024748
	Dca025036	Dca025036
Phalaenopsis aphrodite [3]	PATC043551	PATC043551
	PATC068781	PATC068781
	PATC127710	PATC127710
	PATC133907	PATC133907
	PATC137979	PATC137979

Table A1. Cont.

Species	Gene ID	Accession Number of TPS Gene
	PATC139978	PATC139978
	PATC141250	PATC141250
	PATC144727	PATC144727
	PATC150554	PATC150554
	PATC153230	PATC153230
	PATC155674	PATC155674
	PATC161091	PATC161091
	PATC175129	PATC175129
	PATC183449	PATC183449
	PATC187424	PATC187424
	PATC200022	PATC200022
	PATC208458	PATC208458
Phalaenopsis equestris [1]	Peq006275	Peq006275
	Peq006282	Peq006282
	Peq006283	Peq006283
	Peq006285	Peq006285
	Peq010211	Peq010211
	Peq011221	Peq011221
	Peq011664	Peq011664
	Peq011667	Peq011667
	Peq013045	Peq013045
	Peq013048	Peq013048
	Peq013713	Peq013713
	Peq020239	Peq020239
	Peq020483	Peq020483
	Peq021360	Peq021360
	Peq023325	Peq023325
Phalaenopsis bellina [4]	PbTPS01	CL86.Contig1
	PbTPS02	CL214.Contig2
	PbTPS03	CL376.Contig6
	PbTPS04	CL376.Contig8
	PbTPS05	CL1323.Contig1
	PbTPS06	CL2295.Contig2
	PbTPS07	CL2800.Contig3
	PbTPS08	CL4514.Contig2
	PbTPS09	CL6288.Contig1
	PbTPS10	CL6288.Contig7
	PbTPS11	Unigene4722
Arabidopsis thaliana [2]	AtTPS1	At4g15870
	AtTPS2	At4g16730
	AtTPS3	At4g16740
	AtTPS4	At1g61120
	AtTPS5	At4g20230
	AtTPS6	At1g70080
	AtTPS7	At4g20200
	AtTPS8	At4g20210
	AtTPS9	At2g23230
	AtTPS10	At2g24210
	AtTPS11	At5g44630
	AtTPS12	At4g13280
	AtTPS13	At4g13300
	AtTPS14	At1g61680
	AtTPS15	At3g29190
	AtTPS16	At3g29110
	AtTPS17	At3g14490
	AtTPS18	At3g14520
	AtTPS19	At3g14540

Table A1. Cont.

Species	Gene ID	Accession Number of *TPS* Gene
	AtTPS20	At5g48110
	AtTPS21	At5g23960
	AtTPS22	At1g33750
	AtTPS23	At3g25830
	AtTPS24	At3g25810
	AtTPS25	At3g29410
	AtTPS26	At1g66020
	AtTPS27	At1g48820
	AtTPS28	At1g48800
	AtTPS29	At1g31950
	AtTPS30	At3g32030
	AtTPS31	At4g02780
	AtTPS32	At1g79460
Abies grandis [2]	AAB70707	AGU87910
	AAB70907	AF006193
	AAB71085	U87909
	AAF61454	AF139206
Selaginella moellendorffii [2]	EFJ31965	GL377573
	EFJ37889	GL377565
	J9QS23_SmTPS9	XM_002960304
	IJ9R388_SmTPS10	XM_024672072
	G9MAN7_SmTPS4	XM_024672355.
	G1DGI7_SmTPS7	XM_024689660
	EFJ12417	GL377639
	EFJ37773	GL377565
	EFJ33476	GL377571

[1] OrchidBase 4.0 (http://orchidbase.itps.ncku.edu.tw/est/home2012.aspx (accessed on 9 August 2020)). [2] NCBI database (https://www.ncbi.nlm.nih.gov/ (accessed on 9 August 2020). [3] Orchidstra 2.0 (http://orchidstra2.abrc.sinica.edu.tw/orchidstra2/index.php (accessed on 5 January 2021). [4] *P. bellina* transcriptome database (unpublished).

References

1. Tholl, D. Terpene synthases and the regulation, diversity and biological roles of terpene metabolism. *Curr. Opin. Plant Biol.* **2006**, *9*, 297–304. [CrossRef]
2. Pichersky, E.; Noel, J.P.; Dudareva, N. Biosynthesis of plant volatiles: Nature's diversity and ingenuity. *Science* **2006**, *311*, 808–811. [CrossRef]
3. Yamaguchi, S.; Sun, T.; Kawaide, H.; Kamiya, Y. The GA2 locus of Arabidopsis thaliana encodes ent-kaurene synthase of gibberellin biosynthesis. *Plant Physiol.* **1998**, *116*, 1271–1278. [CrossRef]
4. El Tamer, M.K.; Lücker, J.; Bosch, D.; Verhoeven, H.A.; Verstappen, F.W.; Schwab, W.; van Tunen, A.J.; Voragen, A.G.; De Maagd, R.A.; Bouwmeester, H.J. Domain swapping of Citrus limon monoterpene synthases: Impact on enzymatic activity and product specificity. *Arch. Biochem. Biophys.* **2003**, *411*, 196–203. [CrossRef]
5. Lange, B.M.; Severin, K.; Bechthold, A.; Heide, L. Regulatory role of microsomal 3-hydroxy-3-methylglutaryl-coenzyme A reductase for shikonin biosynthesis in Lithospermum erythrorhizon cell suspension cultures. *Planta* **1998**, *204*, 234–241. [CrossRef] [PubMed]
6. Aubourg, S.; Lecharny, A.; Bohlmann, J. Genomic analysis of the terpenoid synthase (AtTPS) gene family of Arabidopsis thaliana. *Mol. Genet. Genom.* **2002**, *267*, 730–745. [CrossRef] [PubMed]
7. Hayashi, K.; Kawaide, H.; Notomi, M.; Sakigi, Y.; Matsuo, A.; Nozaki, H. Identification and functional analysis of bifunctional ent-kaurene synthase from the moss Physcomitrella patens. *FEBS Lett.* **2006**, *580*, 6175–6181. [CrossRef]
8. Paterson, A.H.; Bowers, J.E.; Bruggmann, R.; Dubchak, I.; Grimwood, J.; Gundlach, H.; Haberer, G.; Hellsten, U.; Mitros, T.; Poliakov, A.; et al. The Sorghum bicolor genome and the diversification of grasses. *Nature* **2009**, *457*, 551–556. [CrossRef]
9. Martin, D.M.; Aubourg, S.; Schouwey, M.B.; Daviet, L.; Schalk, M.; Toub, O.; Lund, S.T.; Bohlmann, J. Functional annotation, genome organization and phylogeny of the grapevine (Vitis vinifera) terpene synthase gene family based on genome assembly, FLcDNA cloning, and enzyme assays. *BMC Plant Biol.* **2010**, *10*, 226. [CrossRef] [PubMed]
10. Falara, V.; Akhtar, T.A.; Nguyen, T.T.; Spyropoulou, E.A.; Bleeker, P.M.; Schauvinhold, I.; Matsuba, Y.; Bonini, M.E.; Schilmiller, A.L.; Last, R.L.; et al. The tomato terpene synthase gene family. *Plant Physiol.* **2011**, *157*, 770–789. [CrossRef]

11. Li, G.; Kollner, T.G.; Yin, Y.; Jiang, Y.; Chen, H.; Xu, Y.; Gershenzon, J.; Pichersky, E.; Chen, F. Nonseed plant Selaginella moellendorffi has both seed plant and microbial types of terpene synthases. *Proc. Natl. Acad. Sci. USA* **2012**, *109*, 14711–14715. [CrossRef] [PubMed]
12. Liu, J.; Huang, F.; Wang, X.; Zhang, M.; Zheng, R.; Wang, J.; Yu, D. Genome-wide analysis of terpene synthases in soybean: Functional characterization of GmTPS3. *Gene* **2014**, *544*, 83–92. [CrossRef] [PubMed]
13. Irmisch, S.; Jiang, Y.; Chen, F.; Gershenzon, J.; Köllner, T.G. Terpene synthases and their contribution to herbivore-induced volatile emission in western balsam poplar (Populus trichocarpa). *BMC Plant Biol.* **2014**, *14*, 270. [CrossRef]
14. Chen, H.; Li, G.; Kollner, T.G.; Jia, Q.; Gershenzon, J.; Chen, F. Positive Darwinian selection is a driving force for the diversification of terpenoid biosynthesis in the genus Oryza. *BMC Plant Biol.* **2014**, *14*, 239. [CrossRef] [PubMed]
15. Yu, Z.; Zhao, C.; Zhang, G.; Teixeira da Silva, J.A.; Duan, J. Genome-Wide Identification and Expression Profile of TPS Gene Family in Dendrobium officinale and the Role of DoTPS10 in Linalool Biosynthesis. *Int. J. Mol. Sci.* **2020**, *21*, 5419. [CrossRef] [PubMed]
16. Chen, F.; Tholl, D.; Bohlmann, J.; Pichersky, E. The family of terpene synthases in plants: A mid-size family of genes for specialized metabolism that is highly diversified throughout the kingdom. *Plant J.* **2011**, *66*, 212–229. [CrossRef]
17. Starks, C.M.; Back, K.; Chappell, J.; Noel, J.P. Structural basis for cyclic terpene biosynthesis by tobacco 5-epi-aristolochene synthase. *Science* **1997**, *277*, 1815–1820. [CrossRef] [PubMed]
18. Rynkiewicz, M.J.; Cane, D.E.; Christianson, D.W. Structure of trichodiene synthase from Fusarium sporotrichioides provides mechanistic inferences on the terpene cyclization cascade. *Proc. Natl. Acad. Sci. USA* **2001**, *98*, 13543–13548. [CrossRef]
19. Whittington, D.A.; Wise, M.L.; Urbansky, M.; Coates, R.M.; Croteau, R.B.; Christianson, D.W. Bornyl diphosphate synthase: Structure and strategy for carbocation manipulation by a terpenoid cyclase. *Proc. Natl. Acad. Sci. USA* **2002**, *99*, 15375–15380. [CrossRef]
20. Christianson, D.W. Structural biology and chemistry of the terpenoid cyclases. *Chem. Rev.* **2006**, *106*, 3412–3442. [CrossRef]
21. Degenhardt, J.; Kollner, T.G.; Gershenzon, J. Monoterpene and sesquiterpene synthases and the origin of terpene skeletal diversity in plants. *Phytochemistry* **2009**, *70*, 1621–1637. [CrossRef]
22. Wise, M.L.; Croteau, R. Monoterpene biosynthesis. In *Comprehensive Natural Products Chemistry*; Elsevier: Amsterdam, The Netherlands, 1999; Volume 2, pp. 97–159.
23. MacMillan, J.; Beale, M.H. Diterpene biosynthesis. In *Comprehensive Natural Products Chemistry*; Elsevier: Amsterdam, The Netherlands, 1999; Volume 2, pp. 217–243.
24. Taniguchi, S.; Hosokawa-Shinonaga, Y.; Tamaoki, D.; Yamada, S.; Akimitsu, K.; Gomi, K. Jasmonate induction of the monoterpene linalool confers resistance to rice bacterial blight and its biosynthesis is regulated by JAZ protein in rice. *Plant Cell Environ.* **2014**, *37*, 451–461. [CrossRef]
25. Zhang, M.; Liu, J.; Li, K.; Yu, D. Identification and characterization of a novel monoterpene synthase from soybean restricted to neryl diphosphate precursor. *PLoS ONE* **2013**, *8*, e75972. [CrossRef] [PubMed]
26. Aharoni, A.; Giri, A.P.; Verstappen, F.W.; Bertea, C.M.; Sevenier, R.; Sun, Z.; Jongsma, M.A.; Schwab, W.; Bouwmeester, H.J. Gain and loss of fruit flavor compounds produced by wild and cultivated strawberry species. *Plant Cell* **2004**, *16*, 3110–3131. [CrossRef]
27. Bohlmann, J.; Crock, J.; Jetter, R.; Croteau, R. Terpenoid-based defenses in conifers: cDNA cloning, characterization, and functional expression of wound-inducible (E)-alpha-bisabolene synthase from grand fir (Abies grandis). *Proc. Natl. Acad. Sci. USA* **1998**, *95*, 6756–6761. [CrossRef] [PubMed]
28. Keeling, C.I.; Weisshaar, S.; Lin, R.P.; Bohlmann, J. Functional plasticity of paralogous diterpene synthases involved in conifer defense. *Proc. Natl. Acad. Sci. USA* **2008**, *105*, 1085–1090. [CrossRef] [PubMed]
29. Yamaguchi, S. Gibberellin metabolism and its regulation. *Annu. Rev. Plant Biol.* **2008**, *59*, 225–251. [CrossRef]
30. Pichersky, E.; Gershenzon, J. The formation and function of plant volatiles: Perfumes for pollinator attraction and defense. *Curr. Opin. Plant Biol.* **2002**, *5*, 237–243. [CrossRef]
31. Unsicker, S.B.; Kunert, G.; Gershenzon, J. Protective perfumes: The role of vegetative volatiles in plant defense against herbivores. *Curr. Opin. Plant Biol.* **2009**, *12*, 479–485. [CrossRef]
32. Weitzel, C.; Simonsen, H.T. Cytochrome P450-enzymes involved in the biosynthesis of mono- and sesquiterpenes. *Phytochem. Rev.* **2015**, *14*, 7–24. [CrossRef]
33. Fay, M.F. Orchid conservation: How can we meet the challenges in the twenty-first century? *Bot. Stud.* **2018**, *59*, 16. [CrossRef] [PubMed]
34. Givnish, T.J.; Spalink, D.; Ames, M.; Lyon, S.P.; Hunter, S.J.; Zuluaga, A.; Doucette, A.; Caro, G.G.; McDaniel, J.; Clements, M.A.; et al. Orchid historical biogeography, diversification, Antarctica and the paradox of orchid dispersal. *J. Biogeogr.* **2016**, *43*, 1905–1916. [CrossRef]
35. Chase, M.W.; Cameron, K.M.; Barrett, R.L.; Freudenstein, J.V. DNA data and Orchidaceae systematics: A new phylogenetic classification. In *Orchid Conservation*; Natural History Publications: Kota Kinabalu, Malaysia, 2003; pp. 69–89.
36. Raguso, R.A.; Levin, R.A.; Foose, S.E.; Holmberg, M.W.; McDade, L.A. Fragrance chemistry, nocturnal rhythms and pollination "syndromes" in Nicotiana. *Phytochemistry* **2003**, *63*, 265–284. [CrossRef]
37. Gregg, K.B. Variation in floral fragrances and morphology: Incipient speciation in Cycnoches? *Bot. Gaz.* **1983**, *144*, 566–576. [CrossRef]

38. Tan, K.H.; Tan, L.T.; Nishida, R. Floral phenylpropanoid cocktail and architecture of Bulbophyllum vinaceum orchid in attracting fruit flies for pollination. *J. Chem. Ecol.* **2006**, *32*, 2429–2441. [CrossRef] [PubMed]
39. Cancino, A.D.M.; Damon, A. Fragrance analysis of euglossine bee pollinated orchids from Soconusco, south-east Mexico. *Plant Species Biol.* **2007**, *22*, 127–132.
40. Salzmann, C.C.; Schiestl, F.P. Odour and colour polymorphism in the food-deceptive orchid Dactylorhiza romana. *Plant Syst. Evol.* **2007**, *267*, 37–45. [CrossRef]
41. Brodmann, J.; Twele, R.; Francke, W.; Luo, Y.B.; Song, X.Q.; Ayasse, M. Orchid Mimics Honey Bee Alarm Pheromone in Order to Attract Hornets for Pollination. *Curr. Biol.* **2009**, *19*, 1368–1372. [CrossRef]
42. Delle-Vedove, R.; Juillet, N.; Bessiere, J.M.; Grison, C.; Barthes, N.; Pailler, T.; Dormont, L.; Schatz, B. Colour-scent associations in a tropical orchid: Three colours but two odours. *Phytochemistry* **2011**, *72*, 735–742. [CrossRef]
43. Wiemer, A.P.; More, M.; Benitez-Vieyra, S.; Cocucci, A.A.; Raguso, R.A.; Sersic, A.N. A simple floral fragrance and unusual osmophore structure in Cyclopogon elatus (Orchidaceae). *Plant Biol.* **2009**, *11*, 506–514. [CrossRef]
44. Nunes, C.E.; Gerlach, G.; Bandeira, K.D.; Gobbo-Neto, L.; Pansarin, E.R.; Sazima, M. Two orchids, one scent? Floral volatiles of Catasetum cernuum and Gongora bufonia suggest convergent evolution to a unique pollination niche. *Flora* **2017**, *232*, 207–216. [CrossRef]
45. Ramya, M.; Park, P.H.; Chuang, Y.-C.; Kwon, O.K.; An, H.R.; Park, P.M.; Baek, Y.S.; Kang, B.-C.; Tsai, W.-C.; Chen, H.-H. RNA sequencing analysis of Cymbidium goeringii identifies floral scent biosynthesis related genes. *BMC Plant Biol.* **2019**, *19*, 337. [CrossRef]
46. Kaiser, R.A. *On the Scent of Orchids*; ACS Publications: Washington, DC, USA, 1993; pp. 240–268.
47. Hsiao, Y.Y.; Pan, Z.J.; Hsu, C.C.; Yang, Y.P.; Hsu, Y.C.; Chuang, Y.C.; Shih, H.H.; Chen, W.H.; Tsai, W.C.; Chen, H.H. Research on orchid biology and biotechnology. *Plant Cell Physiol.* **2011**, *52*, 1467–1486. [CrossRef]
48. Hsiao, Y.Y.; Tsai, W.C.; Kuoh, C.S.; Huang, T.H.; Wang, H.C.; Wu, T.S.; Leu, Y.L.; Chen, W.H.; Chen, H.H. Comparison of transcripts in Phalaenopsis bellina and Phalaenopsis equestris (Orchidaceae) flowers to deduce monoterpene biosynthesis pathway. *BMC Plant Biol.* **2006**, *6*, 14. [CrossRef]
49. Awano, K.; Ichikawa, Y.; Tokuda, K.; Kuraoka, M. Volatile Components of the Flowers of Two Calanthe Species. *Flavour Fragr. J.* **1997**, *12*, 327–333. [CrossRef]
50. Lichtenthaler, H.K. The 1-deoxy-D-xylulose-5-phosphate pathway of isoprenoid biosynthesis in plants. *Annu. Rev. Plant Physiol. Plant Mol. Biol.* **1999**, *50*, 47–65. [CrossRef] [PubMed]
51. Newman, J.D.; Chappell, J. Isoprenoid biosynthesis in plants: Carbon partitioning within the cytoplasmic pathway. *Crit. Rev. Biochem. Mol. Biol.* **1999**, *34*, 95–106. [CrossRef]
52. Sapir-Mir, M.; Mett, A.; Belausov, E.; Tal-Meshulam, S.; Frydman, A.; Gidoni, D.; Eyal, Y. Peroxisomal Localization of Arabidopsis Isopentenyl Diphosphate Isomerases Suggests That Part of the Plant Isoprenoid Mevalonic Acid Pathway Is Compartmentalized to Peroxisomes. *Plant Physiol.* **2008**, *148*, 1219–1228. [CrossRef] [PubMed]
53. Ashour, M.; Wink, M.; Gershenzon, J. Biochemistry of terpenoids: Monoterpenes, sesquiterpenes and diterpenes. In *Annual Plant Reviews: Biochemistry of Plant Secondary Metabolism*; Wiley-Blackwell: Chichester, UK, 2010; pp. 258–303.
54. Zhang, G.Q.; Liu, K.W.; Li, Z.; Lohaus, R.; Hsiao, Y.Y.; Niu, S.C.; Wang, J.Y.; Lin, Y.C.; Xu, Q.; Chen, L.J.; et al. The Apostasia genome and the evolution of orchids. *Nature* **2017**, *549*, 379–383. [CrossRef] [PubMed]
55. Hasing, T.; Tang, H.; Brym, M.; Khazi, F.; Huang, T.; Chambers, A.H. A phased Vanilla planifolia genome enables genetic improvement of flavour and production. *Nat. Food* **2020**, *1*, 811–819. [CrossRef]
56. Zhang, G.-Q.; Xu, Q.; Bian, C.; Tsai, W.-C.; Yeh, C.-M.; Liu, K.-W.; Yoshida, K.; Zhang, L.-S.; Chang, S.-B.; Chen, F.; et al. The Dendrobium catenatum Lindl. genome sequence provides insights into polysaccharide synthase, floral development and adaptive evolution. *Sci. Rep.* **2016**, *6*, 19029. [CrossRef]
57. Cai, J.; Liu, X.; Vanneste, K.; Proost, S.; Tsai, W.C.; Liu, K.W.; Chen, L.J.; He, Y.; Xu, Q.; Bian, C.; et al. The genome sequence of the orchid Phalaenopsis equestris. *Nat. Genet.* **2015**, *47*, 65–72. [CrossRef]
58. Tsai, W.C.; Fu, C.H.; Hsiao, Y.Y.; Huang, Y.M.; Chen, L.J.; Wang, M.; Liu, Z.J.; Chen, H.H. OrchidBase 2.0: Comprehensive collection of Orchidaceae floral transcriptomes. *Plant Cell Physiol.* **2013**, *54*, e7. [CrossRef]
59. Sun, T.P.; Kamiya, Y. The Arabidopsis GA1 locus encodes the cyclase ent-kaurene synthetase A of gibberellin biosynthesis. *Plant Cell* **1994**, *6*, 1509–1518. [PubMed]
60. Tsai, W.C.; Dievart, A.; Hsu, C.C.; Hsiao, Y.Y.; Chiou, S.Y.; Huang, H.; Chen, H.H. Post genomics era for orchid research. *Bot. Stud.* **2017**, *58*, 61. [CrossRef] [PubMed]
61. Finn, R.D.; Coggill, P.; Eberhardt, R.Y.; Eddy, S.R.; Mistry, J.; Mitchell, A.L.; Potter, S.C.; Punta, M.; Qureshi, M.; Sangrador-Vegas, A.; et al. The Pfam protein families database: Towards a more sustainable future. *Nucleic Acids Res.* **2016**, *44*, D279–D285. [CrossRef] [PubMed]
62. Trapp, S.C.; Croteau, R.B. Genomic organization of plant terpene synthases and molecular evolutionary implications. *Genetics* **2001**, *158*, 811–832. [CrossRef] [PubMed]
63. Shalev, T.J.; Yuen, M.M.S.; Gesell, A.; Yuen, A.; Russell, J.H.; Bohlmann, J. An annotated transcriptome of highly inbred Thuja plicata (Cupressaceae) and its utility for gene discovery of terpenoid biosynthesis and conifer defense. *Tree Genet. Genomes* **2018**, *14*, 35. [CrossRef]

64. Williams, D.C.; McGarvey, D.J.; Katahira, E.J.; Croteau, R. Truncation of limonene synthase preprotein provides a fully active 'pseudomature' form of this monoterpene cyclase and reveals the function of the amino-terminal arginine pair. *Biochemistry* **1998**, *37*, 12213–12220. [CrossRef]
65. Hyatt, D.C.; Youn, B.; Zhao, Y.; Santhamma, B.; Coates, R.M.; Croteau, R.B.; Kang, C. Structure of limonene synthase, a simple model for terpenoid cyclase catalysis. *Proc. Natl. Acad. Sci. USA* **2007**, *104*, 5360–5365. [CrossRef]
66. Cao, R.; Zhang, Y.; Mann, F.M.; Huang, C.; Mukkamala, D.; Hudock, M.P.; Mead, M.E.; Prisic, S.; Wang, K.; Lin, F.Y.; et al. Diterpene cyclases and the nature of the isoprene fold. *Proteins Struct. Funct. Bioinform.* **2010**, *78*, 2417–2432. [CrossRef]
67. Keeling, C.I.; Dullat, H.K.; Yuen, M.; Ralph, S.G.; Jancsik, S.; Bohlmann, J. Identification and Functional Characterization of Monofunctional ent-Copalyl Diphosphate and ent-Kaurene Synthases in White Spruce Reveal Different Patterns for Diterpene Synthase Evolution for Primary and Secondary Metabolism in Gymnosperms. *Plant Physiol.* **2010**, *152*, 1197–1208. [CrossRef] [PubMed]
68. Alicandri, E.; Paolacci, A.R.; Osadolor, S.; Sorgona, A.; Badiani, M.; Ciaffi, M. On the Evolution and Functional Diversity of Terpene Synthases in the Pinus Species: A Review. *J. Mol. Evol.* **2020**, *88*, 253–283. [CrossRef]
69. Chen, X.; Kollner, T.G.; Shaulsky, G.; Jia, Q.; Dickschat, J.S.; Gershenzon, J.; Chen, F. Diversity and Functional Evolution of Terpene Synthases in Dictyostelid Social Amoebae. *Sci. Rep.* **2018**, *8*, 14361. [CrossRef]
70. Ker, D.S.; Pang, S.L.; Othman, N.F.; Kumaran, S.; Tan, E.F.; Krishnan, T.; Chan, K.G.; Othman, R.; Hassan, M.; Ng, C.L. Purification and biochemical characterization of recombinant Persicaria minor beta-sesquiphellandrene synthase. *Peer J.* **2017**, *5*, e2961. [CrossRef] [PubMed]
71. Tang, H.; Zhao, T.; Sheng, Y.; Zheng, T.; Fu, L.; Zhang, Y. Dendrobium officinale Kimura et Migo: A Review on Its Ethnopharmacology, Phytochemistry, Pharmacology, and Industrialization. *Evid. Based Complementary Altern. Med.* **2017**, *2017*, 19. [CrossRef] [PubMed]
72. Chuang, Y.C.; Hung, Y.C.; Tsai, W.C.; Chen, W.H.; Chen, H.H. PbbHLH4 regulates floral monoterpene biosynthesis in Phalaenopsis orchids. *J. Exp. Bot.* **2018**, *69*, 4363–4377. [CrossRef] [PubMed]
73. Nieuwenhuizen, N.J.; Wang, M.Y.; Matich, A.J.; Green, S.A.; Chen, X.; Yauk, Y.K.; Beuning, L.L.; Nagegowda, D.A.; Dudareva, N.; Atkinson, R.G. Two terpene synthases are responsible for the major sesquiterpenes emitted from the flowers of kiwifruit (Actinidia deliciosa). *J. Exp. Bot.* **2009**, *60*, 3203–3219. [CrossRef] [PubMed]
74. Chen, X.; Yauk, Y.-K.; Nieuwenhuizen, N.J.; Matich, A.J.; Wang, M.Y.; Perez, R.L.; Atkinson, R.G.; Beuning, L.L. Characterisation of an (S)-linalool synthase from kiwifruit (Actinidia arguta) that catalyses the first committed step in the production of floral lilac compounds. *Funct. Plant Biol.* **2010**, *37*, 232–243. [CrossRef]
75. Krokida, A.; Delis, C.; Geisler, K.; Garagounis, C.; Tsikou, D.; Peña-Rodríguez, L.M.; Katsarou, D.; Field, B.; Osbourn, A.E.; Papadopoulou, K.K. A metabolic gene cluster in Lotus japonicus discloses novel enzyme functions and products in triterpene biosynthesis. *New Phytol.* **2013**, *200*, 675–690. [CrossRef]
76. Jiang, S.Y.; Jin, J.; Sarojam, R.; Ramachandran, S. A Comprehensive Survey on the Terpene Synthase Gene Family Provides New Insight into Its Evolutionary Patterns. *Genome Biol. Evol.* **2019**, *11*, 2078–2098. [CrossRef] [PubMed]
77. Matsuba, Y.; Nguyen, T.T.; Wiegert, K.; Falara, V.; Gonzales-Vigil, E.; Leong, B.; Schafer, P.; Kudrna, D.; Wing, R.A.; Bolger, A.M.; et al. Evolution of a complex locus for terpene biosynthesis in solanum. *Plant Cell* **2013**, *25*, 2022–2036. [CrossRef] [PubMed]
78. Külheim, C.; Padovan, A.; Hefer, C.; Krause, S.T.; Köllner, T.G.; Myburg, A.A.; Degenhardt, J.; Foley, W.J. The Eucalyptus terpene synthase gene family. *BMC Genom.* **2015**, *16*, 450. [CrossRef] [PubMed]
79. Shimura, K.; Okada, A.; Okada, K.; Jikumaru, Y.; Ko, K.-W.; Toyomasu, T.; Sassa, T.; Hasegawa, M.; Kodama, O.; Shibuya, N.; et al. Identification of a Biosynthetic Gene Cluster in Rice for Momilactones. *J. Biol. Chem.* **2007**, *282*, 34013–34018. [CrossRef]
80. Chen, H.; Kollner, T.G.; Li, G.; Wei, G.; Chen, X.; Zeng, D.; Qian, Q.; Chen, F. Combinatorial Evolution of a Terpene Synthase Gene Cluster Explains Terpene Variations in Oryza. *Plant Physiol.* **2020**, *182*, 480–492. [CrossRef]
81. Boutanaev, A.M.; Moses, T.; Zi, J.; Nelson, D.R.; Mugford, S.T.; Peters, R.J.; Osbourn, A. Investigation of terpene diversification across multiple sequenced plant genomes. *Proc. Natl. Acad. Sci. USA* **2015**, *112*, E81–E88. [CrossRef]

Article

Genome-Wide Identification and Analysis of the APETALA2 (AP2) Transcription Factor in *Dendrobium officinale*

Danqi Zeng [1,2], Jaime A. Teixeira da Silva [3], Mingze Zhang [1,2], Zhenming Yu [1], Can Si [1], Conghui Zhao [1,2], Guangyi Dai [4], Chunmei He [1,*] and Juan Duan [1,5,*]

1. Key Laboratory of South China Agricultural Plant Molecular Analysis and Genetic Improvement, Provincial Key Laboratory of Applied Botany, South China Botanical Garden, Chinese Academy of Sciences, Guangzhou 510650, China; zengdanqi20@scbg.ac.cn (D.Z.); zhangmingze@scbg.ac.cn (M.Z.); zhenming311@scbg.ac.cn (Z.Y.); cans2013@163.com (C.S.); zhaoconghui@scbg.ac.cn (C.Z.)
2. College of Life Sciences, University of the Chinese Academy of Sciences, No. 19A Yuquan Road, Shijingshan District, Beijing 100049, China
3. Independent Researcher, P. O. Box 7, Ikenobe 3011-2, Miki-cho, Kagawa-ken 761-0799, Japan; jaimetex@yahoo.com
4. Opening Public Laboratory, Chinese Academy of Sciences, Guangzhou 510650, China; daigy@scbg.ac.cn
5. Center of Economic Botany, Core Botanical Gardens, Chinese Academy of Sciences, Guangzhou 510650, China
* Correspondence: hechunmei2012@scbg.ac.cn (C.H.); duanj@scib.ac.cn (J.D.); Tel.: +86-20-37252993 (J.D.); Fax: +86-20-37252978 (J.D.)

Abstract: The APETALA2 (AP2) transcription factors (TFs) play crucial roles in regulating development in plants. However, a comprehensive analysis of the AP2 family members in a valuable Chinese herbal orchid, *Dendrobium officinale*, or in other orchids, is limited. In this study, the 14 DoAP2 TFs that were identified from the *D. officinale* genome and named DoAP2-1 to DoAP2-14 were divided into three clades: euAP2, euANT, and basalANT. The promoters of all *DoAP2* genes contained cis-regulatory elements related to plant development and also responsive to plant hormones and stress. qRT-PCR analysis showed the abundant expression of *DoAP2-2*, *DoAP2-5*, *DoAP2-7*, *DoAP2-8* and *DoAP2-12* genes in protocorm-like bodies (PLBs), while *DoAP2-3*, *DoAP2-4*, *DoAP2-6*, *DoAP2-9*, *DoAP2-10* and *DoAP2-11* expression was strong in plantlets. In addition, the expression of some *DoAP2* genes was down-regulated during flower development. These results suggest that *DoAP2* genes may play roles in plant regeneration and flower development in *D. officinale*. Four *DoAP2* genes (*DoAP2-1* from euAP2, *DoAP2-2* from euANT, and *DoAP2-6* and *DoAP2-11* from basal ANT) were selected for further analyses. The transcriptional activation of DoAP2-1, DoAP2-2, DoAP2-6 and DoAP2-11 proteins, which were localized in the nucleus of *Arabidopsis thaliana* mesophyll protoplasts, was further analyzed by a dual-luciferase reporter gene system in *Nicotiana benthamiana* leaves. Our data showed that pBD-DoAP2-1, pBD-DoAP2-2, pBD-DoAP2-6 and pBD-DoAP2-11 significantly repressed the expression of the LUC reporter compared with the negative control (pBD), suggesting that these DoAP2 proteins may act as transcriptional repressors in the nucleus of plant cells. Our findings on *AP2* genes in *D. officinale* shed light on the function of *AP2* genes in this orchid and other plant species.

Keywords: *Dendrobium officinale*; AP2 transcription factor; development; dual-luciferase reporter gene system; gene expression

1. Introduction

As relatively static organisms, biochemical and genetic mechanisms in plants tend to be sophisticated, including delicate networks involved in regulatory mechanisms that allow plants to adapt to varying environments and resist biotic and abiotic stresses. Transcription factors (TFs) are proteins that can bind to DNA in a sequence-specific manner to regulate transcription. The regulation of gene transcription by TFs is an extremely complicated process [1], and is vital to plant growth and environmental responses. APETALA2

(AP2) belongs to the APETALA2/Ethylene Response Factor (AP2/ERF) superfamily [2,3], which participates in the regulation of various biological processes in plants, such as growth and development (flower development, somatic embryogenesis, meristem and leaf growth, etc.), hormones and stress responses [4–13].

Historically, the AP2/ERF superfamily has been divided into four separate families, namely the ERF, AP2, RAV and Soloist families [3,10]. The AP2 protein is exclusive to plants and contains two AP2 domains that have also been found in ancient plants such as gymnosperms, mosses, and *Chlamydomonas*, indicating that AP2 and EREBP (ethylene responsive element binding protein) families differentiated before Chlorophyta and Streptophyta lineages differentiated [14]. There are also similarities, including structures and conserved motifs, among the four families, suggesting that they have similar properties. For example, three *A. thaliana* AP2 genes (*AtAP2-6*, *AtAP2-7* and *AtAP2-11*) are involved in the regulation of seed development [15].

TFs recognize target DNA sequences with different DBDs, thereby controlling the expression of target gene promoters at the transcriptional level [16]. The AP2 domain not only plays a key role in transcriptional regulation [7], but also serves as the basis for family classification. In particular, the AP2 family, which contains two AP2 domains and a small number of proteins with a single AP2 domain [2,3,10,17], was further subdivided into the euAP2 (which is characterized by the miR172 binding motif) and AINTEGUMENTA (ANT) (which is characterized by signature amino acid insertions in the AP2 domain) clades [18,19] based on the amino acid sequences and nuclear localization of the two AP2 domains. Among them, the ANT clade was further divided into the basalANT and euANT clades. euANT proteins are defined by a long pre-domain region and four conserved motifs, which is the main difference between basalANT and euANT clades [19,20]. In total, based on differences in gene structure, plant AP2 proteins are divided into three subfamilies: euAP2, euANT and basalANT in the model plant *Arabidopsis thaliana* [20].

Early studies found that in *A. thaliana*, AP2 was a homeotic gene with a profound effect on floral organs that could determine the identity and fate of floral organs [21–24]. In rapeseed (*Brassica napus*), the *AP2*-like gene *BABY BOOM* regulates somatic embryogenesis [5]. In addition, the ANT clade is involved in the development of various plant organs, such as vegetative organs [25] and ovule development [26] in *A. thaliana*, fruit development in apple [27], berry size in grapevine [28], and many other examples. In addition, the rice *AP2*-like gene *SNB*, which belongs to the euAP2 clade, regulates the development of grains, such as seed shattering [29], *Eriobotrya japonica* EjAP2-1 interacts with EjMYB to induce fruit lignification [30], while the maize (*Zea mays*) *AP2* genes *ids1* and *sid1* regulate the initiation of corn flower meristems to determine the fate of meristem cells [31]. These studies demonstrate that the AP2 family is involved in the regulation of processes associated with plant development, such as flower development, embryonic development, meristem and leaf growth, among others.

Even though the AP2 family in plants has been extensively studied, AP2 family members in the medicinal (herbal) orchid, *Dendrobium officinale*, i.e., DoAP2s, have not been analyzed. The identification and analysis of DoAP2 family members will not only provide clues for revealing the function of AP2s in *D. officinale*, but also provide a theoretical basis for studying the functional conservation of AP2/ERF TFs, such as identifying new conserved protein domains and motifs, and enriching the functions of TF families in plants. In this study, 14 *DoAP2* genes were identified from *D. officinale*. They were systematically analyzed, a phylogenetic tree was constructed, protein interactions were predicted, and promoter *cis*-acting elements were analyzed, including their subcellular localization and using a dual-luciferase reporter assay. Moreover, using quantitative real-time polymerase chain reaction (qRT-PCR), their expression patterns were analyzed at different developmental stages, including flower development, and in response to different stress treatments, to lay a theoretical foundation for further analyzing the functions of DoAP2s in *D. officinale*.

2. Results

2.1. Identification and Analysis of AP2 Gene Family in D. officinale

A total of 14 genes (*DoAP2-1* to *DoAP2-14*) annotated as AP2 TFs were identified from the *D. officinale* genome. All 14 AP2 proteins from *D. officinale*, 16 AP2 proteins from *Oryza sativa*, and 18 AP2 proteins from *A. thaliana* were used to perform a phylogenetic analysis. The DoAP2 proteins were classified into three clades: euAP2, euANT, and basalANT (Figure 1A). The euANT clade contained the most (seven) AP2 proteins (DoAP2-2, DoAP2-4, DoAP2-5, DoAP2-7, DoAP2-9, DoAP2-13 and DoAP2-14), followed by the euAP2 clade with four members (DoAP2-1, DoAP2-3, DoAP2-8 and DoAP2-10), while only three AP2 proteins (DoAP2-6, DoAP2-11 and DoAP2-12) were found in the basalANT clade. The AP2 domain is responsible for DNA binding and protein complex formation of AP2 proteins [32]. Two AP2 domains were present in 13 of the 14 DoAP2 proteins, whereas DoAP2-14 only contained one AP2 domain (Figure 1B and Figure S1).

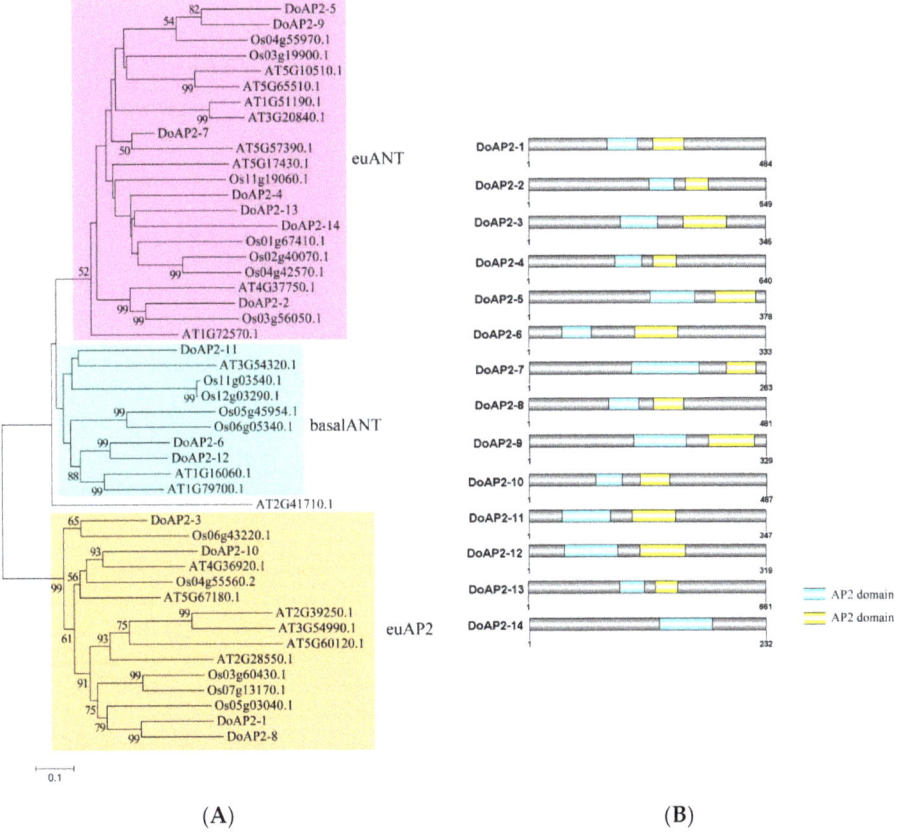

(A) (B)

Figure 1. Phylogenetic analysis and conserved domain of the AP2 proteins from *D. officinale*, *O. sativa* and *A. thaliana*. (**A**) Phylogenetic tree of DoAP2 proteins. A total of 14 AP2 proteins from *D. officinale*, 16 from *O. sativa* and 18 from *A. thaliana* were aligned using ClustalX to generate a FASTA alignment file. The phylogenetic tree was constructed using the MEGA 7.0 program and the neighbor-joining (NJ) method with 1000 bootstrap replications based on the alignment file; (**B**) Conserved domain of DoAP2 proteins. The location and size of AP2 domains are shown by different colors.

2.2. Prediction of Protein–Protein Interaction Network of AP2 Proteins

Protein–protein interactions play a role in transcriptional activation/repression and serve crucial functions in cellular regulation and biological processes in plants. Hence, we analyzed the protein–protein interaction network of the 14 DoAP2 proteins by STRING 11 and found that they were analogous to the interactions displayed by the corresponding *A. thaliana* orthologous proteins (Figure 2). These results show that several members of the DoAP2 family may have a certain connection to LEC1 and LEC2 of the LEC protein family, which is involved in embryonic development [33]. Interestingly, DoAP2-2, DoAP2-3, DoAP2-6, DoAP2-12, and DoAP2-14 were not linked to, nor did they interact with, any other DoAP2s based on the protein–protein interaction network of DoAP2s. Some DoAP2 proteins interacted with other DoAP2 proteins, such as DoAP2-4 and DoAP2-13 while others interacted with TFs involved in plant growth and development, such as DoAP2-1 and b-ZIP. These interactions suggest that DoAP2s play a broad role in plant growth and development.

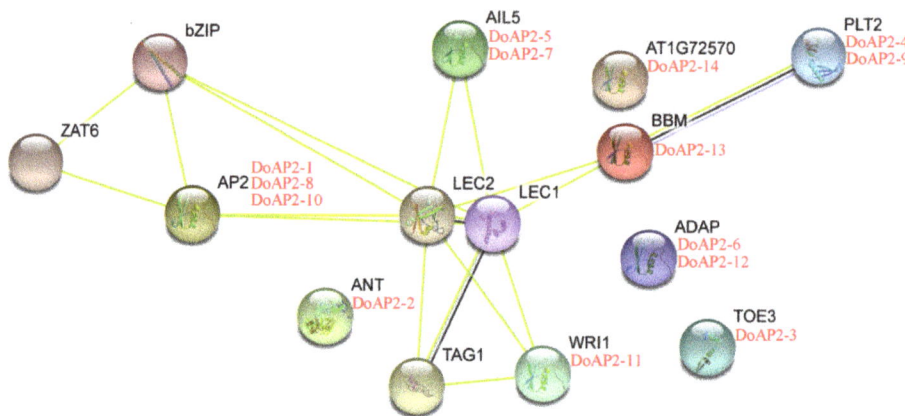

Figure 2. Protein–protein interaction network analysis of DoAP2 proteins using STRING 11. Cyan line represents data from curated databases, green lines indicate gene neighborhoods, and black lines indicate co-expression. Red text = DoAP2s; black text = the proteins in *A. thaliana*.

2.3. Analysis of Cis-Regulatory Elements in the Promoters of DoAP2 Genes

AP2 genes that are involved in plant growth and development are regulated by different factors. In order to investigate the *cis*-regulatory elements in the promoters of *DoAP2* genes, we isolated the 2000-bp upstream section according to the *D. officinale* genome and analyzed the *cis*-regulatory elements using the PlantCARE web site. The *cis*-regulatory elements of the promoters of *DoAP2* genes were related to growth and development (meristem expression and specific to the endosperm), plant hormones (auxin, abscisic acid, methyl jasmonate (MeJA), gibberellin, and salicylic acid) and stress (drought inducibility, low temperature responsiveness, anaerobic induction, and defense and stress responsiveness). As depicted in Figure 3, more than half of the *DoAP2* genes harbored a total of seven meristem expression-responsive elements and eight endosperm specific-responsive elements, indicating that *DoAP2* genes may play a vital role in meristem growth and embryonic development of *D. officinale*. In addition, an abundance of elements responsive to plant hormones was present in the promoters of all *DoAP2* genes, demonstrating the response of these genes to these hormones. Interestingly, MeJA-responsive elements formed the largest group of elements among the promoters of *DoAP2* genes, indicating that *DoAP2* genes are MeJA-responsive genes (Figure 3). As a vital cellular regulator, MeJA plays a crucial role in mediating various developmental processes and defense responses against biotic and abiotic stresses [34]. Furthermore, except for *DoAP2-10*, the remaining 13 genes contained a total of 45 abiotic stress-responsive elements, not only suggesting that the expression of 13 *DoAP2* genes was associated with these abiotic stresses, but also that they played a role in various stress regulatory networks. Collectively, these results indicate that *AP2* family members participate in embryonic development, meristem growth and environmental stress regulation during the growth and development of *D. officinale*.

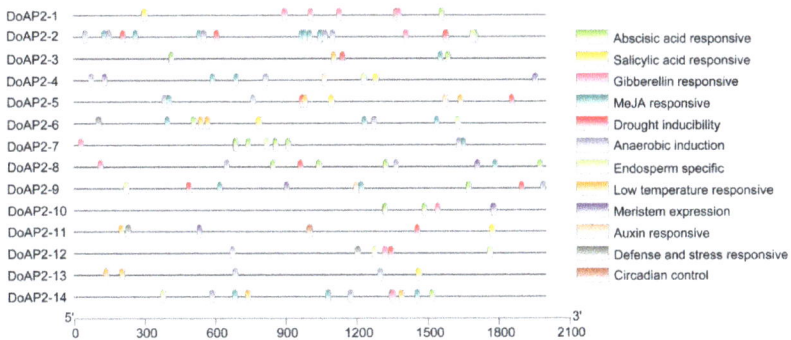

Figure 3. Prediction of *cis*-responsive elements in the 2-kbp upstream region of the initiation codon of 14 *DoAP2* genes. Different colored boxes indicated different *cis*-responsive elements. MeJA, methyl jasmonate.

2.4. Expression Analysis of DoAP2 Genes at Different Developmental Stages

AP2 TFs are regarded as factors that are primarily responsible for the regulation of developmental programs [10]. Protocorm-like bodies (PLBs), which form during the in vitro culture of orchid plants, can proliferate or develop into a complete plant. We analyzed the changes in expression of *DoAP2* genes during the development of PLBs to plantlets (PLBs, multiple shoots and plantlet). All *DoAP2* genes were detected at all three developmental stages, except for *DoAP2-4*, which showed the highest expression in PLBs and the lowest expression in plantlets (Figure 4). The decrease in expression from PLBs to plantlets suggests that *DoAP2-4* may play a role in PLB development. In addition, *DoAP2-7* and *DoAP2-8* were strongly detected in PLBs (Figure 4). *DoAP2-3*, *DoAP2-4*, *DoAP2-6*, *DoAP2-9*, *DoAP2-10*, and *DoAP2-11*, but especially *DoAP-6*, were abundant in plantlets (Figure 4). The expression of *DoAP2-1* was not different among the three developmental stages. The *DoAP2* genes displayed different expression patterns, even within the same clade. For example, in the euANT clade, *DoAP2-2* and *DoAP2-7* were highly expressed in PLBs while *DoAP2-4* and *DoAP2-9* were highly expressed in plantlets (Figure 4).

Flowers are important functional organs of plants. AP2 is involved in flower development [23]. For example, two genes, *ANT* and *ANT-LIKE6*, regulate *A. thaliana* floral growth and patterning [35]. We detected the expression of *DoAP2* genes during three stages of *D. officinale* flower development, in small flower buds (FB1), medium flower buds (FB2), and fully bloomed flowers (FBF) (Figure 5). As shown in Figure 5, *DoAP2-8* and *DoAP2-10* exhibited a similar expression pattern, showing relatively high expression levels in FB1, decreasing as the flower developed further. The expression of *DoAP2-3*, which was in the same clade as *DoAP2-8* and *DoAP2-10*, was up-regulated during flower development, and most expressed in FBF. A similar pattern was found in *DoAP2-2*, but a different expression pattern in *DoAP2-4* and *DoAP2-5*, all from the euANT clade (Figure 5). Four out of seven euANT genes were abundant during FB1 (Figure 5). In particular, *DoAP2-11* was specifically expressed in FBF, the expression of *DoAP2-11* in FBF was about 193.56- and 2225.64-fold higher than in FB1 and FB2 while its expression level was much higher than that of other *DoAP2* genes. These findings show the specificity of expression of different *DoAP2* genes in floral development. Moreover, seven *DoAP2* genes (*DoAP2-2*, *DoAP2-7*, *DoAP2-8*, *DoAP2-10*, *DoAP2-12*, *DoAP2-13*, and *DoAP2-14*) had the highest expression in FB1 compared to FB2 and FBF, suggesting that they might play an essential role in the pre-flowering developmental state where differentiation is not yet complete. These results suggest that *DoAP2* family members play a role in the development of *D. officinale* flowers.

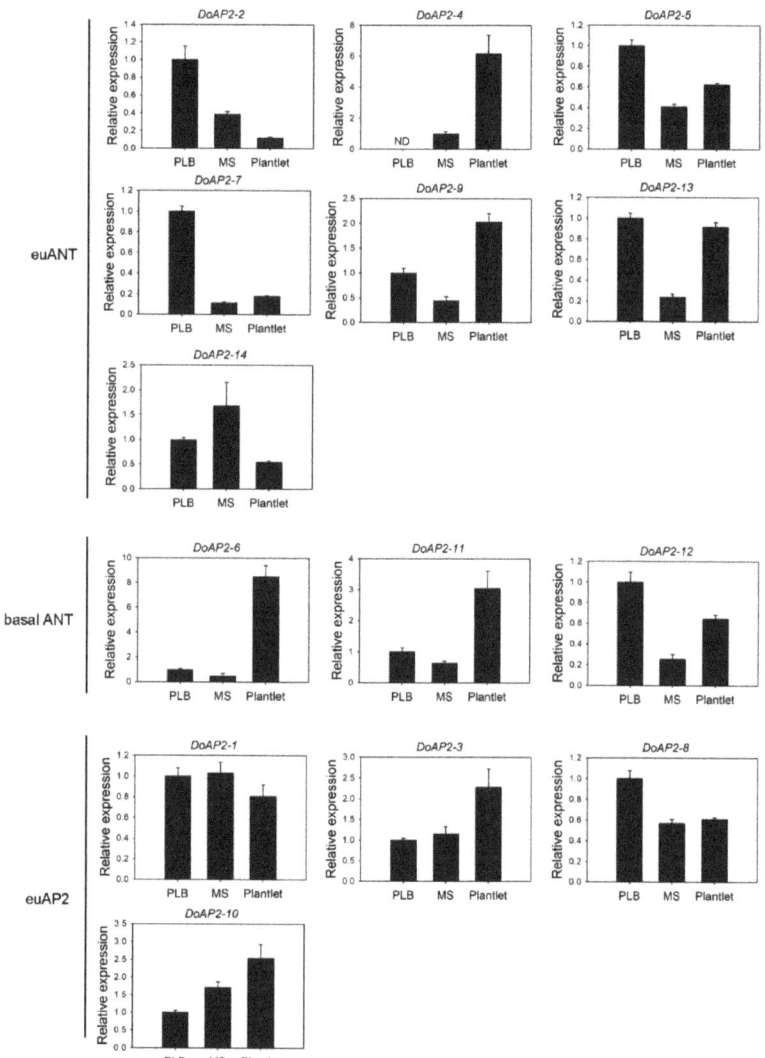

Figure 4. Expression analysis of *DoAP2* genes during plant regeneration from PLBs by qRT-PCR. PLB, protocorm-like bodies; MS, multiple shoots. Each data bar represents the mean ± standard deviation (SD) of three biological replicates ($n = 3$). ND, not detected.

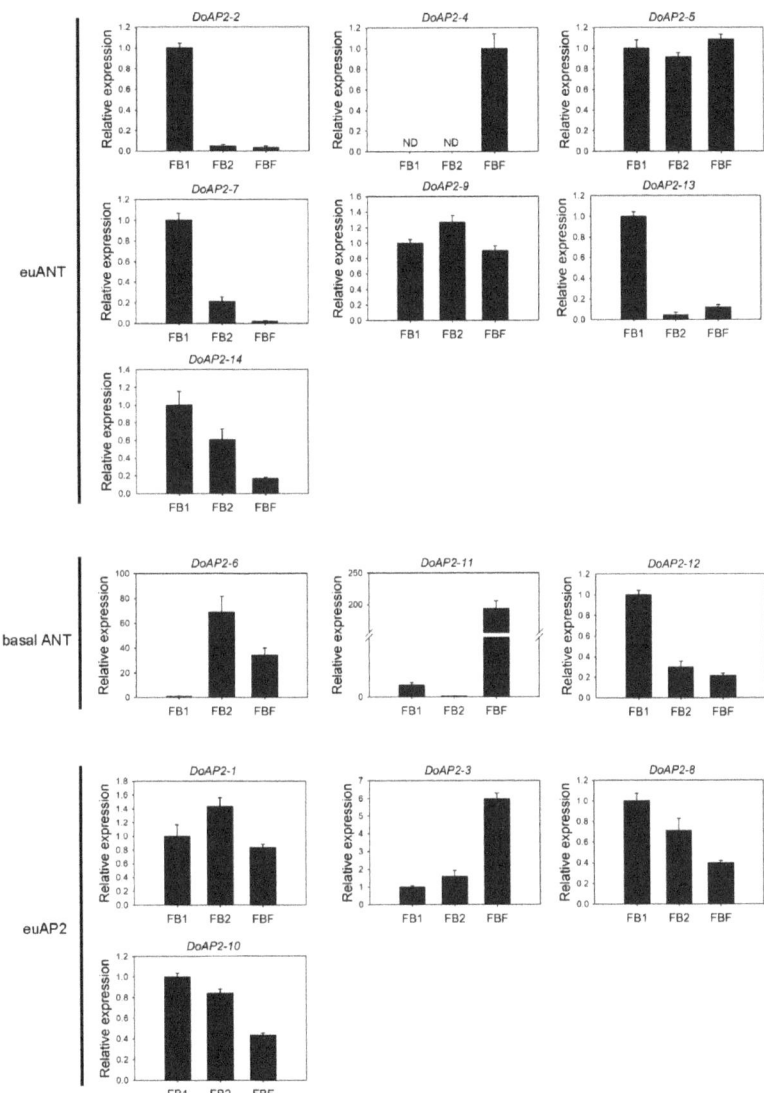

Figure 5. Expression analysis of *DoAP2* genes during flower development by qRT-PCR. FB1, small flower buds (about 5 mm long); FB2, medium flower buds (about 10 mm long); FBF, fully bloomed flowers. Each data bar represents the mean ± standard deviation (SD) of three biological replicates (n = 3). ND, not detected.

The unique flower shape of orchids gives them high ornamental value [36]. We then analyzed the expression of *DoAP2* genes in different floral tissues (sepal, petal, lip, and column) at the FBF stage (Figure S2). The expression of *DoAP2-4* and *DoAP2-13* was not detected in any of the four tissues of FBFs, and *DoAP2-2*, *DoAP2-7*, and *DoAP2-14* were expressed only in the column and might be closely related to flower morphogenesis. Interestingly, the aforementioned genes that were specifically expressed, or not expressed, belong to the euANT clade. Remarkably, the remaining nine *DoAP2* genes were expressed in different tissues of FBFs. Among them, *DoAP2-1*, *DoAP2-3*, and *DoAP2-11* were more highly expressed in petals, and the expression trends of *DoAP2-1* and *DoAP2-3* of the same clade were consistent; *DoAP2-6* was the most highly expressed in the column, the expression of the column was about 12.88-, 5.88- and 7.72-fold higher than in the sepal, petal and lip, respectively; the expression of *DoAP2-5*, *DoAP2-9*, *DoAP2-10*, and *DoAP2-12* were all abundant and the variations were small among the four tissues of FBF. Based on these findings, *DoAP2* genes had unique expression patterns, indicating that they probably played diverse roles in different *D. officinale* FBF tissues.

2.5. Expression Analysis of DoAP2 Genes in Response to Abiotic Stresses

AP2 responds to abiotic stresses [37]. To further investigate the changes in expression of *DoAP2* genes to different abiotic stress treatments (cold, PEG and NaCl), we examined the expression levels of 14 *DoAP2* genes under abiotic stress using qRT-PCR, while samples from untreated plantlets served as the control (Figure 6). According to their expression profiles, the expression of *DoAP2-2*, *DoAP2-4*, *DoAP2-5*, *DoAP2-12*, *DoAP2-13*, and *DoAP2-14* were reduced to varying degrees. In addition, the differences in expression of *DoAP2-11* in different treatments were slight, indicating that the above treatments had little effect on *DoAP2-11*. The expression of *DoAP2-3* and *DoAP2-7* in the NaCl treatment was about 3.49- and 2.56-fold higher than in the control, demonstrating that they played a role in the mechanism of response to stress in the face of adversity, especially high salt stress. The expression of *DoAP2-9* in the cold treatment was more abundant than in the other three treatments. Compared to the control, the expression of at least one of the remaining *DoAP2* genes was up-regulated in response to these abiotic stress treatments. In particular, the expression of *DoAP2-6* was 1.93-, 1.95- and 1.21-fold higher than the control in the cold, PEG and NaCl treatments, respectively, implying its potential importance in the adaptation of this orchid to adverse growth conditions experiencing abiotic stresses.

2.6. Subcellular Localization of Selected DoAP2 Proteins

To explore the localization of DoAP2 proteins, *A. thaliana* protoplasts were PEG-mediated transformed with a transient expression vector containing YFP. According to the phylogenetic tree (Figure 1), AP2 proteins of *D. officinale* and *A. thaliana* were classified into three clades. We selected four representative genes, namely DoAP2-2 of the euANT clade, DoAP2-6 and DoAP2-11 of the basalANT clade, and DoAP2-1 of the euAP2 clade, for further analysis. Subcellular localization was observed using a Zeiss LSM 510 Meta confocal microscope. Yellow fluorescence signals of the positive control (empty YFP vector) were detected in the cytoplasm and plasma membrane (Figure S3). As expected, the yellow fluorescent signals of four YFP-DoAP2-fused proteins were localized in the nucleus (Figure 7), conforming to their transcriptional regulatory function in the nucleus.

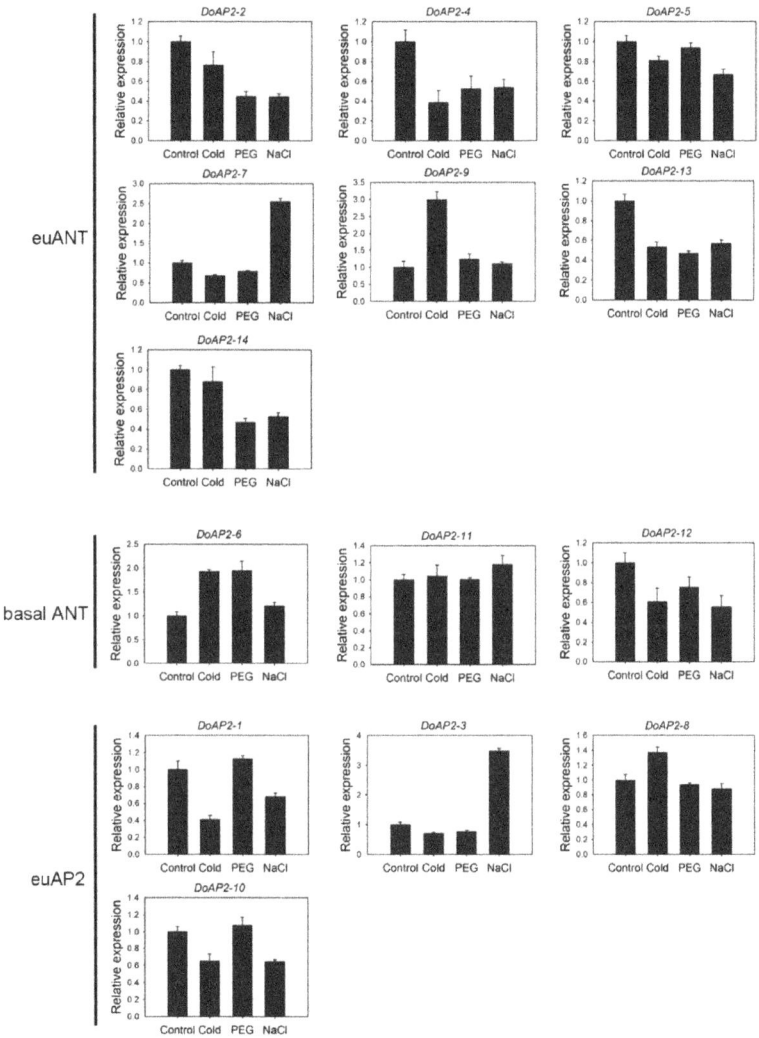

Figure 6. Expression analysis of *DoAP2* genes in response to abiotic stresses (cold, PEG and NaCl) by qRT-PCR. PEG treatment, 15% polyethylene glycol (PEG)-6000; NaCl treatment, 250 mM NaCl; cold treatment, 4 °C. Each data bar represents the mean ± standard deviation (SD) of three biological replicates (n = 3).

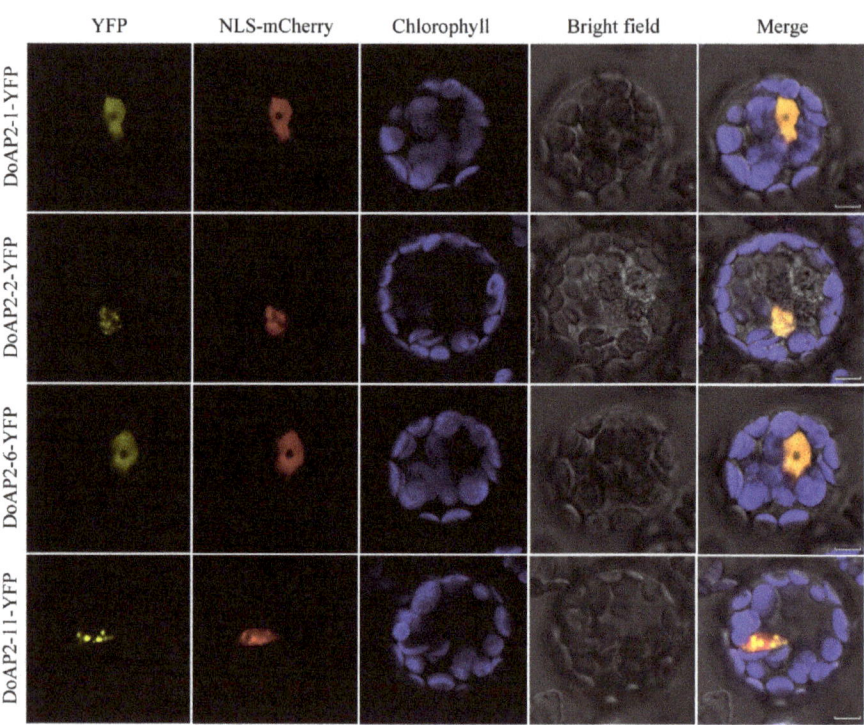

Figure 7. Subcellular localization of four DoAP2 proteins (DoAP2-1, DoAP2-2, DoAP2-6 and DoAP2-11) in *A. thaliana* protoplasts. Bars = 5 μm.

2.7. Four DoAP2 Proteins Displayed Transcriptional Repression in Tobacco Leaves

AP2 family members are considered to be TFs, and a defining feature of a TF is its transactivation activity [38]. Hence, we investigated the transcriptional activity of four *DoAP2* genes (*DoAP2-1, DoAP2-2, DoAP2-6* and *DoAP2-11*) using a dual-luciferase reporter gene system in tobacco (*Nicotiana benthamiana*) leaves. These representative genes are the same as those that were used for subcellular localization. The constructed vectors are shown in Figure 8A. We used vectors containing the CaMV35S-driven pBD and fusion protein vectors pBD-VP16, pBD-DoAP2s (DoAP2-1, DoAP2-2, DoAP2-6 and DoAP2-11) as the effector, and the CaMV35S-driven LUC and TATA cassette-driven REN as reporters. pBD-EMPTY and pBD-VP16 were used as negative and positive controls, respectively. In this system, the reporter vector was generated by fusing the *firefly luciferase* (*LUC*) gene after five GAL4 binding sites, and a *renilla luciferase* (*REN*) gene driven by a *CaMV35S* promoter in the reporter vector was used as the internal control (Figure 8A). The ORF of the four *DoAP2* genes was cloned into the site of the pBD vector, which is after the GAL4 binding domain (Figure 8A). The ratio of the two luciferases (LUC and REN) was detected using a dual fluorescent reporter gene system assay (Figure 8B). The LUC/REN ratio of the positive control pBD-VP16 was 27-fold higher than the negative control pBD, while the LUC/REN ratio of the four pBD-DoAP2 proteins was significantly lower than that of pBD (Figure 8B). These results show that *DoAP2-1, DoAP2-2, DoAP2-6,* and *DoAP2-11* genes had transcriptional repression activity in tobacco plants, i.e., they are transcriptional repressors.

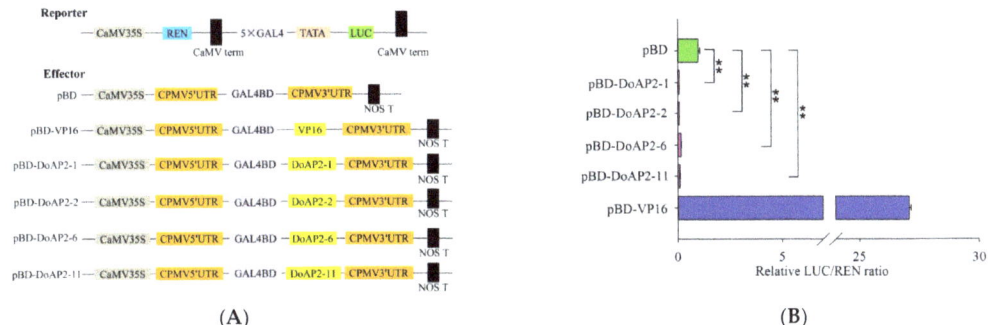

Figure 8. Transcriptional activity assay of *DoAP2* genes in tobacco leaves. (**A**) Schematic presentation of the reporter and effector vectors; (**B**) Transcriptional repression ability of DoAP2 proteins in tobacco leaves. Double asterisks (**) indicate significant differences in two-treatment comparisons ($p < 0.01$) using Dunnett's test, compared with the negative control (pBD). Each value represents the means of six biological replicates.

3. Discussion

3.1. Bioinformatics Analysis of DoAP2 TFs

In this study, we identified 14 *AP2* genes in the *D. officinale* genome. The AP2 family is a small TF family with fewer members in plants than the ERF TF family. For example, 18 *AP2* genes were found in model plant Arabidopsis [3], 26 in *Indica* rice [39], and 62 in wheat [40]. The AP2 proteins from plants are divided into three subfamilies: euAP2, euANT and basalANT [10,14,20,41]. In this study, DoAP2 proteins were classified into euAP2, euANT and basalANT subfamilies, similar to other plants. All AP2 proteins contain two AP2 domains except for DoAP2-14, which has a single AP2 domain (Figure 1B and Figure S1). This is consistent with a prior finding that AP2 family members contain one or two AP2 domains [10].

A study of protein interactions can enrich the characteristics of TFs, such as localization, transcription activity, target specificity and function. Based on the STRING 11 tool, all protein–protein interaction networks of DoAP2 proteins were predicted. Interestingly, some DoAP2 proteins may interact with TFs involved in plant growth and development such as b-ZIP. Additionally, there are many reports on the interaction of proteins with members of the AP2/ERF superfamily. For example, rice OsERF3 interacted with WOX11 to regulate rice crown root development [42], AP2 TF HaDREB2 in sunflower (*Helianthus annuus* L.) interacted with another TF HaHSFA9 to regulate zygotic embryogenesis [43], and AtERF5 in *A. thaliana* interacted with AtERF6, AtERF8, SCL13 and other proteins to exert a wide range of regulatory effects, such as defense against phytopathogenic fungi [44,45]. The results of interaction protein prediction in this paper (Figure 2) provide a notion about the possible binding nature of DoAP2 proteins, although further verification is needed with yeast two hybrid, co-immunoprecipitation and bimolecular fluorescence complementation assays to determine the DoAP2 protein interaction network and its protein interaction under different conditions.

Increasing lines of evidence have shown that the AP2/ERF superfamily is mainly involved in development and abiotic stress responses [5,7,8,13,46,47]. In the present study, many growth and development, hormone and stress *cis*-acting elements were detected in the promoter regions of *DoAP2* genes (Figure 3). In addition, *cis*-acting elements regulate the expression of stress-inducible genes, leading them to be referred to as molecular switches that regulate various biological processes [48]. The promoters of *DoAP2* genes contain endosperm-specific *cis*-acting elements with expression in the meristem in response to growth and developmental processes, and that respond to multiple stress signals, which might regulate various biological processes. This is consistent with the basic functions of AP2 TFs [37,49]. Different DoAP2 TFs play a variety of roles that may be related to their specific and/or differential binding to different *cis*-acting elements or other proteins, suggesting their involvement in different regulatory processes [50–52].

3.2. The AP2 TF Play Important Roles in Plant Regeneration and Flower Development

An increasing number of studies provide evidence for the involvement of *AP2* genes in plant regeneration and flower development.

In plant regeneration, AP2/ERF superfamily TFs promoted callus induction and proliferation, shoot differentiation, root differentiation and differentiation of somatic cells [53]. Several examples are provided next. Overexpression of the AP2 gene *ZmBBM2* promoted callus induction and proliferation in maize [54]. In cacao (*Theobroma cacao*), overexpressed of the *TcBBM* gene induced embryo formation; moreover, the *TcBBM* gene can be used as an embryogenesis biomarker in cacao [11]. *HbAP2-3* and *HbAP2-7* genes are marker genes of somatic embryogenesis during callus proliferation in rubber tree (*Hevea brasiliensis*) [55]. An AP2/EREBP-type transcription activator NtCEF1 regulated gene expression in tobacco callus [56]. In rapeseed and *A. thaliana*, BBM, which shares similarities with the AP2/ERF superfamily of TFs, led to the differentiation of somatic cells, inducing embryonic development [5]. An AP2/ERF TF WOUND INDUCED DEDIFFERENTIATION1 (WIND1) promoted shoot regeneration in *A. thaliana* [57]. In *A. thaliana*, AP2/ERF TFs play an important role in root regeneration [58]. The AP2/ERF gene *GmRAV1* regulated the regeneration of roots and adventitious buds in soybean [59]. *D. officinale* PLBs are considered to be somatic embryos that can proliferate and also differentiate into complete plants [60]. The induction, proliferation and regeneration of PLBs is an advantageous method for the large-scale production of *D. officinale* [61]. In addition, AP2 TFs specify the identity of floral organs and regulate the expression of genes related to flower development [20–24,62,63]. For example, *D. officinale* flowers contain three petalized sepals, two petals, one lip and one column [64], the unique floral patterning gives this orchid its ornamental value, it provides advantages to pollination, and promotes normal plant development [65]. In this study, there were five *DoAP2* genes (*DoAP2-2*, *DoAP2-5*, *DoAP2-7*, *DoAP2-8* and *DoAP2-12*) that showed abundant expression in PLBs from among the three development stages of *D. officinale*. In particular, it is important to emphasize that we found that *DoAP2-2* and *DoAP2-7* were specifically expressed in PLBs while *DoAP2-2* was down-regulated during the development of *D. officinale*. Seven *DoAP2* genes (*DoAP2-2*, *DoAP2-7*, *DoAP2-8*, *DoAP2-10*, *DoAP2-12*, *DoAP2-13* and *DoAP2-14*) were strongly expressed in the early flower buds, and were down-regulated as flowers developed (Figure 5), suggesting that *DoAP2* genes play a role in *D. officinale* flower development. Among them, the expression levels of *DoAP2-2* and *DoAP2-7* were highest in PLBs and FB1, and both strongly expressed in the column (Figure S2), indicating that *DoAP2-2* and *DoAP2-7* are highly specific genes that may play an important role in immature tissues of *D. officinale*. These findings imply that *DoAP2* and *DoAP2-7* genes are involved in regulating the maintenance of immature tissues and flower development, supporting the view that AP2 TFs have important functions in regulating plant growth and development. The expression levels of *DoAP2-6* and *DoAP2-11* of the same clade basal ANT were similar at different stages (Figure 4), and their expression continued to increase during development, peaking in plantlets and FBF. However, they also displayed some differences. Among the expression levels of different tissues in FBF, *DoAP2-6* showed abundant expression in the column, while *DoAP2-11* expression was abundant in sepals and petals. In *D. officinale*, *DoAP2* genes have diverse roles in flowers, similar to the expression pattern of the *NsAP2* gene in floral organs of water lily (*Nymphaea* sp. cv. 'Yellow Prince') [49]. Despite these similarities, flower development is a very complicated process, and *AP2* genes may be directly or indirectly involved in a certain regulatory role. Therefore, it is necessary to conduct more in-depth and detailed research on these genes.

3.3. DoAP2 Genes May Play a Role in Abiotic Stress Response

The expression of *DoAP2-6* increased to varying degrees, especially in response to cold stress and PEG treatment (Figure 6). The expression of *DoAP2-11* also increased, but its amplitude was much lower than that of *DoAP2-6* (Figure 6). This shows that the basal clade members *DoAP2-6* and *DoAP2-11* may have important regulatory effects under adverse abiotic stresses, allowing for a response to salt and drought stress during the growth and development of *D. officinale* plants. *DoAP2-6* and *DoAP2-11* contained a large number of stress-related *cis*-acting elements related to drought and low temperature (Figure 3), which may be closely related to their increased expression levels under different stress treatments (Figure 6), indicating that they are involved in the regulation of adverse abiotic stresses. Collectively, the above results suggest that genes with similar *cis*-acting elements among genes of the same clade may perform similar functions [66].

3.4. The DoAP2 Proteins Are Localized in the Nucleus and Display Transcription Activity

Nuclear localization is a key regulatory mechanism of TFs [67]. Our subcellular localization analysis of DoAP2 proteins indicated that, like many other AP2 and ERF TFs such as PsAP2 [68], OsDREBL [69] and GsERF71 [70], DoAP2-1, DoAP2-2, DoAP2-6 and DoAP2-11 were localized in the nucleus (Figure 7). These findings demonstrate that DoAP2 proteins have the basic characteristics of TFs, performing functions in the nucleus. However, since we selected representative genes of each clade from among the 14 DoAP2 TFs, this does not mean that all DoAP2 proteins have the above characteristics, and the related characteristics of the remaining proteins still needs additional research.

Based on their functions, TFs can be divided into either activators or repressors. Repressors play an important role in the regulation of gene expression by inhibiting the expression of certain genes by combining with DNA elements, transcription activators or promoter sequences [71], enabling plants to save energy under normal (non-adverse) conditions [72]. The expression of a repressor is also closely related to growth and development, and overexpression of repressors can lead to abnormal plant development [73,74]. However, compared with activators, there is less research on suppressors, especially in non-model plants. Previous studies showed that *AP2* genes can negatively regulate the expression of certain genes to achieve corresponding functions, and this has been well studied in *A. thaliana*. For example, early studies found that *A. thaliana* AP2 is a negative regulator of the AGAMOUS gene, both of which were involved in flower development, and the regulatory mechanism of their interaction established the expression pattern of floral homologous genes in *A. thaliana* to some extent [75]. An *A. thaliana AP2* gene (*At4g36920*) negatively regulated the *REPLUMLESS* (*RPL*) gene that controls fruit dehiscence to achieve the function of controlling fruit development [62]. The *AP2* gene negatively regulates the size and number of embryonic cells, thus achieving the function of affecting seed mass and seed yield in *A. thaliana* [15,76]. Moreover, the AP2-like TF mutant of rice showed enlarged grains and increased grain weight [29], and *AP2* genes also negatively regulated the formation of the abscission layer [77], thereby affecting the development of rice grains. In this study, we further tested the transcription activity of DoAP2 TFs using the dual-luciferase assay. As shown in Figure 8, *DoAP2* genes had a strong repressive effect. This is likely to be related to the AP2/EREBP domain, which belongs to the DBD of a plant transcription repressor [32]. Such a repressor needs to bind to DNA to repress transcription [78]. In short, studies of repressors have important biological significance, and they can not only enrich our understanding of the negative regulatory role of plants in response to external environmental stresses, but also provide new theoretical guidance for genetic improvement of plant resistance to stresses in adverse growth conditions. Therefore, further in-depth studies on DoAP2 TFs will be of great significance for breeding and trait improvement.

4. Materials and Methods

4.1. Plant Material and Growth Conditions

The *D. officinale* plants used in this study were grown and maintained in the South China Botanical Garden, Chinese Academy of Sciences, Guangzhou, China. The expression patterns of *DoAP2* genes were performed on different *D. officinale* tissues (see below). The sampling method is also described below. PLBs, multiple shoots (MS, i.e., without roots) and plantlets (about 5 cm high) of *D. officinale* were grown on half-strength Murashige and Skoog (1/2MS) [79] medium supplemented with 20 g/L sucrose, 6 g/L agar and 1 g/L activated carbon (pH 5.4) in a growth chamber. We collected *D. officinale* material treated with PEG, NaCl and cold stress. Among them, concentrations were selected based on a relevant previous study [80]. First, *D. officinale* plants under normal growth conditions were cultured on 1/2MS with 0.1% activated carbon, 2% sucrose, and 0.6% agar medium (pH 5.4). *D. officinale* plants were then separately exposed to one of several abiotic stresses (on the basis of the above growth conditions): 15% polyethylene glycol (PEG)-6000 (Sigma-Aldrich, Shanghai, China; the PEG treatment), 250 mM NaCl (Guangzhou Chemical Reagent Factory, Guangzhou, China; the NaCl treatment) and 4 °C (the cold treatment). The culture conditions in controlled-climate chambers were: 26 ± 1 °C, 86.86 µmol·m^{-2}·s^{-1}, a 12-h photoperiod, and about 60% relative humidity. In addition, MS and plantlets were derived from PLBs, as PLBs can grow into MS and plantlets after culture in the above medium. Each treatment was conducted as three replications and five *D. officinale* PLBs, MS and plantlets were used for each treatment. All samples were instantaneously frozen in liquid nitrogen for 15 min then stored at -80 °C for later use.

4.2. Identification of DoAP2 Genes from the D. officinale Genome

From the NCBI (https://ftp.ncbi.nlm.nih.gov, accessed on 27 September 2020) genome database, we selected *D. officinale* and downloaded the *D. officinale* genome file. The AP2 protein sequences from *O. sativa* and *A. thaliana* were obtained from Plant Transcription Factor Database (http://planttfdb.gao-lab.org/index.php, accessed on 9 October 2020). Genes were identified by a hidden Markov model (HMM) search based on the AP2 domain using the Pfam protein domain database (http://pfam.xfam.org/, accessed on 10 October 2020). The HMM file was assessed by the HMMER3 software package under default parameters (http://hmmer.janelia.org/, accessed on 10 October 2020). Subsequently, DoAP2 proteins were verified via a local HMM-based search program (E-value $\leq 1 \times 10^{-10}$). Furthermore, the identified sequences were confirmed to be AP2 proteins by annotation as an AP2 protein, either in the Uniprot database (https://www.uniprot.org/, accessed on 11 October 2020) or in the NCBI database. Retrieved AP2 protein sequences were compared with the *A. thaliana* AP2 protein sequences and a phylogenetic tree was constructed using the MEGA version 7 program [81]. Lastly, the remaining 14 proteins were considered to be *D. officinale* DoAP2 proteins.

4.3. Bioinformatics Analysis of DoAP2 Proteins

DNAMAN version 8.0 software (Lynnon Biosoft, Foster City, CA, USA) was used to generate multiple sequence alignments of full-length amino acid sequences of the DoAP2 proteins. In addition, we also used Clustal X 2.0 [82] to perform multiple alignments of DoAP2 proteins to further verify the results of sequence alignment. For the phylogenetic analysis, based on the alignment of AP2 proteins, we used MEGA version 7 [81] to perform phylogenetic and molecular evolutionary analyses of AP2 proteins. Initially, the amino acid sequences of the AP2 proteins from *D. officinale* and *A. thaliana* (*FASTA* format) were arrayed with Clustal X 2.0 [82], and the UniProt BLAST online website (http://www.uniprot.org/blast/, accessed on 11 October 2020) was used to calculate sequence identity based on the neighbor-joining method [83] with 1000 bootstrap replicates. Thus, a phylogenetic tree of *D. officinale* and *A. thaliana* AP2s was constructed. *D. officinale* AP2 proteins were categorized by their phylogenetic relationships with the corresponding *A. thaliana* AP2 proteins. Additionally, we used NCBI's conserved domain database (CDD) [84] to identify the conserved domains of DoAP2 proteins, and to calculate the conserved domain start sites and lengths. Finally, DOG2.0 software (http://dog.biocuckoo.org/, accessed on 22 October 2020) was used to map the distribution of conserved domains.

We obtained the promoter sequences of *DoAP2* genes (Table S1) from the *D. officinale* whole genome sequencing files. Subsequently, the upstream 2000 bp sequence relative to the translation initiation codon (ATG) of the promoter of each *DoAP2* gene was selected as the promoter region, and the PlantCare online software (http://bioinformatics.psb.ugent.be/webtools/plantcare/html/, accessed on 19 November 2020) was used to predict the *cis*-acting elements in the promoters of *DoAP2* genes. Finally, the prediction map of *DoAP2* genes' promoters were drawn by TBtools [85].

Based on the association model of *A. thaliana*, the STRING 11 tool (https://string-db.org, accessed on 2 December 2020) [86] was used to predict the protein–protein interaction network between DoAP2 proteins and other proteins.

4.4. RNA Extraction, cDNA Synthesis and qRT-PCR

The RNA extraction kit, RNAout2.0 reagent (Tiandz Inc., Beijing, China) was used to extract total RNA from the aforementioned *D. officinale* materials according to the operation manual. We used RNase-free DNase I (Takara Bio Inc., Kyoto, Japan) to purify RNA. After the extracted RNA was digested with DNase, 2 μL was applied to agarose gel electrophoresis, and the integrity of the RNA was detected by the Clinx GenoSens gel documentation system (Clinx Science Instruments, Shanghai, China). Total RNA was reverse transcribed using the GoScript™ Reverse Transcription System (Promega, Madison, WI, USA) in accordance with the manufacturer's protocol, and 4 μg of purified total RNA was used for reverse transcription. The reaction system was 20 μL to synthesize first strand cDNA. The obtained cDNA was diluted in ddH$_2$O to 1:50 and applied as a template for qRT-PCR analysis. The LightCycler 480 system (Roche, Basel, Switzerland) that uses the Aptamer™ qPCR SYBR® Green Master Mix (Tianjin Novogene Bioinformatics Technology Co. Ltd., Tianjin, China) was used to perform qRT-PCR. Reaction conditions were: 95 °C for 5 min, and 40 subsequent cycles of 95 °C for 10 s and 60 °C for 1 min. *D. officinale* *ACTIN* (NCBI accession number: JX294908) was used as the internal reference gene [87] to standardize cDNA concentration. Relative gene expression was calculated using the $2^{-\Delta\Delta CT}$ method [88]. Supplementary Table S2 lists the specific primer sequences for the *DoAP2* genes. Three independent biological replicates were performed for each sample.

4.5. Subcellular Localization Analysis

The transient gene expression system that uses *A. thaliana* mesophyll protoplasts is an advantageous tool, and is often used for subcellular localization. First, we inserted the entire coding sequence of the four *DoAP2* genes (*DoAP2-1*, *DoAP2-2*, *DoAP2-6* and *DoAP2-11*) without stop codons into the *Eco*RI site of the pSAT6-EYFP-N1 vector [89]. In addition, this study used an additional method [90] to isolate protoplasts from *A. thaliana* leaves at the 4-weeks-old stage. Since TFs are generally located in the nucleus, for further verification, the recombinant protein was combined with the NLS location marker to transform *A. thaliana* mesophyll protoplasts using a PEG-mediated method [90]. After incubation for 12–18 h in the dark, a Leica TCS SP8 STED 3× microscope (Wetzlar, Hesse, Germany) was used to excite the YFP fluorescence signal at 514 nm to observe the yellow fluorescence signal of *A. thaliana* mesophyll protoplasts. The primers used to construct the four YFP-DoAP2 fused proteins are listed in Supplementary Table S3.

4.6. Dual-Luciferase Reporter (DLR) Assay

The DLR assay, which was used to investigate the transcriptional activation of TFs, was performed according to a previous report [91]. Briefly, the coding sequences of *DoAP2* (*DoAP2-1*, *DoAP2-2*, *DoAP2-6* and *DoAP2-11*) genes without the stop codon were inserted into the constructed pBD vector driven by the 35S promoter as effector, and the double-reporter vector as reporter, which includes a GAL4-LUC and an internal control REN driven by the 35S promoter. The effector and reporter were genetically transformed into tobacco (about 5–6 weeks, young non-flowering plants) leaves using *Agrobacterium tumefaciens* strain GV3101 (Weidi, Shanghai, China) and tobacco plants were cultured in the dark for 3 days at 25 °C. Finally, according to the manufacturer's instructions, the activities ratio of the two luciferases (LUC and REN) was carried out using the DLR assay (Promega Corp., Madison, WI, USA) and measured using a GloMax 20/20 luminometer (Promega Corp.). The results were calculated as the ratio of LUC to REN. Six independent biological replicates were performed. At least six transient assay measurements were performed for each assay. The primers used to construct the four pBD-DoAP2 fusion constructs are listed in Supplementary Table S4.

4.7. Statistical Analysis

In figures, data have been plotted as means ± standard deviation (SD). Analysis of variance (ANOVA) followed by the Dunnet test was used to determine significant differences at $p < 0.05$ and $p < 0.01$. Analyses were conducted with SPSS v. 22.0 software (IBM) for Windows (IBM Corp., Armonk, NY, USA).

5. Conclusions

We identified 14 DoAP2 TFs from a precious Chinese herbal medicinal orchid, *D. officinale*. We analyzed the expression of *DoAP2* genes in different tissues of *D. officinale* and provided evidence for the specific expression and important regulatory roles at different developmental stages. Promoter analysis of *DoAP2* genes showed that they contained a large number of *cis*-acting elements related to development and abiotic stress, supporting the diversity of their regulatory functions. Results of the protein interaction prediction helped to find putative functions for the DoAP2 TFs, providing a basis for further analysis and verification. Moreover, qRT-PCR analysis showed that *DoAP2* genes are involved in regulating many biological processes such as floral development, embryonic development and adversity to stress, thus play a variety of roles in *D. officinale*. Importantly, use of CLSM to observe the subcellular localization of DoAP2-1, DoAP2-2, DoAP2-6, and DoAP2-11 allowed for the verification that all were localized in the nucleus. Furthermore, the DLR assay demonstrated that DoAP2-1, DoAP2-2, DoAP2-6, and DoAP2-11 proteins displayed strong transcription inhibitory activity in *Nicotiana benthamiana*, indicating that they are transcriptional repressors that inhibit expression. Overall, our study shows not only that DoAP2 TFs have transcriptional inhibitory activity, but also that they were mainly involved in regulating different growth and development stages of *D. officinale*, especially flower development. Our results have relevance to genetically modified resistant breeding since *DoAP2* genes were expressed in PLBs, flowers, and plantlets, and were involved in biological processes such as stress response.

Supplementary Materials: The following are available online at https://www.mdpi.com/article/10.3390/ijms22105221/s1, Figure S1. Multiple sequence alignment of all DoAP2 proteins using ClustalX 2.1, Figure S2. Expression analysis of *DoAP2* genes in different tissues of FBF by qRT-PCR, Figure S3. Subcellular localization of positive control (empty YFP vector) in *A. thaliana* protoplasts, Table S1. Promoter sequences of *DoAP2* genes, Table S2. Primers used for qRT-PCR, Table S3. Primers used for subcellular localization analysis, Table S4. Primers used for the dual-luciferase reporter gene system.

Author Contributions: Conceptualization, J.D. and C.H.; methodology, D.Z. and C.H.; software, D.Z., C.H. and G.D.; validation, D.Z., M.Z., Z.Y., C.S. and C.Z.; formal analysis, D.Z., C.H. and J.A.T.d.S.; investigation, D.Z.; resources, J.D.; data curation, D.Z. and C.H.; writing—original draft preparation, D.Z., C.H. and J.A.T.d.S.; writing—review and editing, D.Z., J.A.T.d.S. and C.H.; visualization, D.Z. and C.H.; supervision, J.D.; and project administration, J.D. All authors have read and agreed to the published version of the manuscript.

Funding: This research was supported by the Foundation of the Key Laboratory of South China Agricultural Plant Molecular Analysis and Genetic Improvement, South China Botanical Garden,

Chinese Academy of Sciences, grant number KF202008, and the Natural Science Foundation of Guangdong, China, grant number 2021A1515012170.

Institutional Review Board Statement: Not applicable.

Informed Consent Statement: Not applicable.

Data Availability Statement: Not applicable.

Conflicts of Interest: The authors declare no conflict of interest.

Abbreviations

1/2MS	half-strength Murashige and Skoog (1962) medium
AD	transcriptional activation domain
AP2	APETALA2
AP2/ERF	APETALA2/Ethylene Response Factor
BD	DNA-binding domain
CDD	conserved domain database
Co	column
DBD	DNA binding domain
EREB	ethylene response element binding factor
FB1	small flower bud
FB2	medium flower bud
FBF	fully bloomed flower
HMM	Hidden Markov mode
Li	lip
LUC	*firefly luciferase*
MEGA	Molecular Evolutionary Genetics Analysis
MS	multiple shoots
PEG	polyethylene glycol
PLB	protocorm-like body
Pe	petal
qRT-PCR	quantificational real-time polymerase chain reaction
REN	*renilla luciferase*
Se	sepal
TF	transcription factor
YFP	yellow fluorescent protein

References

1. Lambert, S.A.; Jolma, A.; Campitelli, L.F.; Das, P.K.; Weirauch, M.T. The human transcription factors. *Cell* **2018**, *172*, 650–665. [CrossRef] [PubMed]
2. Sakuma, Y.; Liu, Q.; Dubouzet, J.G.; Abe, H.; Shinozaki, K.; Yamaguchi-Shinozaki, K. DNA-binding specificity of the ERF/AP2 domain of Arabidopsis DREBs, transcription factors involved in dehydration- and cold-inducible gene expression. *Biochem. Biophys. Res. Commun.* **2002**, *290*, 998–1009. [CrossRef]
3. Nakano, T.; Suzuki, K.; Fujimura, T.; Shinshi, H. Genome-wide analysis of the ERF gene family in Arabidopsis and rice. *Plant Physiol.* **2006**, *140*, 411–432. [CrossRef] [PubMed]
4. Elliott, R.C.; Betzner, A.S.; Huttner, E.; Oakes, M.P.; Tucker, W.; Gerentes, D.; Smyth, P. *AINTEGUMENTA*, an *APETALA2*-like gene of Arabidopsis with pleiotropic roles in ovule development and floral organ growth. *Plant Cell* **1996**, *8*, 155–168. [CrossRef]
5. Boutilier, K.; Offringa, R.; Sharma, V.K.; Kieft, H.; Campagne, M.M.V.L. Ectopic expression of BABY BOOM triggers a conversion from vegetative to embryonic growth. *Plant Cell* **2002**, *14*, 1737–1749. [CrossRef]
6. Graaff, E. A new role for the Arabidopsis AP2 transcription factor, LEAFY PETIOLE, in gibberellin-induced germination is revealed by the misexpression of a homologous gene, *SOB2/DRN-LIKE*. *Plant Cell* **2006**, *18*, 29–39. [CrossRef]
7. Mizoi, J.; Shinozaki, K.; Yamaguchi-Shinozaki, K. AP2/ERF family transcription factors in plant abiotic stress responses. *BBA Gene Regul. Mech.* **2012**, *1819*, 86–96. [CrossRef] [PubMed]
8. Koichiro, A.; Tokunori, H.; Kanna, S.I.; Miyako, U.T.; Hidemi, K.; Makoto, M. A novel AP2-type transcription factor, SMALL ORGAN SIZE1, controls organ size downstream of an auxin signaling pathway. *Plant Cell Physiol.* **2014**, 897–912. [CrossRef]
9. Chuck, G.; Meeley, R.B.; Hake, S. The control of maize spikelet meristem fate by the APETALA2-like gene indeterminate spikelet1. *Genes Dev.* **1998**, *12*, 1145. [CrossRef] [PubMed]

10. Licausi, F.; Ohme-Takagi, M.; Perata, P. APETALA/Ethylene Responsive Factor (AP2/ERF) transcription factors: Mediators of stress responses and developmental programs. *New Phytol.* **2013**, *199*, 639–649. [CrossRef] [PubMed]
11. Florez, S.L.; Erwin, R.L.; Maximova, S.N.; Guiltinan, M.J.; Curtis, W.R. Enhanced somatic embryogenesis in *Theobroma cacao* using the homologous BABY BOOM transcription factor. *BMC Plant Biol.* **2015**, *15*, 1–12. [CrossRef] [PubMed]
12. Lowe, K.; Wu, E.; Wang, N.; Hoerster, G.; Hastings, C.; Cho, M.J.; Scelonge, C.; Lenderts, B.; Chamberlin, M.; Cushatt, J. Morphogenic regulators *Baby boom* and *Wuschel* improve monocot transformation. *Plant Cell* **2016**, *28*, 1998–2015. [CrossRef]
13. Chandler, J.W. Class VIIIb APETALA2 ethylene response factors in plant development. *Trends Plant Sci.* **2017**, *23*, 151–162. [CrossRef]
14. Shigyo, M.; Hasebe, M.; Ito, M. Molecular evolution of the AP2 subfamily. *Gene* **2006**, *366*, 256–265. [CrossRef] [PubMed]
15. Ohto, M.A.; Fischer, R.L.; Goldberg, R.B.; Nakamura, K.; Harada, J.J. Control of seed mass by APETALA2. *Proc. Natl. Acad. Sci. USA* **2005**, *102*, 3123–3128. [CrossRef] [PubMed]
16. Yamasaki, K.; Kigawa, T.; Seki, M.; Shinozaki, K.; Yokoyama, S. DNA-binding domains of plant-specific transcription factors: Structure, function, and evolution. *Trends Plant Sci.* **2013**, *18*, 267–276. [CrossRef]
17. Zhuang, J.; Cai, B.; Peng, R.-H.; Zhu, B.; Jin, X.-F.; Xue, Y.; Gao, F.; Fu, X.-Y.; Tian, Y.-S.; Zhao, W.; et al. Genome-wide analysis of the AP2/ERF gene family in *Populus trichocarpa*. *Biochem. Biophys. Res. Commun.* **2008**, *371*, 468–474. [CrossRef] [PubMed]
18. Shigyo, M.; Ito, M. Analysis of gymnosperm two-AP2-domain-containing genes. *Dev. Genes Evol.* **2004**, *214*, 105–114. [CrossRef] [PubMed]
19. Horstman, A.; Willemsen, V.; Boutilier, K.; Heidstra, R. AINTEGUMENTA-LIKE proteins: Hubs in a plethora of networks. *Trends Plant Sci.* **2014**, *19*, 146–157. [CrossRef] [PubMed]
20. Kim, S.; Soltis, P.S.; Wall, K.; Soltis, D.E. Phylogeny and domain Evolution in the *APETALA2*-like gene family. *Mol. Biol. Evol.* **2006**, *23*, 107–120. [CrossRef] [PubMed]
21. Komaki, M.K.; Okada, K.; Nishino, E.; Shimura, Y. Isolation and characterization of novel mutants of *Arabidopsis thaliana* defective in flower development. *Development* **1988**, *104*, 1536–1546. [CrossRef]
22. Bowman, J.L.; Smyth, D.R.; Meyerowitz, E.M. Genes directing flower development in *Arabidopsis*. *Plant Cell Online* **1989**, *1*, 37–52. [CrossRef]
23. Kunst, L.; Klenz, J.E.; Haughn, M. AP2 gene determines the identity of perianth organs in flowers of *Arabidopsis thaliana*. *Plant Cell* **1989**, *1*, 1195–1208. [CrossRef]
24. Bowman, J.L.; Smyth, D.R.; Meyerowitz, E.M. Genetic interactions among floral homeotic genes of *Arabidopsis*. *Development* **1991**, *112*, 1–20. [CrossRef]
25. Krizek, B.A. *AINTEGUMENTA-LIKE* genes have partly overlapping functions with *AINTEGUMENTA* but make distinct contributions to *Arabidopsis thaliana* flower development. *J. Exp. Bot.* **2015**, *66*, 4537–4549. [CrossRef] [PubMed]
26. Klucher, K.M.; Chow, H.; Fischer, R. The *AINTEGUMENTA* gene of *Arabidopsis* required for ovule and female gametophyte development is related to the floral homeotic gene *APETALA2*. *Plant Cell Online* **1996**, *8*, 137–153. [CrossRef]
27. Dash, M.; Malladi, A. The *AINTEGUMENTA* genes, *MdANT1* and *MdANT2*, are associated with the regulation of cell production during fruit growth in apple (*Malus* × *domestica* Borkh.). *BMC Plant Biol.* **2012**, *12*, 1–15. [CrossRef] [PubMed]
28. Chialva, C.; Eichler, E.; Grissi, C.; Mu?Oz, C.; Gomez-Talquenca, S.; Martínez-Zapater, J.; Lijavetzky, D. Expression of grapevine *AINTEGUMENTA*-like genes is associated with variation in ovary and berry size. *Plant Mol. Biol.* **2016**, *91*, 67–80. [CrossRef] [PubMed]
29. Jiang, L.; Ma, X.; Zhao, S.; Tang, Y.; Liu, F.; Gu, P.; Fu, Y.; Zhu, Z.; Cai, H.; Sun, C. The APETALA2-like transcription factor SUPERNUMERARY BRACT controls rice seed shattering and seed size. *Plant Cell* **2019**, *31*, 17–36. [CrossRef] [PubMed]
30. Zeng, J.K.; Li, X.; Xu, Q.; Chen, J.Y.; Yin, X.R.; Ferguson, I.B.; Chen, K.S. *EjAP2-1*, an *AP2/ERF* gene, is a novel regulator of fruit lignification induced by chilling injury, via interaction with *EjMYB* transcription factors. *Plant Biotechnol. J.* **2016**, *13*, 1325–1334. [CrossRef] [PubMed]
31. Chuck, G.; Meeley, R.; Hake, S. Floral meristem initiation and meristem cell fate are regulated by the maize *AP2* genes *ids1* and *sid1*. *Development* **2008**, *135*, 3013–3019. [CrossRef] [PubMed]
32. Jofuku, K.D.; Boer, B.G.d.; Montagu, M.V.; Okamuro, J.K. Control of Arabidopsis flower and seed development by the homeotic gene *APETALA2*. *Plant Cell* **1994**, *6*, 1211–1225. [CrossRef]
33. Stone, S.L.; Kwong, L.W.; Yee, K.M.; Pelletier, J.; Lepiniec, L.; Fischer, R.L.; Goldberg, R.B.; Harada, J.J. LEAFY COTYLEDON2 encodes a B3 domain transcription factor that induces embryo development. *Proc. Natl. Acad. Sci. USA* **2001**, *98*, 11806–11811. [CrossRef] [PubMed]
34. Cheong, J.J.; Choi, Y.D. Methyl jasmonate as a vital substance in plants. *Trends Genet.* **2003**, *19*, 409–413. [CrossRef]
35. Krizek, B.A. AINTEGUMENTA and AINTEGUMENTA-like6 act redundantly to regulate arabidopsis floral growth and patterning. *Plant Physiol.* **2009**, *150*, 1916–1929. [CrossRef]
36. Rudall, P.J.; Bateman, R.M. Roles of synorganisation, zygomorphy and heterotopy in floral evolution: The gynostemium and labellum of orchids and other lilioid monocots. *Biol. Rev.* **2010**, *77*, 403–441. [CrossRef] [PubMed]
37. Hawku, M.D.; Goher, F.; Islam, M.A.; Guo, J.; He, F.; Bai, X.; Yuan, P.; Kang, Z.; Guo, J. *TaAP2-15*, An AP2/ERF transcription factor, is positively involved in wheat resistance to *Puccinia striiformis* f. sp. *tritici*. *Int. J. Mol. Sci.* **2021**, *22*, 2080. [CrossRef]
38. Xie, Z.; Nolan, T.M.; Jiang, H.; Yin, Y. AP2/ERF transcription factor regulatory networks in hormone and abiotic stress responses in *Arabidopsis*. *Front. Plant Sci.* **2019**, *10*, 228. [CrossRef] [PubMed]

39. Ahmeda, S.; Rashid, M.A.R.; Zafar, S.A.; Azhar, M.T.; Waqas, M.; Uzair, M.; Rana, I.A.; Azeem, F.; Chung, G.; Ali, Z.; et al. Genome-wide investigation and expression analysis of APETALA-2 transcription factor subfamily reveals its evolution, expansion and regulatory role in abiotic stress responses in Indica rice (*Oryza sativa* L. ssp. *indica*). *Genomics* **2020**, *113*, 1029–1043. [CrossRef]
40. Zhao, Y.; Ma, R.; Xu, D.; Bi, H.; Xia, Z.; Peng, H. Genome-wide identification and analysis of the AP2 transcription factor gene family in wheat (*Triticum aestivum* L.). *Front. Plant Sci.* **2019**, *10*, 1286. [CrossRef] [PubMed]
41. Dipp-Álvarez, M.; Cruz-Ramírez, A. A phylogenetic study of the ANT family points to a preANT gene as the ancestor of basal and euANT transcription factors in land plants. *Front. Plant Sci.* **2019**, *10*. [CrossRef]
42. Zhao, Y.; Cheng, S.; Song, Y.; Huang, Y.; Zhou, S.; Liu, X.; Zhou, D.X. The interaction between rice ERF3 and WOX11 promotes crown root development by regulating gene expression involved in cytokinin signaling. *Plant Cell* **2015**, *27*, 2469–2483. [CrossRef]
43. Diaz-Martin, J.; Almoguera, C.N.; Prieto-Dapena, P.; Espinosa, J.M.; Jordano, J. Functional interaction between two transcription factors involved in the developmental regulation of a small heat stress protein gene promoter. *Plant Physiol.* **2005**, *139*, 1483–1494. [CrossRef] [PubMed]
44. Son, G.H.; Wan, J.; Kim, H.J.; Xuan Canh, N.; Chung, W.S.; Hong, J.C.; Stacey, G. Ethylene-responsive element-binding factor 5, ERF5, is involved in chitin-induced innate immunity response. *Mol. Plant Microbe Interact.* **2012**, *25*, 48–60. [CrossRef] [PubMed]
45. Moffat, C.S.; Ingle, R.A.; Wathugala, D.L.; Saunders, N.J.; Knight, H.; Knight, M.R. ERF5 and ERF6 play redundant roles as positive regulators of JA/Et-mediated defense against *botrytis cinerea* in *Arabidopsis*. *PLoS ONE* **2012**, *7*, e35995. [CrossRef]
46. Riechmann, J.L.; Meyerowitz, E.M. The AP2/EREBP family of plant transcription factors. *Biol. Chem.* **1998**, *379*, 633–646. [CrossRef]
47. Shukla, R.K.; Raha, S.; Tripathi, V.; Chattopadhyay, D. Expression of CAP2, an APETALA2-family transcription factor from chickpea, enhances growth and tolerance to dehydration and salt stress in transgenic tobacco. *Plant Physiol.* **2006**, *142*, 113–123. [CrossRef]
48. Yamaguchi-Shinozaki, K.; Shinozaki, K. Organization of cis-acting regulatory elements in osmotic- and cold-stress-responsive promoters. *Trends Plant Sci.* **2005**, *10*, 88–94. [CrossRef] [PubMed]
49. Luo, H.; Chen, S.; Jiang, J.; Teng, N.; Chen, Y.; Chen, F. The AP2-like gene *NsAP2* from water lily is involved in floral organogenesis and plant height. *J. Plant Physiol.* **2012**, *169*, 992–998. [CrossRef]
50. Shoji, T.; Mishima, M.; Hashimoto, T. Divergent DNA-binding specificities of a group of ETHYLENE RESPONSE FACTOR transcription factors involved in plant defense. *Plant Physiol.* **2013**, *162*, 977–990. [CrossRef] [PubMed]
51. Wang, L.; Qin, L.; Liu, W.; Zhang, D.; Wang, Y. A novel ethylene-responsive factor from Tamarix hispida, ThERF1, is a GCC-box- and DRE-motif binding protein that negatively modulates abiotic stress tolerance in *Arabidopsis*. *Physiol. Plant.* **2014**, *152*, 84–97. [CrossRef]
52. Phukan, U.J.; Jeena, G.S.; Tripathi, V.; Shukla, R.K. Regulation of Apetala2/Ethylene response factors in plants. *Front. Plant Sci.* **2017**, *8*. [CrossRef]
53. Che, P.; Lall, S.; Nettleton, D.; Howell, S.H. Gene expression programs during shoot, root and callus development in *Arabidopsis* tissue culture. *Plant Biol.* **2006**, *2006*, 282–283. [CrossRef]
54. Du, X.; Fang, T.; Liu, Y.; Huang, L.; Zang, M.; Wang, G.; Liu, Y.; Fu, J. Transcriptome profiling predicts new genes to promote maize callus formation and transformation. *Front. Plant Sci.* **2019**, *10*. [CrossRef]
55. Piyatrakul, P.; Putranto, R.-A.; Martin, F.; Rio, M.; Dessailly, F.; Leclercq, J.; Dufayard, J.-F.; Lardet, L.; Montoro, P. Some ethylene biosynthesis and AP2/ERF genes reveal a specific pattern of expression during somatic embryogenesis in *Hevea brasiliensis*. *BMC Plant Biol.* **2012**, *12*, 1–20. [CrossRef]
56. Lee, J.H.; Kim, D.M.; Lee, J.H.; Kim, J.; Bang, J.W.; Kim, W.T.; Pai, H.S. Functional characterization of NtCEF1, an AP2/EREBP-type transcriptional activator highly expressed in tobacco callus. *Planta* **2005**, *222*, 211–224. [CrossRef] [PubMed]
57. Iwase, A.; Harashima, H.; Ikeuchi, M.; Rymen, B.; Ohnuma, M.; Komaki, S.; Morohashi, K.; Kurata, T.; Nakata, M.; Ohme-Takagi, M.; et al. WIND1 promotes shoot regeneration through transcriptional activation of ENHANCER OF SHOOT REGENERATION1 in *Arabidopsis*. *Plant Cell* **2017**, *29*, 54–69. [CrossRef] [PubMed]
58. Ye, B.-B.; Shang, G.-D.; Pan, Y.; Xu, Z.-G.; Zhou, C.-M.; Mao, Y.-B.; Bao, N.; Sun, L.; Xu, T.; Wang, J.-W. AP2/ERF transcription factors integrate age and wound signals for root regeneration. *Plant Cell* **2020**, *32*, 226–241. [CrossRef] [PubMed]
59. Zhang, K.; Zhao, L.; Yang, X.; Li, M.; Sun, J.; Wang, K.; Li, Y.; Zheng, Y.; Yao, Y.; Li, W. GmRAV1 regulates regeneration of roots and adventitious buds by the cytokinin signaling pathway in Arabidopsis and soybean. *Physiol. Plant.* **2019**, *165*, 814–829. [CrossRef] [PubMed]
60. Lee, Y.; Hsu, S.; Yeung, E.C. Orchid protocorm-like bodies are somatic embryos. *Am. J. Bot.* **2013**, *100*, 2121–2131. [CrossRef] [PubMed]
61. Teixeira da Silva, J.A.; Cardoso, J.C.; Dobránszki, J.; Zeng, S. *Dendrobium* micropropagation: A review. *Plant Cell Rep.* **2015**, *34*, 671–704. [CrossRef]
62. Ripoll, J.J.; Roeder, A.; Ditta, G.S.; Yanofsky, M.F. A novel role for the floral homeotic gene *APETALA2* during *Arabidopsis* fruit development. *Development* **2011**, *138*, 5167–5176. [CrossRef]
63. Samad, A.; Muhammad, S.; Nazaruddin, N.; Fauzi, I.A.; Murad, A.; Zamri, Z.; Ismanizan, I. MicroRNA and transcription factor: Key players in plant regulatory network. *Front. Plant Sci.* **2017**, *8*. [CrossRef] [PubMed]

64. He, C.; Si, C.; Teixeira da Silva, J.A.; Li, M.; Duan, J. Genome-wide identification and classification of MIKC-type MADS-box genes in Streptophyte lineages and expression analyses to reveal their role in seed germination of orchid. *BMC Plant Biol.* **2019**, *19*, 1–15. [CrossRef]
65. Yu, H.; Goh, C.J. Molecular genetics of reproductive biology in orchids. *Plant Physiol.* **2001**, *127*, 1390–1393. [CrossRef]
66. Cao, Y.; Song, F.; Goodman, R.M.; Zheng, Z. Molecular characterization of four rice genes encoding ethylene-responsive transcriptional factors and their expressions in response to biotic and abiotic stress. *J. Plant Physiol.* **2006**, *163*, 1167–1178. [CrossRef] [PubMed]
67. Igarashi, D.; Ishida, S.; Fukazawa, J.; Takahashi, Y. 14-3-3 proteins regulate intracellular localization of the bZIP transcriptional activator RSG. *Plant Cell* **2001**, *13*, 2483–2497. [CrossRef] [PubMed]
68. Mishra, S.; Phukan, U.J.; Tripathi, V.; Singh, D.K.; Luqman, S.; Shukla, R.K. PsAP2 an AP2/ERF family transcription factor from *Papaver somniferum* enhances abiotic and biotic stress tolerance in transgenic tobacco. *Plant Mol. Biol.* **2015**, *89*, 173–186. [CrossRef]
69. Chen, J.-Q.; Dong, Y.; Wang, Y.-J.; Liu, Q.; Zhang, J.-S.; Chen, S.-Y. An AP2/EREBP-type transcription-factor gene from rice is cold-inducible and encodes a nuclear-localized protein. *Theor. Appl. Genet.* **2003**, *107*, 972–979. [CrossRef] [PubMed]
70. Yu, Y.; Duan, X.; Ding, X.; Chen, C.; Zhu, D.; Yin, K.; Cao, L.; Song, X.; Zhu, P.; Li, Q.; et al. A novel AP2/ERF family transcription factor from *Glycine soja*, GsERF71, is a DNA binding protein that positively regulates alkaline stress tolerance in *Arabidopsis*. *Plant Mol. Biol.* **2017**, *94*, 509–530. [CrossRef]
71. Thiel, G.; Lietz, M.; Hohl, M. How mammalian transcriptional repressors work. *Eur. J. Biochem.* **2004**, *271*, 2855–2862. [CrossRef] [PubMed]
72. Kazan, K. Negative regulation of defence and stress genes by EAR-motif-containing repressors. *Trends Plant Sci.* **2006**, *11*, 109–112. [CrossRef]
73. Yaish, M.W.; El-kereamy, A.; Zhu, T.; Beatty, P.H.; Good, A.G.; Bi, Y.-M.; Rothstein, S.J. The APETALA-2-Like transcription factor OsAP2-39 controls key interactions between abscisic acid and gibberellin in rice. *PLoS Genet.* **2010**, *6*, e1001098. [CrossRef] [PubMed]
74. Pan, I.C.; Li, C.-W.; Su, R.-C.; Cheng, C.-P.; Lin, C.-S.; Chan, M.-T. Ectopic expression of an EAR motif deletion mutant of *SlERF3* enhances tolerance to salt stress and *Ralstonia solanacearum* in tomato. *Planta* **2010**, *232*, 1075–1086. [CrossRef]
75. Drews, G.N.; Bowman, J.L.; Meyerowitz, E.M. Negative regulation of the Arabidopsis homeotic gene AGAMOUS by the APETALA2 product. *Cell* **1991**, *65*, 991–1002. [CrossRef]
76. Jofuku, K.D.; Omidyar, P.K.; Gee, Z.; Okamuro, J.K. Control of seed mass and seed yield by the floral homeotic gene APETALA2. *Proc. Natl. Acad. Sci. USA* **2005**, *102*, 3117–3122. [CrossRef]
77. Zhou, Y.; Lu, D.; Li, C.; Luo, J.; Zhu, B.-F.; Zhu, J.; Shangguan, Y.; Wang, Z.; Sang, T.; Zhou, B.; et al. Genetic control of seed shattering in rice by the APETALA2 transcription factor SHATTERING ABORTION1. *Plant Cell* **2012**, *24*, 1034–1048. [CrossRef] [PubMed]
78. Liu, L.S.; White, M.J.; MacRae, T.H. Transcription factors and their genes in higher plants functional domains, evolution and regulation. *Eur. J. Biochem.* **1999**, *262*, 247–257. [CrossRef] [PubMed]
79. Murashige, T.; Skoog, F. A revised medium for rapid growth and bio assays with tobacco tissue cultures. *Physiol. Plant.* **1962**, *15*, 473–497. [CrossRef]
80. He, C.; Zeng, S.; Teixeira da Silva, J.A.; Yu, Z.; Tan, J.; Duan, J. Molecular cloning and functional analysis of the *phosphomannomutase* (*PMM*) gene from *Dendrobium officinale* and evidence for the involvement of an abiotic stress response during germination. *Protoplasma* **2017**, *254*, 1693–1704. [CrossRef]
81. Kumar, S.; Stecher, G.; Tamura, K. MEGA7: Molecular Evolutionary Genetics Analysis Version 7.0 for bigger datasets. *Mol. Biol. Evol.* **2016**, *33*, 1870–1874. [CrossRef]
82. Larkin, M.A.; Blackshields, G.; Brown, N.P.; Chenna, R.; McGettigan, P.A.; McWilliam, H.; Valentin, F.; Wallace, I.M.; Wilm, A.; Lopez, R.; et al. Clustal W and clustal X version 2.0. *Bioinformatics* **2007**, *23*, 2947–2948. [CrossRef]
83. Saitou, N.; Nei, M. The neighbor-joining method - a new method for reconstructing phylogenetic trees. *Mol. Biol. Evol.* **1987**, *4*, 406–425. [CrossRef] [PubMed]
84. Lu, S.; Wang, J.; Chitsaz, F.; Derbyshire, M.K.; Geer, R.C.; Gonzales, N.R.; Gwadz, M.; Hurwitz, D.I.; Marchler, G.H.; Song, J.S.; et al. CDD/SPARCLE: The conserved domain database in 2020. *Nucleic Acids Res.* **2020**, *48*, D265–D268. [CrossRef] [PubMed]
85. Chen, C.; Chen, H.; Zhang, Y.; Thomas, H.R.; Frank, M.H.; He, Y.; Xia, R. TBtools: An integrative toolkit developed for interactive analyses of big biological data. *Mol. Plant* **2020**, *13*, 1194–1202. [CrossRef] [PubMed]
86. Szklarczyk, D.; Gable, A.L.; Lyon, D.; Junge, A.; Wyder, S.; Huerta-Cepas, J.; Simonovic, M.; Doncheva, N.T.; Morris, J.H.; Bork, P.; et al. STRING v11: Protein-protein association networks with increased coverage, supporting functional discovery in genome-wide experimental datasets. *Nucleic Acids Res.* **2019**, *47*, D607–D613. [CrossRef] [PubMed]
87. He, C.; Zhang, J.; Liu, X.; Zeng, S.; Wu, K.; Yu, Z.; Wang, X.; Teixeira da Silva, J.A.; Lin, Z.; Duan, J. Identification of genes involved in biosynthesis of mannan polysaccharides in *Dendrobium officinale* by RNA-seq analysis. *Plant Mol. Biol.* **2015**, *88*, 219–231. [CrossRef] [PubMed]
88. Livak, K.J.; Schmittgen, T.D. Analysis of relative gene expression data using real-time quantitative PCR and the $2^{-\Delta\Delta CT}$ method. *Methods* **2001**, *25*, 402–408. [CrossRef]
89. Citovsky, V.; Lee, L.-Y.; Vyas, S.; Glick, E.; Chen, M.-H.; Vainstein, A.; Gafni, Y.; Gelvin, S.B.; Tzfira, T. Subcellular localization of interacting proteins by bimolecular fluorescence complementation in planta. *J. Mol. Biol.* **2006**, *362*, 1120–1131. [CrossRef]

90. Yoo, S.-D.; Cho, Y.-H.; Sheen, J. *Arabidopsis* mesophyll protoplasts: A versatile cell system for transient gene expression analysis. *Nat. Protoc.* **2007**, *2*, 1565–1572. [CrossRef] [PubMed]
91. Han, Y.-C.; Kuang, J.-F.; Chen, J.-Y.; Liu, X.-C.; Xiao, Y.-Y.; Fu, C.-C.; Wang, J.-N.; Wu, K.-Q.; Lu, W.-J. Banana transcription factor MaERF11 recruits histone deacetylase MaHDA1 and Represses the expression of *MaACO1* and *Expansins* during fruit ripening. *Plant Physiol.* **2016**, *171*, 1070–1084. [CrossRef] [PubMed]

Article

How Are the Flower Structure and Nectar Composition of the Generalistic Orchid *Neottia ovata* Adapted to a Wide Range of Pollinators?

Emilia Brzosko *, Andrzej Bajguz *, Magdalena Chmur, Justyna Burzyńska, Edyta Jermakowicz, Paweł Mirski and Piotr Zieliński

Faculty of Biology, University of Bialystok, Ciolkowskiego 1J, 15-245 Bialystok, Poland; m.chmur@uwb.edu.pl (M.C.); j.burzynska@uwb.edu.pl (J.B.); edytabot@uwb.edu.pl (E.J.); p.mirski@uwb.edu.pl (P.M.); p.zielinski@uwb.edu.pl (P.Z.)
* Correspondence: emilka@uwb.edu.pl (E.B.); abajguz@uwb.edu.pl (A.B.); Tel.: +48-85-738-8424 (E.B.); +48-85-738-8361 (A.B.)

Abstract: Plant-pollinator interactions significantly influence reproductive success (RS) and drive the evolution of pollination syndromes. In the context of RS, mainly the role of flower morphology is touched. The importance of nectar properties is less studied, despite its significance in pollination effectiveness. Therefore, the aim of this study was to test selection on flower morphology and nectar chemistry in the generalistic orchid *Neottia ovata*. In 2019–2020, we measured three floral displays and six flower traits, pollinaria removal (PR), female reproductive success (FRS), and determined the soil properties. The sugars and amino acids (AAs) were analyzed using the HPLC method. Data were analyzed using multiple statistical methods (boxplots, ternary plot, one-way ANOVA, Kruskal-Wallis test, and PCA). Variation of flower structure and nectar chemistry and their weak correlation with RS confirms the generalistic character of *N. ovata*. In particular populations, different traits were under selection. PR was high and similar in all populations in both years, while FRS was lower and varied among populations. Nectar was dominated by glucose, fructose, and included 28 AAs (Ala and Glu have the highest content). Sugars and AAs influenced mainly FRS. Among soil parameters, carbon and carbon:nitrogen ratio seems to be the most important in shaping flower structure and nectar chemistry.

Keywords: amino acids; female reproductive success; pollinaria removal; natural selection; orchids; plant-pollinator interactions; sugars

1. Introduction

Plants dependent on animals in the pollination process evolved different strategies to attract pollinators, thereby increasing reproductive success. The main parts of these strategies are flower traits (the size, shape, color, scent, and nectar) adapted to a given pollinator or their whole group. Pollinator-mediated selection on floral traits is well documented, and adaptation of plants to the most effective pollinators drives the evolution of pollination syndromes [1]. The flagship example of the unusual diversity of flowers and equally differentiated pollination mechanisms is Orchidaceae, which is one of the biggest families among flowering plants [2]. About one-third of its representatives deceive pollinators through sexual or food deception [2–4]. Other groups of orchids reward pollinators in a different way, producing oils, nectar, resin, wax, and fragrances [5,6]. Among rewards offered by orchids, nectar is the most effective [2,7,8]. Fruiting in nectariferous orchids is significantly higher than in nectarless [2,8]. Although nectariferous orchids constitute a large part of the family, and the role of nectar in highly effective pollination is indisputable, information on its chemical composition in Orchidaceae is very scarce. Moreover, many data derive from studies using less sensitive methods in comparison to those applied recently. Importantly, more data on nectar chemistry provide results of studies on plants from other families [9–15], but they often focus on cultivars and the feeding needs of their pollinators, mainly bees.

Although available data document a great variability of nectar properties at different levels (species, population, and even individual), some patterns are outlined. In flower nectar, three main sugar components dominate, i.e., sucrose, glucose, and fructose, with different ratios between them. Nectar of the majority species is sucrose dominated [12,13,16], but some papers document domination of hexoses over sucrose [14,17–19]. The concentration of sugars also shows a great variation (from about 7–70%, [20,21]) and is connected with pollinator types [9,10,22,23], especially with the adaptation of their mouthparts to use nectar of a given viscosity. For example, bees prefer the highest concentration of sugars in nectar (35% on average), while bats and hawkmoths can suck nectar with a 17–19% concentration of sugars [23–26]. In orchids, nectar sugar concentrations range from a few to about 50% [17,18,27–30]. The preferences of pollinators also concern other components of nectar: amino acids (AAs). They are present in nectar at a lower amount than sugars but play a significant role as a source of nutrition and in attracting pollinators, thereby affecting reproductive success and survival of nectar-feeding animals [14,16,31–33]. Some authors suggest that taste function is even more important than a nutritive value [34,35]. Nectar of plants adapted to pollination by butterflies is characterized by high AA concentration, while those pollinated by birds or flies are characterized by their lower concentration [34]. In the nectar of different species, some AAs dominate, and others are present in low concentrations or are absent [17,28,36,37].

Apart from nectar quantity and quality, its accessibility also influences plant-pollinator interaction, thereby affects plant reproductive success. If nectar is secreted inside the corolla or in a spur, it is protected against evaporation and is available for specific, restricted groups of pollinators. On the other hand, exposed nectar may be collected by pollinators representing different morphological and ecological types and is more vulnerable to evaporation and robbery [38]. Moreover, nectar in flowers with concealed nectaries tends to be dominated by sucrose, while in more open flowers, it is dominated by glucose and fructose [29,39].

In papers dedicated to plant-pollinator interaction, the role of flower structure in attracting pollinators was studied more often than nectar properties [40–43]. In particular, phenotypic selection and its dependence on the mutual match between pollinator and flower traits are well documented [44]. This match is one of the most important evolutionary mechanisms [2,4] and is an effect of the potential for adaptation to the local partners. Many researchers have shown that pollinators act as selection agents on floral morphology and contribute to plant fitness [45–47]. Van der Niet, et al. [46] stated that when pollinators' fitness is strongly influenced by an ability to access the reward in flowers of a given species, the adaptation of pollinators to flowers, rather than flowers to pollinators, takes place. In the case of plants, in which flowers are arranged as the inflorescences, floral display (the length of inflorescence and number of flowers) may also contribute to reproductive success. Plants with larger inflorescences often set more fruit, due to attracting more pollinators, which visit more flowers on larger inflorescences [48–52]. However, in cases in which larger inflorescences suffer from factors that decrease fitness, such as a higher probability of geitonogamy or intense herbivore activity, smaller inflorescences are favored by natural selection [52–54].

Both floral characters and pollinator assemblages vary in space [19,41,55,56]. Variation of floral traits in the geographic range of plant species is often an adaptation to the locally most-effective pollinators, being an answer for requirements of their specific assemblages present in a given environment [14,28,46,56,57]. The shift of floral traits and pollinators assemblages in space translate into differentiated direction and strength of selection and variation of the level of reproductive success [58].

Reproductive success depends on more than an evolutionary match between plants and pollinators. Environmental factors, both biotic (co-occurring plants) and abiotic (soil resources, weather conditions), in places where populations exist may also importantly shape plant-pollinator interactions. The composition of local pollinators is strictly connected to the diversity of the plant community because more plant species accumulate a wider spectrum of resources for flower-visiting animals [59]. Plant species richness, blossom cover, and especially the presence of attractive plant species influence assemblages of pollinators and the frequency of their visits [60,61]. It seems especially important in the case of generalist plants, which depend on many species in the pollination process. On the one hand, the presence of other flowering plants may facilitate the visitation rate, and as a consequence, increase the reproductive success of a given species [62,63]. On the other hand, a higher diversity of plant species may increase competition for pollinators when species share pollinators [62–65], especially when populations of pollinators are not abundant. Competition for pollination resources can also include intraspecific competition, which may be stronger than interspecies competition, according to niche theory [66]. The richness of the plant community, and the growth and flowering of particular species, strictly depend on soil conditions. For example, David, et al. [67] found that a high level of N in soil and a low pH decrease species diversity and the abundance of nectariferous plants. In effect, nectar and pollen resources decline, causing a decrease in pollinators' assemblages [68]. Soil properties also shape other plant traits, which influence the level of reproductive success, e.g., the flowering [69] or quantity and quality of nectar [27].

Due to the unusual richness of orchids' flowers and the wide variation of relations with pollinators, orchids are often considered a model system to study plant-pollinator interactions and evolutionary processes. The majority of orchid species are specialists and are connected to only one pollinator species (67% of all orchids; [70]) or a single functional group [71–74]. Others are generalists, and a wide range of animals may pollinate them. For example, *Epipactis palustris* is pollinated by more than 100 species [75], and in *Neottia ovata* almost 300 different species were noted as visitors, with about 50 species carrying pollinia [76]. Specialist orchids are more frequent objects of studies on selection/coevolution between plant and pollinators than generalists. Therefore, it seems interesting to choose the generalist orchid *N. ovata* as a model species to test in which way flower traits are adapted to pollination by a wide range of pollinators. *N. ovata* was the object of studies on pollination mechanism [76–78], demographic processes [79–81], genetic variation [82,83], and flower anatomy [84]. So far, there are no published data on nectar composition and floral structure in this orchid and their role in the effectiveness of reproduction. Therefore, the main aim of our study was to determine the floral traits of the generalist orchid *N. ovata* and to test for selection on floral morphology and nectar chemistry in populations existing in different habitats. Such studies enrich knowledge about evolutionary factors and processes that underlie the generalization or specialization and consequences at the population and species levels.

2. Results

2.1. Floral Display

We found statistically significant differences between populations in the height of the flowering shoots in both years (F = 9.390/3.422, $p < 0.0001/0.01$), while the inflorescence length differed only in 2019 (F = 14.740, $p < 0.0001$), and the number of flowers per inflorescence in 2020 (F = 2.510, $p < 0.05$) (Table 1). The highest shoots and the longest inflorescences were noted in TUR in both years. The number of flowers developed on *N. ovata* shoots was the lowest in ZAB1 in 2019 and in ZAB2 in 2020. Higher values of floral display traits were found in 2020 in four out of five cases, where statistically significant differences between years were noted. Soil parameters did not influence floral display traits [Personal communication].

Table 1. Variation of floral display and flower structure in *Neottia ovata* populations. Data show the mean ± standard deviation. Dark blue values in bold indicate statistically significant differences between years.

Population	Year	N	Shoot Height (cm)	Inflorescence Length (cm)	Number of Flowers	Cavity Length (mm)	Cavity Width (mm)	Labellum Length (mm)	Groove Length (mm)	Labellum Width (mm)	Flower Width (mm)
OPA	2019	38	51.02 ± 5.56	17.75 ± 3.22	22.75 ± 7.22	0.98 ± 0.10	1.34 ± 0.11	8.60 ± 1.11	3.70 ± 0.60	3.29 ± 0.45	7.96 ± 0.95
	2020	26	44.50 ± 8.58	16.28 ± 4.25	28.56 ± 12.94	0.94 ± 0.09	1.41 ± 0.08	10.05 ± 1.17	3.94 ± 0.35	3.88 ± 0.49	8.21 ± 0.53
LUB	2019	38	53.84 ± 10.62	17.71 ± 5.83	24.27 ± 7.43	1.03 ± 0.10	1.37 ± 0.16	9.03 ± 1.01	3.99 ± 0.38	3.35 ± 0.38	8.69 ± 0.53
	2020	28	46.63 ± 13.03	17.75 ± 6.41	26.43 ± 8.04	0.92 ± 0.11	1.29 ± 0.10	9.63 ± 1.02	3.98 ± 0.44	3.46 ± 0.33	8.54 ± 0.74
POG	2019	44	47.45 ± 6.53	14.06 ± 3.53	26.55 ± 4.78	0.91 ± 0.09	1.39 ± 0.13	9.12 ± 0.91	4.03 ± 0.36	3.42 ± 0.36	8.77 ± 0.57
	2020	44	53.31 ± 8.49	20.15 ± 4.14	25.75 ± 7.56	1.03 ± 0.11	1.44 ± 0.12	10.78 ± 1.17	4.12 ± 0.38	3.93 ± 0.46	8.96 ± 0.60
ZAB1	2019	40	51.70 ± 9.21	19.63 ± 4.14	22.86 ± 9.13	0.97 ± 0.15	1.35 ± 0.11	8.59 ± 1.33	4.13 ± 0.37	3.59 ± 0.41	8.61 ± 0.81
	2020	30	52.73 ± 8.82	18.82 ± 2.33	29.25 ± 9.45	0.99 ± 0.12	1.39 ± 0.11	9.39 ± 1.26	3.85 ± 0.32	3.59 ± 0.34	8.39 ± 0.40
ZAB2	2019	46	52.09 ± 6.74	14.86 ± 4.15	28.38 ± 9.25	1.02 ± 0.11	1.46 ± 0.15	8.67 ± 1.38	4.14 ± 0.39	3.59 ± 0.43	8.81 ± 0.69
	2020	33	51.04 ± 8.55	17.58 ± 3.94	24.46 ± 6.42	1.03 ± 0.10	1.35 ± 0.09	9.71 ± 1.11	3.94 ± 0.42	3.48 ± 0.54	8.64 ± 0.71
TUR	2019	44	61.91 ± 8.49	24.34 ± 4.73	27.33 ± 8.45	1.04 ± 0.10	1.38 ± 0.07	7.91 ± 0.92	3.70 ± 0.37	3.22 ± 0.41	8.30 ± 0.61
	2020	40	59.32 ±8.50	20.00 ± 4.61	35.78 ± 10.49	1.07 ± 0.12	1.40 ± 0.09	10.17 ± 0.94	3.98 ± 0.37	3.78 ± 0.47	8.72 ± 0.69
WIS	2019	nd	nd	nd	nd	nd	nd	nd	nd	nd	nd
	2020	12	51.67 ± 6.55	18.25 ± 4.55	29.83 ± 11.30	1.06 ± 0.11	1.32 ± 0.05	10.92 ± 0.75	4.18 ± 0.51	3.62 ± 0.33	8.63 ± 0.75
SKA	2019	42	45.88 ± 7.09	15.48 ± 3.92	26.22 ± 6.08	1.05 ± 0.12	1.42 ± 0.12	8.85 ± 1.28	4.13 ± 0.50	3.55 ± 0.46	8.86 ± 0.60
	2020	44	46.32 ± 10.46	19.73 ± 5.60	25.27 ± 5.95	1.00 ± 0.09	1.38 ± 0.09	9.96 ± 1.30	4.04 ± 0.43	3.70 ± 0.33	8.61 ± 0.72
IPD	2019		$p < 0.001$	$p < 0.01$	NSD	$p < 0.05$	$p < 0.05$	$p < 0.05$	$p < 0.05$	$p < 0.05$	$p < 0.05$
	2020		$p < 0.01$	NSD	$p < 0.05$	$p < 0.01$	$p < 0.01$	$p < 0.01$	NSD	$p < 0.05$	NSD

nd—no data. IPD—inter-population differentiation of particular traits. N—number of analyzed flowers. NSD—nonsignificant differences.

2.2. Flower Structure

All measured flower traits were differentiated between *N. ovata* populations in 2019, while in 2020 populations differed only in labellum length and width as well as in cavity length and width (Table 1). The width of the flower was the lowest in OPA in both years and the highest in ZAB2, POG, and SKA. Labellum length was the most intra-population differentiated flower trait (CV to 1.9); it also showed the largest variability between years (in 5 among 7 populations, significant differences were noted). The labellum was the widest in ZAB1 and ZAB2 in 2019 and in OPA and POG in 2020. The size of the groove with nectar along the labellum was the shortest in OPA and TUR populations, and in the remaining cases in both years, the values of this trait reached about or even above 4 mm. The minimal and maximal values of size of the cavity with nectar were noted in different populations in both years (Table 1). In 11 out of 14 cases, where year-to-year statistically significant changes were noted, values of floral traits were higher in 2020. In populations with a higher concentration of P in the soil (Table 2), we observed correlated with shorter labellum ($r_s = -0.71$) and groove length ($r_s = -0.83$). On the other hand, in populations where a higher C:P ratio was noted (Table 2), labellum length was longer ($r_s = 0.86$).

Table 2. Soil parameters for *Neottia ovata* populations.

Population	% C	% N	% P	C:N	C:P	N:P	pH Water	pH KCl	CaCO$_3$
OPA	14.28	0.32	0.08	44.1	185.90	4.20	7.94	7.63	7.53
LUB	16.08	0.11	0.15	150.3	110.10	0.70	8.00	7.77	12.96
POG	15.27	0.06	0.05	260.00	309.40	1.20	7.36	7.21	0
ZAB1	0.01	0.32	0.20	0.03	0.06	1.60	8.04	7.77	13.22
ZAB2	37.41	0.70	0.60	53.60	61.90	1.20	6.81	6.73	0
TUR	44.26	0.65	0.16	67.70	278.20	4.10	7.13	6.89	0
WIS	13.13	0.15	0.05	85.70	253.80	3.00	7.78	7.65	13.52
SKA	5.31	0.19	0.06	27.40	95.10	3.50	7.04	6.35	0

2.3. Nectar Chemistry

2.3.1. Sugars

In *N. ovata* nectar, three main sugars, i.e., glucose, fructose, and sucrose, were detected (Table 3). Generally, in POG, the highest concentration of all three sugars was reported (Q_3 = 192.96 µM for glucose, 113.68 for µM fructose, and 28.75 µM for sucrose). Other populations had either significantly lower or statistically equal concentration of sugars (in terms of mean and median). *N. ovata* nectar was dominated by glucose and fructose — sucrose concentration was about 3–5 times lower than the other two sugars. Distribution of individual sugar amounts significantly varied between populations (PermANOVA, F = 5.862, R^2 = 0.277, $p < 0.001$) (Table 3, Figure 1).

Table 3. The concentration of sugars (µM) in *Neottia ovata* nectar. Data represent the mean ($\bar{x}$) ± standard error (SE), lower quartile (Q1), median (Q2), upper quartile (Q3), interquartile range (IQR). The same letters indicate statistically nonsignificant differences according to the pairwise Wilcoxon Rank Sum test with Benjamini-Hochberg adjustment ($p \geq 0.05$).

Sugar	Statistic	OPA (n = 12)	LUB (n = 13)	POG (n = 22)	ZAB1 (n = 12)	ZAB2 (n = 14)	SKA (n = 17)	TUR (n = 18)	WIS (n = 4)
Glucose	$\bar{x}$ ± SE	64.95 ± 6.80	85.23 ± 9.05	160.85 ± 7.80	59.48 ± 3.78	92.99 ± 7.82	57.86 ± 3.85	73.94 ± 2.40	80.02 ± 3.61
	Q_1	48.16	65.92	148.83	50.29	73.02	55.10	67.42	75.75
	Q_2 (IQR)	69.90 (30.10) [ab]	75.31 (16.17) [ab]	168.67 (44.14) [c]	56.15 (18.03) [a]	93.96 (33.46) [c]	64.29 (10.31) [d]	73.77 (14.64) [ab]	80.13 (8.66) [bc]
	Q_3	78.25	82.09	192.96	68.32	106.48	65.41	82.06	84.40
Fructose	$\bar{x}$ ± SE	54.10 ± 5.21	78.77 ± 7.21	99.12 ± 5.84	61.69 ± 4.25	71.64 ± 4.96	61.86 ± 2.55	70.77 ± 2.58	63.52 ± 1.18
	Q_1	43.57	64.21	85.04	54.79	66.30	55.36	63.50	61.93
	Q_2 (IQR)	59.90 (22.77) [ab]	68.75 (11.97) [a]	100.31 (28.64) [ab]	57.18 (18.83) [ab]	70.16 (12.33) [ab]	65.93 (13.40) [c]	71.77 (10.85) [a]	63.30 (2.96) [b]
	Q_3	66.34	76.18	113.68	73.62	78.63	68.76	74.35	64.89
Sucrose	$\bar{x}$ ± SE	11.84 ± 2.03	18.88 ± 2.46	19.51 ± 2.19	19.28 ± 2.98	15.96 ± 2.46	10.99 ± 2.08	23.17 ± 2.62	21.87 ± 2.00
	Q_1	9.26	13.84	11.76	14.53	10.53	6.54	12.69	20.98
	Q_2 (IQR)	11.08 (4.37) [ab]	16.27 (4.82) [a]	18.89 (16.98) [b]	17.67 (5.99) [ab]	15.48 (12.18) [ab]	6.68 (6.40) [ab]	24.19 (20.70) [ab]	22.61 (2.51) [ab]
	Q_3	13.63	18.66	28.75	20.52	22.70	12.94	33.39	23.49

The amount of sugars varied between populations (F = 16.294, $p < 0.001$) and ranged from 25.33 ± 6.79 mg/mL and 25.50 ± 8.79 mg/mL in SKA and OPA populations to 53.51 ± 12.90 mg/mL in POG (Table 4, Figure 1). Nectar was dominated by hexoses—the sucrose:hexoses ratio shaped from 0.14 (POG) to 0.30 (ZAB1, TUR, and WIS) and varied significantly between populations (F = 5.897, $p < 0.001$) (Table 4). Populations also differed in the fructose:glucose ratio (F = 19.011, $p < 0.001$), which was close to 1 in four populations (LUB, ZAB1, SKA, and TUR). The sum of sugars as well as the amount of both hexoses positively correlated with the C:N ratio (r_s = 0.83 in all cases), while the N:P ratio negatively correlated with the sum of sugar (r_s = −0.71) and fructose content (r_s = −0.71).

Table 4. The amount of sugars in *Neottia ovata* nectar. The same letters indicate statistically nonsignificant differences according to Tukey's post-hoc test.

	OPA	LUB	POG	ZAB1	ZAB2	SKA	TUR	WIS
Mean ± SD of total sugars (mg/mL)	25.50 ± 8.79 [b]	36.01 ± 12.80 [b]	53.51 ± 12.90 [a]	28.43 ± 7.82 [b]	35.12 ± 10.08 [b]	25.33 ± 6.79 [b]	34.00 ± 6.63 [b]	33.34 ± 2.77 [b]
Glucose content in nectar (w/v) (%)	1.17	1.54	2.89	1.07	1.68	1.04	1.33	1.44
Fructose content in nectar (w/v) (%)	0.97	1.42	1.78	1.11	1.29	1.11	1.27	1.14
Sucrose content in nectar (w/v) (%)	0.41	0.65	0.67	0.66	0.55	0.38	0.79	0.75
Sugar content in nectar (w/v) (%)	2.55	3.60	5.35	2.84	3.51	2.53	3.40	3.33
Fructose:glucose	0.83	0.92	0.62	1.04	0.77	1.07	0.96	0.79
Sucrose/(fructose + glucose)	0.19	0.22	0.14	0.30	0.18	0.17	0.30	0.29

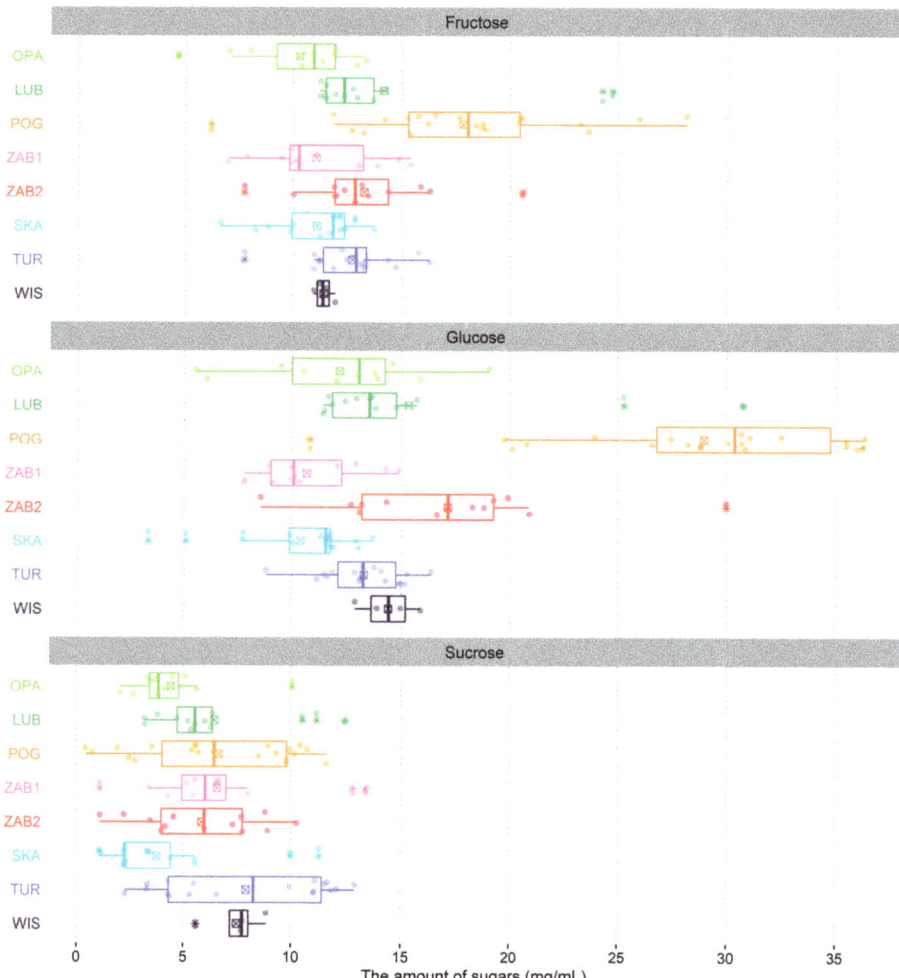

Figure 1. Boxplots of sugar amounts for *Neottia ovata* populations. Colored dots are individual samples. The crossed square shows the mean. The lower and upper hinges correspond to the lower (Q1) and upper (Q3) quartiles. Thus box length shows the interquartile range (IQR). The thicker line inside boxes corresponds to the median. The lower whisker extends from the hinge to the smallest value at most Q1 − 1.5 × IQR of the hinge. The upper whisker extends from the hinge to the largest value no further than Q3 + 1.5 × IQR. Data beyond the end of the whiskers, indicated with an asterisk symbol, are outliers.

2.3.2. Amino Acids

The content of all AAs differed between *N. ovata* populations (PermANOVA, F = 8.228, R^2 = 0.474, $p < 0.001$) and was the lowest in OPA and SKA, and the highest in POG (about 3–8 times higher than in other populations (16,662.1 ± 655.4 µg/mL, Table 5). The content of non-proteogenic AAs was also the highest in POG (735.1 ± 54.2 µg/mL). The percentage of this group of AAs differed between populations (F = 4.525, $p < 0.001$) and was the highest in OPA and SKA (about 10%) and the lowest in POG and LUB (a little above 4%, Table 5).

Table 5. The concentration of amino acids (μM) in *Neottia ovata* nectar. Data represent the mean (x̄) ± standard error (SE), lower quartile (Q1), median (Q2), upper quartile (Q3), interquartile range (IQR). The same letters indicate statistically nonsignificant differences according to the pairwise Wilcoxon Rank Sum test with Benjamini-Hochberg adjustment ($p \geq 0.05$). The number of classes represents the effect of amino acids on insect chemoreceptors: I—no effect; II—inhibition of chemoreceptors; III—stimulate the salt cell; IV—the ability to stimulate the sugar cell [12].

Amino Acid	Class	Statistic	OPA (n = 12)	LUB (n = 13)	POG (n = 22)	ZAB1 (n = 12)	ZAB2 (n = 14)	SKA (n = 17)	TUR (n = 18)	WIS (n = 4)
					Proteogenic amino acids					
Asp	I	x̄ ± SE	114.90 ± 34.78	258.79 ± 29.19	923.49 ± 97.10	154.67 ± 39.38	259.12 ± 65.73	44.14 ± 8.32	405.13 ± 43.23	296.39 ± 34.05
		Q1	39.94	176.71	736.74	39.52	92.58	18.26	297.02	275.96
		Q2 (IQR)	73.93 (131.69) [ab]	259.49 (136.37) [c]	873.31 (352.41) [d]	127.18 (223.98) [abc]	167.22 (269.40) [ac]	26.50 (57.86) [b]	397.49 (216.17) [e]	298.75 (43.22) [ace]
		Q3	171.63	313.08	1089.16	263.50	361.98	76.12	513.19	319.18
Glu	I	x̄ ± SE	339.81 ± 56.75	755.19 ± 51.04	2395.91 ± 154.04	434.54 ± 57.93	672.67 ± 148.76	216.76 ± 37.74	800.82 ± 110.67	510.73 ± 69.55
		Q1	212.65	635.19	1945.63	236.66	329.39	89.08	478.09	397.38
		Q2 (IQR)	403.58 (273.74) [ab]	785.68 (195.29) [c]	2284.82 (751.22) [d]	518.04 (371.77) [a]	443.10 (382.86) [ac]	175.38 (212.96) [c]	798.38 (559.64) [c]	501.67 (217.64) [ac]
		Q3	486.39	830.48	2696.85	608.43	712.25	301.99	1037.73	615.02
Ala	I	x̄ ± SE	402.29 ± 79.45	558.92 ± 61.82	3324.72 ± 348.13	591.94 ± 149.41	573.48 ± 88.64	391.30 ± 95.09	715.04 ± 71.29	481.91 ± 80.45
		Q1	247.36	439.60	1978.72	218.60	408.74	123.56	439.03	403.61
		Q2 (IQR)	361.57 (241.01) [a]	516.32 (300.81) [ab]	3111.70 (2612.50) [c]	385.60 (766.26) [ab]	510.81 (207.66) [ab]	177.95 (459.66) [a]	716.84 (460.90) [b]	518.95 (193.64) [ab]
		Q3	488.38	740.41	4591.21	984.85	616.40	583.23	899.93	597.25
Cys	I	x̄ ± SE	170.11 ± 30.27	173.11 ± 22.06	732.41 ± 62.49	194.07 ± 53.75	183.14 ± 33.14	124.34 ± 23.34	250.40 ± 30.48	195.32 ± 60.22
		Q1	92.36	124.85	541.84	71.35	80.17	70.33	149.91	118.68
		Q2 (IQR)	137.97 (149.36) [ab]	161.86 (71.71) [ab]	682.44 (224.80) [c]	126.02 (193.01) [ab]	157.57 (155.67) [ab]	111.26 (65.35) [a]	254.05 (172.38) [b]	161.43 (119.38) [ab]
		Q3	241.72	196.56	766.23	264.37	235.84	135.68	322.29	238.06
Gly	I	x̄ ± SE	139.61 ± 28.14	121.60 ± 15.01	506.45 ± 46.47	126.32 ± 27.03	192.83 ± 43.85	95.29 ± 19.02	163.19 ± 14.15	115.21 ± 23.62
		Q1	98.38	87.58	328.59	63.14	108.77	37.49	114.31	80.47
		Q2 (IQR)	109.77 (85.36) [ab]	113.70 (45.14) [ab]	513.24 (294.99) [c]	98.99 (114.81) [ab]	130.81 (97.77) [a]	74.43 (115.47) [b]	160.19 (90.30) [a]	114.07 (68.34) [ab]
		Q3	183.74	132.73	623.58	177.95	206.55	152.96	204.61	148.81
Ser	I	x̄ ± SE	53.79 ± 15.35	206.09 ± 17.32	745.52 ± 94.33	153.20 ± 51.36	212.98 ± 48.44	82.30 ± 18.78	322.11 ± 35.99	177.59 ± 26.57
		Q1	19.16	186.77	448.70	25.91	132.32	25.94	197.66	135.41
		Q2 (IQR)	29.34 (52.48) [a]	212.33 (50.33) [bc]	715.29 (568.49) [d]	76.13 (228.27) [abe]	166.96 (101.26) [b]	31.23 (130.75) [c]	316.74 (230.35) [c]	168.08 (74.85) [bce]
		Q3	71.64	237.10	1017.19	254.18	233.58	156.69	428.01	210.26

Table 5. Cont.

Amino Acid	Class	Statistic	Population								
			OPA (n = 12)	LUB (n = 13)	POG (n = 22)	ZAB1 (n = 12)	ZAB2 (n = 14)	SKA (n = 17)	TUR (n = 18)	WIS (n = 4)
			Proteogenic amino acids								
Thr	I	$\bar{x} \pm SE$	131.63 ± 26.07	128.20 ± 12.19	351.60 ± 50.49	149.27 ± 35.45	140.66 ± 25.20	101.54 ± 18.39	168.98 ± 14.22	118.27 ± 12.90
		Q_1	77.30	106.53	108.57	51.90	85.46	44.89	128.16	99.85
		Q_2 (IQR)	105.86 (88.10) [ab]	116.73 (33.95) [ab]	345.52 (403.22) [a]	111.56 (225.62) [ab]	117.61 (78.14) [ab]	75.12 (79.41) [b]	167.10 (85.02) [a]	117.13 (35.70) [ab]
		Q_3	165.40	140.48	511.79	277.53	163.60	124.31	213.17	135.55
Tyr	I	$\bar{x} \pm SE$	10.24 ± 4.35	30.30 ± 9.86	101.24 ± 18.59	19.73 ± 7.27	18.79 ± 8.64	4.56 ± 3.28	2.48 ± 1.41	4.81 ± 4.81
		Q_1	0.00	0.00	29.35	0.00	0.00	0.00	0.00	0.00
		Q_2 (IQR)	0.00 (18.82) [abc]	15.09 (42.50) [a]	74.45 (122.45) [d]	4.74 (37.97) [ab]	0.00 (32.27) [abc]	0.00 (0.00) [c]	0.00 (0.00) [c]	0.00 (4.81) [abc]
		Q_3	18.82	42.50	151.80	37.97	32.27	0.00	0.00	4.81
Arg	II	$\bar{x} \pm SE$	17.58 ± 4.57	24.62 ± 4.69	78.84 ± 12.67	26.18 ± 5.95	43.45 ± 12.93	15.94 ± 4.28	24.32 ± 4.07	14.61 ± 6.08
		Q_1	8.18	13.41	22.48	8.48	8.69	0.00	13.17	8.00
		Q_2 (IQR)	15.44 (12.65) [a]	29.55 (27.28) [ab]	85.17 (103.93) [b]	19.20 (39.33) [ab]	27.41 (61.44) [ab]	13.86 (21.80) [a]	17.65 (23.61) [a]	14.95 (13.55) [ab]
		Q_3	20.83	40.69	126.41	47.81	70.14	21.80	36.78	21.55
Asn	II	$\bar{x} \pm SE$	60.01 ± 22.28	344.62 ± 59.19	1448.64 ± 236.17	212.85 ± 78.93	252.49 ± 111.30	46.00 ± 12.32	280.67 ± 43.12	178.70 ± 36.53
		Q_1	10.88	200.50	798.45	19.63	17.64	13.43	182.66	133.28
		Q_2 (IQR)	17.96 (72.58) [ab]	345.78 (259.59) [c]	1169.15 (856.17) [d]	82.46 (282.10) [abc]	55.56 (253.22) [abc]	24.67 (39.87) [a]	229.73 (154.64) [c]	151.80 (63.94) [bc]
		Q_3	83.46	460.09	1654.63	301.73	270.86	53.30	337.30	197.22
Gln	II	$\bar{x} \pm SE$	92.01 ± 37.32	338.85 ± 59.08	1011.30 ± 218.53	384.08 ± 109.89	373.92 ± 132.31	119.67 ± 22.68	441.51 ± 63.80	208.35 ± 67.29
		Q_1	33.21	223.92	446.99	28.86	126.66	55.79	248.54	133.84
		Q_2 (IQR)	53.19 (58.73) [a]	294.41 (220.74) [b]	711.60 (604.83) [c]	342.75 (609.76) [abd]	203.08 (293.26) [bd]	65.23 (158.28) [ad]	447.78 (331.45) [b]	224.13 (164.80) [abd]
		Q_3	91.93	444.66	1051.82	638.62	419.92	214.07	579.99	298.64
His	II	$\bar{x} \pm SE$	17.43 ± 6.28	155.67 ± 22.72	397.86 ± 38.95	15.70 ± 4.37	228.64 ± 70.16	48.83 ± 9.41	109.24 ± 15.75	38.14 ± 12.43
		Q_1	2.83	101.98	245.04	4.79	56.16	21.27	64.36	25.16
		Q_2 (IQR)	6.14 (26.03) [a]	149.72 (76.04) [b]	351.39 (302.51) [c]	13.76 (15.28) [a]	130.01 (151.81) [d]	48.83 (44.11) [d]	93.45 (88.48) [b]	41.44 (29.25) [ad]
		Q_3	28.86	178.01	547.56	20.07	207.97	65.38	152.84	54.41
Lys	II	$\bar{x} \pm SE$	37.15 ± 13.63	58.05 ± 8.24	389.39 ± 51.81	82.89 ± 27.14	87.32 ± 24.46	61.59 ± 16.32	110.20 ± 15.52	64.39 ± 18.46
		Q_1	11.21	28.22	215.21	10.43	40.43	17.31	57.70	35.61
		Q_2 (IQR)	19.28 (26.07) [a]	61.79 (51.87) [abc]	335.74 (333.13) [d]	33.51 (137.00) [abc]	64.20 (59.81) [bc]	35.58 (76.61) [ab]	84.39 (97.23) [c]	61.18 (54.35) [abc]
		Q_3	37.29	80.09	548.34	147.42	100.24	93.92	154.93	89.96

Table 5. *Cont.*

Amino Acid	Class	Statistic	OPA (n = 12)	LUB (n = 13)	POG (n = 22)	ZAB1 (n = 12)	ZAB2 (n = 14)	SKA (n = 17)	TUR (n = 18)	WIS (n = 4)
					Proteogenic amino acids					
Pro	III	$\bar{x} \pm SE$	72.45 ± 16.97	303.90 ± 37.63	1159.30 ± 99.85	106.13 ± 24.51	383.73 ± 92.20	69.36 ± 12.31	336.08 ± 61.83	80.50 ± 18.19
		Q_1	22.18	218.12	800.62	35.06	170.61	32.87	200.48	58.47
		Q_2 (IQR)	75.19 (90.30) [a]	328.10 (133.69) [b]	1013.74 (675.10) [c]	112.77 (113.32) [a]	296.34 (166.64) [b]	59.87 (52.86) [a]	336.27 (203.78) [b]	67.35 (30.92) [a]
		Q_3	112.48	351.81	1475.72	148.38	337.26	85.74	404.26	89.38
Ile	IV	$\bar{x} \pm SE$	101.47 ± 17.98	139.14 ± 14.56	452.06 ± 39.40	161.12 ± 53.32	142.63 ± 29.08	81.21 ± 19.36	239.01 ± 31.25	146.79 ± 17.76
		Q_1	59.14	109.40	334.50	26.89	76.93	30.90	126.10	123.77
		Q_2 (IQR)	97.27 (98.19) [ab]	138.68 (43.90) [ac]	439.52 (155.53) [d]	100.01 (174.61) [abc]	116.72 (99.30) [ab]	62.65 (73.69) [b]	207.61 (165.25) [c]	149.76 (49.01) [ac]
		Q_3	157.33	153.31	490.03	201.50	176.22	104.59	291.35	172.78
Leu	IV	$\bar{x} \pm SE$	220.90 ± 34.29	255.87 ± 23.30	934.70 ± 60.70	258.17 ± 57.32	260.36 ± 56.28	132.27 ± 24.78	385.27 ± 38.22	256.47 ± 33.52
		Q_1	138.39	217.03	758.48	35.19	120.63	44.87	266.85	233.41
		Q_2 (IQR)	211.82 (193.35) [ab]	250.99 (67.78) [a]	888.18 (279.22) [c]	277.10 (402.64) [abd]	221.51 (151.65) [abd]	109.11 (163.00) [b]	420.63 (229.31) [d]	260.92 (50.56) [abd]
		Q_3	331.75	284.80	1037.70	437.83	272.28	207.87	496.16	283.98
Met	IV	$\bar{x} \pm SE$	22.33 ± 3.92	15.81 ± 2.13	30.42 ± 2.86	31.41 ± 2.96	30.74 ± 6.98	7.62 ± 3.01	24.30 ± 3.78	7.58 ± 5.50
		Q_1	11.50	10.44	20.30	25.72	2.99	0.00	13.84	0.00
		Q_2 (IQR)	28.28 (20.85) [ab]	17.89 (11.74) [ac]	27.73 (18.47) [b]	28.41 (7.33) [b]	32.70 (42.63) [ab]	0.00 (12.32) [c]	24.44 (18.37) [ab]	3.49 (11.07) [ac]
		Q_3	32.35	22.17	38.77	33.06	45.63	12.32	32.21	11.07
Phe	IV	$\bar{x} \pm SE$	110.48 ± 17.14	132.80 ± 17.35	461.23 ± 26.52	156.89 ± 41.20	118.70 ± 20.79	77.69 ± 13.22	164.46 ± 18.43	97.28 ± 13.26
		Q_1	80.67	114.91	32.06	32.06	74.64	37.48	93.62	77.60
		Q_2 (IQR)	106.12 (63.57) [ab]	129.69 (32.66) [a]	448.66 (149.68) [c]	141.74 (197.18) [ab]	101.08 (59.09) [ab]	70.90 (53.69) [b]	159.20 (108.91) [a]	97.12 (39.19) [ab]
		Q_3	144.24	147.57	536.63	229.24	133.73	91.17	202.54	116.79
Trp	IV	$\bar{x} \pm SE$	62.87 ± 10.81	103.53 ± 10.00	181.22 ± 19.70	65.93 ± 13.74	72.15 ± 14.17	55.42 ± 9.41	102.43 ± 15.29	92.35 ± 28.59
		Q_1	41.92	63.64	127.52	29.45	31.33	31.26	49.03	62.00
		Q_2 (IQR)	56.85 (41.94) [ab]	119.90 (67.55) [c]	151.57 (68.77) [d]	62.45 (72.09) [abc]	55.52 (61.12) [abc]	49.10 (35.97) [a]	86.06 (82.37) [bc]	69.82 (38.17) [abcd]
		Q_3	83.85	131.19	196.29	101.54	92.45	67.23	131.40	100.17
Val	IV	$\bar{x} \pm SE$	49.48 ± 22.50	80.96 ± 27.94	116.49 ± 15.57	95.92 ± 52.29	25.57 ± 17.67	197.81 ± 43.41	108.59 ± 33.25	248.59 ± 66.92
		Q_1	8.80	22.65	64.26	0.00	0.00	84.84	22.18	229.28
		Q_2 (IQR)	16.74 (40.02) [ab]	26.08 (83.53) [a]	94.40 (87.82) [cd]	13.11 (86.12) [abe]	0.66 (12.39) [b]	183.90 (130.31) [c]	50.82 (78.80) [ade]	289.59 (79.62) [cde]
		Q_3	48.82	106.18	152.08	86.12	12.39	215.15	100.98	308.90

Table 5. Cont.

Amino Acid	Class	Statistic	OPA (n = 12)	LUB (n = 13)	POG (n = 22)	ZAB1 (n = 12)	ZAB2 (n = 14)	SKA (n = 17)	TUR (n = 18)	WIS (n = 4)
					Non-proteogenic amino acids					
Orn		$\bar{x} \pm SE$	153.64 ± 88.35	126.73 ± 10.87	134.27 ± 14.46	96.86 ± 27.01	55.19 ± 9.62	37.66 ± 7.55	134.56 ± 11.75	84.90 ± 11.86
		Q_1	39.32	99.23	76.40	7.81	32.90	11.24	89.02	72.34
		Q_2 (IQR)	72.35 (71.66) [ab]	134.77 (45.44) [a]	103.38 (109.81) [a]	78.80 (171.73) [b]	53.61 (30.28) [b]	23.75 (53.99) [b]	142.82 (87.98) [a]	88.30 (28.52) [ab]
		Q_3	110.67	144.66	186.22	179.54	63.18	65.24	177.00	100.86
Cit		$\bar{x} \pm SE$	10.81 ± 2.35	15.84 ± 2.27	58.45 ± 13.10	7.64 ± 2.72	25.72 ± 8.72	4.14 ± 2.25	10.21 ± 2.77	7.53 ± 4.43
		Q_1	4.47	10.84	28.44	0.00	0.00	0.00	0.00	0.00
		Q_2 (IQR)	11.67 (10.63) [a]	17.85 (9.46) [a]	36.65 (27.01) [ac]	3.81 (13.89) [ac]	11.12 (35.77) [a]	0.00 (0.00) [c]	4.64 (18.33) [ac]	6.49 (14.02) [ac]
		Q_3	15.10	20.29	55.46	13.89	35.77	0.00	18.33	14.02
Tau		$\bar{x} \pm SE$	59.85 ± 18.63	63.83 ± 17.30	121.95 ± 17.04	55.39 ± 15.94	135.37 ± 32.55	62.97 ± 10.85	82.06 ± 14.64	64.33 ± 44.18
		Q_1	2.93	17.10	67.07	0.00	28.06	37.00	47.28	15.29
		Q_2 (IQR)	44.09 (93.96) [a]	39.37 (59.25) [a]	102.23 (108.07) [a]	55.81 (101.43) [a]	111.77 (204.77) [a]	61.84 (51.65) [a]	62.94 (57.95) [a]	31.54 (65.29) [a]
		Q_3	96.89	76.35	175.13	101.43	232.82	88.64	105.23	80.58
AABA		$\bar{x} \pm SE$	16.04 ± 5.27	22.01 ± 5.50	3.64 ± 2.93	26.22 ± 11.74	29.80 ± 8.95	13.91 ± 3.45	50.51 ± 7.12	57.26 ± 29.22
		Q_1	0.00	11.38	0.00	0.00	1.65	0.00	25.64	21.07
		Q_2 (IQR)	9.24 (29.53) [a]	16.41 (19.36) [a]	0.00 (0.00) [b]	5.68 (32.69) [a]	27.32 (43.50) [ac]	13.18 (15.72) [c]	43.22 (42.48) [c]	42.17 (57.30) [ac]
		Q_3	29.53	30.74	0.00	32.69	45.15	15.72	68.12	78.37
BABA		$\bar{x} \pm SE$	19.90 ± 7.25	6.84 ± 2.80	38.33 ± 7.54	12.48 ± 6.94	14.99 ± 9.62	18.28 ± 4.29	8.94 ± 3.23	18.46 ± 12.21
		Q_1	0.00	0.00	12.44	0.00	0.00	0.00	0.00	0.00
		Q_2 (IQR)	10.17 (30.44) [ab]	0.00 (10.99) [a]	31.21 (35.59) [b]	0.00 (11.02) [ab]	0.00 (5.96) [a]	20.29 (29.93) [ab]	0.00 (19.74) [a]	11.17 (29.63) [ab]
		Q_3	30.44	10.99	48.03	11.02	5.96	29.93	19.74	29.63
GABA		$\bar{x} \pm SE$	94.97 ± 30.51	41.58 ± 10.22	500.42 ± 77.81	138.96 ± 44.84	161.11 ± 35.08	116.97 ± 24.03	62.50 ± 19.01	118.27 ± 73.87
		Q_1	21.78	14.23	194.27	0.00	59.13	22.42	0.00	48.23
		Q_2 (IQR)	60.89 (89.58) [ab]	41.48 (54.03) [a]	455.26 (427.76) [c]	87.63 (235.14) [ab]	129.89 (222.39) [b]	102.89 (163.50) [ab]	0.00 (144.08) [ab]	69.39 (91.20) [ab]
		Q_3	111.36	68.26	622.03	235.14	281.52	185.92	144.08	139.43
β-Ala		$\bar{x} \pm SE$	11.42 ± 8.46	21.52 ± 9.13	0.00 ± 0.00	7.87 ± 4.37	0.00 ± 0.00	28.31 ± 19.34	7.58 ± 6.42	0.00 ± 0.00
		Q_1	0.00	0.00	0.00	0.00	0.00	0.00	0.00	0.00
		Q_2 (IQR)	0.00 (2.02)	0.00 (30.49)	0.00 (0.00)	0.00 (6.28)	0.00 (0.00)	0.00 (20.83)	0.00 (0.00)	0.00 (0.00)
		Q_3	2.02	30.49	0.00	6.28	0.00	20.83	0.00	0.00
Nva		$\bar{x} \pm SE$	6.95 ± 3.33	4.05 ± 1.75	62.30 ± 8.17	4.06 ± 2.23	1.71 ± 0.92	11.44 ± 2.52	0.00 ± 0.00	0.00 ± 0.00
		Q_1	0.00	0.00	42.07	0.00	0.00	0.00	0.00	0.00
		Q_2 (IQR)	0.76 (8.41)	0.00 (6.52)	55.42 (31.73)	0.00 (5.27)	0.00 (0.00)	15.18 (19.73)	0.00 (0.00)	0.00 (0.00)
		Q_3	8.41	6.52	73.81	5.27	0.00	19.73	0.00	0.00

Among all *N. ovata* populations, 28 AAs were detected (20 proteogenic and eight non-proteogenic), with the lowest detected in populations from SLP (TUR-25 and WIS-26). Twenty-six AAs among 28 were present in all populations (Table 5). In some populations, β-Ala and Nva were absent. In each population, five non-proteogenic AAs were found. The highest participation in all populations had Ala (12.4–19.3%) and Glu (12.4–17.3%), and nine others (Leu, Gln, Asp, Asn, Cys, Pro, Val, Ser, and GABA) were noted with 5–10% frequency, although most of them reached such a frequency only in some populations (Table 5, Figure S1). Populations differed in the ratio of sugars to AAs—from 27.8 in POG to 94.6 in SKA (Table 6).

Table 6. Sugar and amino acid ratio in *Neottia ovata* populations.

Population	Total Sugars [mg/mL]	Total AAs [mg/mL]	Total Sugars/Total AAs
OPA	306.01	3.71	82.41
LUB	468.12	6.93	67.54
POG	1177.32	42.37	27.79
ZAB1	341.13	5.42	62.89
ZAB2	491.74	7.87	62.48
SKA	430.61	4.55	94.56
TUR	612.02	11.76	52.02
WIS	133.38	1.74	76.68

Amino acids responsible for nectar taste were divided into four classes. Possible simulation of insect chemoreceptors by AAs in nectars have AAs from classes II (chemoreceptor inhibitors), III (stimulation of salt cells), and IV (stimulation of sugar cells). The percentage share of class II ranged between 10.0–82.5%, while that for class III ranged between 0–19.7% and that for class IV ranged between 11.8–86.3%; the mean percentage shares were 49.9%, 11.19%, and 39.0% for classes II-IV, respectively. Fifty percent of samples had a percentage share in the range of app. 35–78% for class II, 10–22% for class III, and 22–58% for class IV (Figure 2).

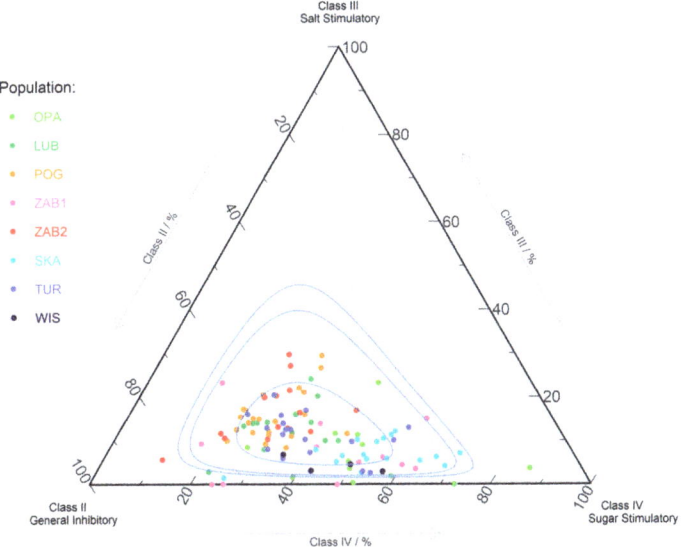

Figure 2. Ternary plot of amino acid classes for *Neottia ovata* populations: II (Asp, Glu, His, Arg, Lys), III (Hyp, Pro), and IV (Val, Met, Trp, Phe, Ile, Leu). Blue lines show 50%, 90%, and 95% confidence intervals via the Mahalanobis Distance and use of the Log-Ratio Transformation. The first class of AAs (Asn, Gln, Ala, Cys, Gly, Ser, Thr, Tyr) does not affect the chemoreceptors of fly (data not shown).

PCA (especially when considering only Dim1) and UMAP gave similar results in clustering individual samples and studying the underlying relations of AAs. However, it should be noted that, contrary to PCA, in the UMAP model the size of clusters relative to each other is essentially meaningless, and the distances between clusters are likely to be meaningless (Figure 3 and Figure S4). Positive scores for the first principal component (Dim1) generally indicate higher values of Gly, Ala, Cys, Glu, Lys, Phe, Asp, Leu, Pro, Ile, Ser, Trp, Cit, and BABA than mean values. Moreover, positive scores for the second principal component (Dim2) indicate values for GABA, Arg, Tau, Met, Nva, and Thr that are higher than mean values for all the populations, while a negative score shows higher values of AABA, Val, and Gln (Figure 3). The AAs show a very good quality of representation on the created model. POG population is the unique one because it has the largest amount of AAs, but it is differentiated by the levels of, e.g., AABA, Val, Tau, GABA, Arg, and Met. SKA population is very similar to OPA, while TUR is similar to LUB and possibly WIS. Samples of ZAB1 and ZAB2 populations vary; some of them are similar to SKA, OPA, and WIS (e.g., 48, 54, 55, and 83), but others to TUR (e.g., 60, 69, 70, 72, 85, and 86) and even POG (e.g., 49 and 66).

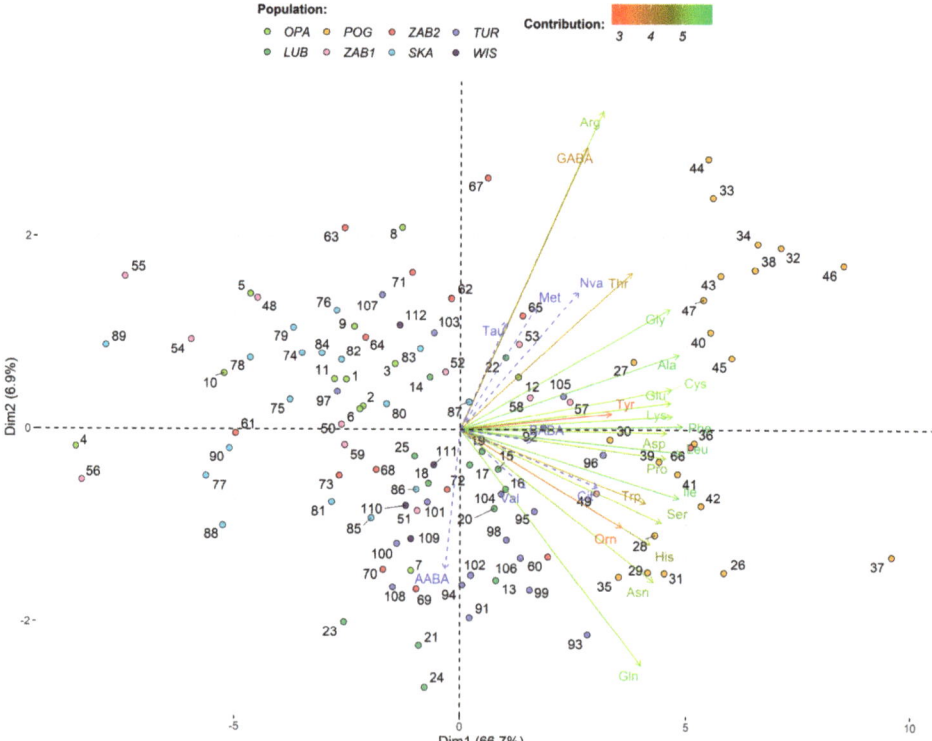

Figure 3. Biplot of amino acid profiles for *Neottia ovata* populations, showing the first two dimensions/factors (Dim1-2) of PCA that together explain 73.5% of the variance. Biplot vectors indicate the strength and direction of factor loading for the first two factors. Vectors of supplementary variables are in blue. Individuals (populations) are color-coded and labeled with a number corresponding to Id used in Table S3.

No correlation between soil parameters and the total amount of AAs was found, but the concentration of some AAs was correlated with soil properties. Different soil parameters were correlated with the amount and percentage of different AAs, although negative correlations dominated. Among soil traits, the concentration of C and the C:N ratio were most often correlated with AAs in *N. ovata* nectar (data not shown).

2.4. Reproductive Success

The pollinaria removal (PR) in *N. ovata* populations was shaped at a high level—above 90% in all populations in 2019, and from 81.6% (TUR) to 95% (ZAB1) in 2020 (Table 7). No statistically significant differences between populations in PR (F = 1.318, p = 0.28 in 2019 and F = 0.628, p = 0.71 in 2020) were found. Fruiting was more differentiated in both years (F = 15.430 in 2019 and F = 10.971 in 2020, p < 0.001) and significantly lower than PR, excluding LUB in 2019, where female reproductive success (FRS) was slightly higher than PR. In 2019, the ratio of flowers that developed into fruits was lowest in TUR and SKA populations (40.7% and 44.4%, respectively), while it was the highest in LUB (93.7%). In 2020, in most cases, the level of fruiting was lower and ranged from 14.9% (SKA) to 86.8% (POG). The efficiency of pollination varied between populations and was higher in populations from Biebrza Valley, while from other regions, the PR was about 5–9 times higher than that of FRS.

Table 7. Spatial and temporal variation of FRS and PR in *Neottia ovata* populations. Data show the mean ± standard deviation.

Population	Year	FRS (%)	PR (%)	PR:FRS
OPA	2019	72.41 ± 20.12	98.44 ± 4.42	1.19 ± 0.22
	2020	63.00 ± 29.56	92.75 ± 11.98	1.44 ± 0.52
LUB	2019	93.74 ± 9.58	90.10 ± 16.59	0.97 ± 0.19
	2020	73.65 ± 18.52	92.11 ± 6.69	1.36 ± 45
POG	2019	nd	nd	nd
	2020	86.76 ± 14.83	86.97 ± 33.98	1.04 ± 0.49
ZAB1	2019	69.59 ± 25.92	97.09 ± 6.50	1.40 ± 0.63
	2020	75.24 ± 21.10	95.00 ± 6.21	1.42 ± 0.68
ZAB2	2019	86.63 ± 14.69	93.45 ± 7.76	1.12 ± 0.23
	2020	44.77 ± 24.78	94.80 ± 7.90	4.65 ± 6.83
TUR	2019	40.73 ± 16.45	nd	nd
	2020	41.02 ± 27.86	81.61 ± 16.33	5.55 ± 8.36
WIS	2019	nd	nd	nd
	2020	38.69 ± 34.35	88.35 ± 18.58	4.81 ± 4.67
SKA	2019	44.35 ± 14.16	nd	nd
	2020	14.85 ± 8.34	85.48 ± 35.56	9.11 ± 5.13

nd—no data.

2.5. Factors Influencing Reproductive Success

Both PR and FRS in *N. ovata* populations were weakly correlated with flower traits. Only in five cases among 84 analyzed were statistically significant correlations between flower traits and RS parameters noted, and in particular populations, different flower traits were under selection. PR was positively correlated with flower width in TUR in 2019 (r_s = 0.74) and with cave width in ZAB2 in 2020 (r_s = 0.82), while in WIS in 2020 it was negatively correlated with groove length (r_s = −0.94). On the other hand, in 2020, we noted correlations between FRS and groove length in OPA (r_s = 0.85) as well as cave length in SKA (r_s = 0.76).

The statistically significant relationship between the amount of sugar and its participation and PR or FRS was noted only in three populations (ZAB1, SKA, and TUR). ZAB1 was the only population where PR, was negatively correlated with sugars: fructose (r_s = −0.68), sucrose (r_s = −0.64), and the sum of sugars (r_s = −0.68). The amount of glucose and the sum of sugars negatively correlated with FRS in SKA and TUR (r_s = −0.63 and r_s = −0.94, respectively). Moreover, FRS in TUR decreased with the increasing amount of sucrose and its percentage (r_s = −0.56 and r_s = −0.50, respectively) as well as the sucrose:hexose ratio (r_s = −0.50) and increased with the increase of fructose participation (r_s = 0.55).

In three populations (OPA, LUB, and WIS), AAs were not correlated with RS in any way, and in two others, only single statistically significant correlations were noted. In TUR, only the percentage of Cys positively correlated with FRS ($r_s = 0.57$), and in SKA only the percentage of Cys negatively correlated with FRS ($r_s = -0.83$) and an increased amount of Gln benefited FRS ($r_s = 0.83$). In the remaining three populations (POG, ZAB1, and ZAB2), we noted more statistically significant correlations between AAs and FRS (Table 6). Pollinia removal correlated only with Pro in POG ($r_s = 0.53$), with GABA and Cit in ZAB1 ($r_s = -0.65$ and $r_s = 0.85$, respectively), and with Val and Gly in ZAB2 ($r_s = -0.83$ and $r_s = -0.72$, respectively). Only in two populations (POG and ZAB1) did we find correlations between the amount of AAs from a particular taste group and RS. In POG, AAs from taste group IV positively correlated with PR ($r_s = 0.56$), while taste group I negatively correlated with FRS ($r_s = -0.59$). In ZAB1, the sum of AAs from taste groups I, II, and IV negatively correlated with FRS ($r_s = -0.64$, $r_s = -0.78$, and $r_s = -0.79$, respectively).

3. Discussion

Plants evolved different strategies to achieve reproductive success. In animal pollinated species, the level of RS depends, first of all, on the presence and abundance of pollinators [85]. Their deficiency is recognized as the main cause of low RS in orchids [2]. Assemblages of pollinators are strictly connected to the character of vegetation [60,61,86–88]. Plants being hosts of *N. ovata* pollinators are common (e.g., species from the Apiaceae family, and *Alnus, Crataegus, Betula, Salix, Corylus, Vaccinium* genera) [76] and were present, more or less frequently, in plant communities in which the studied populations exist. However, vegetation in populations and surrounding areas also showed differences, which certainly influenced insects' assemblages. Nilsson [76] found that saw-flies, one of the most important *N. ovata* pollinators, were present only in the population near marsh vegetation. This could partially explain the higher level of RS, especially FRS, in populations that existed in BNP on mineral islands among peat bogs in comparison to others (SKA, TUR, and WIS) surrounded by a distinct type of vegetation. Other plants may also decrease RS, competing successfully for pollinators, offering them more and/or better food [62–65], especially when populations of pollinators are not abundant. Nilsson [76] found differentiation of the presence and abundance of visitors and pollinators in distinct Swedish *N. ovata* populations. Variability in insect assemblages, and their abundance was probably one of the main factors shaping the levels of RS in populations of this orchid in northeast Poland. Insect assemblages also fluctuate from year to year [76,89], which may explain the temporary variation of RS in some *N. ovata* populations.

According to Nilsson [76], *N. ovata* may be visited by almost 300 species, representing different systematic groups with a wide spectrum of body sizes, mouth apparatus, and nutritional preferences. Undoubtedly, the main role in the attraction of these insects is played by the scent bouquet, comprised of compounds that are known as general attractants of a wide range of insects [71,76]. Numerous insects capable of pollinating flowers of this orchid, together with the easily available nectar on the labellum, create a chance for a high level of RS. In the majority of populations, we observed a higher level of fruiting than those found by Brzosko [80] and Brys, et al. [79]. PR in all *N. ovata* populations in both years (always above 80% or even above 90% in many cases) suggests that a large number of insects penetrated flowers. On the other hand, FRS was more differentiated (similar to the seven-year studies of Brzosko [80]). The higher efficiency of pollination we noted in populations from Biebrza National Park, and in remaining FRS was 5–11 times lower than that of PR. This indicates that not all insects that visited flowers (even able to collect pollinia) were effective pollinators. Probably some of the visitors, especially the smallest or the weakest, may suck nectar only from the groove along the labellum and do not penetrate flowers in-depth, omitting in this way the cave at the labellum base, which decreases the probability of contact with the column. Moreover, pollinaria may be attached to different parts of the insect's body [76], and the position of the visitor sucking nectar may sometimes be unsuitable for the collection of pollinia and/or to place them on the stigma. The low efficiency of pollination could be explained by Nilsson [76]. The author found behavioral disturbances of smaller ichneumons (dominant pollinators) if they have big loads of pollinia. Insects that do not penetrate the flowers correctly may occasionally contribute to the pollination of *N. ovata* flowers. This indicates that pollination in this species has a haphazard character. The disparity between PR and FRS indicates that pollinia are often lost, as observed by Brys, et al. [79].

Incorrect flower penetration, causing ineffective pollination, is an effect of mismatch between flower and pollinator. The mechanical fit between partners is one of the essential preconditions of successful pollination [45–47] and one of the most important evolutionary mechanisms [2,4,44]. Such a match is generally stronger in specialized systems [90], which confirms, for example, the results of studies on long-spurred orchids [41–43,51]. Our results suggest the best fit between flowers and pollinators in POG and LUB populations. PR:FRS in LUB in 2019 and in POG in 2020 was close to 1, and additionally, in LUB in 2019, FRS was higher than PR, indicating the presence of effective pollinators and their high efficiency. The high PR:FRS ratio in other populations and the relatively low FRS in some of them denote a mismatch between flowers and visitors and, as a result, a larger loss of pollinia. Weak correlations between flower traits and PR and FRS (five cases among 84 analyzed in both years) confirm this mismatch. In these single cases, distinct flower traits were under selection in particular populations. Nevertheless, four among five flower traits correlated with PR or FRS concerned the sizes of structures (groove and cage), in which nectar is secreted and accumulated. Because we supposed that groove and cavity sizes are the measures of nectar quantity, it could indicate that the amount of nectar is the most important trait influencing RS in *N. ovata*. We expected that the labellum in this orchid, as a landing platform and flower part, should be adapted to pollinators' sizes. Although we did not find an influence of labellum on RS parameters, its length was the most differentiated flower structure between populations, which suggests that it reacts on local insect assemblages. The disparity between the level of PR and FRS in most populations, probably being an effect of structural mismatch, may be explained through the great variation in body sizes of *N. ovata* visitors and the differentiation of their behaviors as nectar consumers. Even the main group of pollinators of this species (ichneumonids) includes representatives with a wide range of sizes [76]. In populations with lower FRS, these were probably predominant insects that more accidentally remove pollinia and less often place them on stigmas of other flowers. Because their main dietary sources are connected to other plant species, and *N. ovata* is a marginal part of the food (if only because of small population sizes), they do not need to adapt to its flowers. This suggests that the level of *N. ovata* RS depends on accompanying plant species, their diversity, and their abundance. Contrary to Brys, et al. [79] results, we did not find an influence of floral display on RS in *N. ovata*.

In nectariferous plants, the amount and composition of nectar are known to affect plant-pollinator interactions [9,10,12,13,16,19,22,23,27,28,34]. Our studies document that *N. ovata* is characterized by exceptionally diluted nectar with the lowest sugar concentration among orchids [17,18,27–30]. To our knowledge, these values are comparable only to the concentration of nectar used by some hummingbirds [26,91]. The relatively low sugar concentration was noted for plants pollinated by moths and flies [17,23,24,26,92,93], and only fly-pollinated species have extremely low volume and sugar concentration but high amino acids and hexose content [12,34]. Some fly species are also known to pollinate *N. ovata* [76]. The nectar of *N. ovata* is dominated by hexoses, which is in agreement with the statements of Gottsberger, et al. [39] and Pais, et al. [29] that nectar in flowers with concealed nectaries tends to be dominated by sucrose, while in more open flowers by glucose and fructose. Hexose solution has a higher osmolarity, and therefore lower evaporation rates, than sucrose solution, which can explain the high proportion of hexoses in shallow flowers [12]. However, the prevalence of hexoses was noted in the nectar of some long-spurred orchids [17,18]. On the other hand, contrary to our results, Galetto, et al. [18], studying nectar in five orchid species, found that nectar located in the spurs in two *Habenaria* species was copious and less concentrated (<20%), while in species in which nectar was accumulated in the basal lateral parts of the labellum, it was more concentrated (ca. 50%). Our results are in accordance with the studies of Johnson and Nicolson [94], who documented a clear distinction between nectar sucrose content of specialized (40–60%) and generalized (0–5%) bird-pollinated species. Nonspecialized insects, i.e., syrphids, flies, and beetles (insects from these groups are *N. ovata* pollinators) [76] preferred monosaccharide nectar of plants from phryganic communities [22]. Hexose-rich nectar, which is taken up more easily than sucrose, may be an adaptation and advantage for attracting a wide range of nonspecialized pollinators. It is worth noting that low sucrose nectar is also characteristic for species from the Apiaceae family [22,95,96], which are pollinated by the same systematic groups of insects as *N. ovata*. The lack of influence of nectar sugars on RS in five populations confirms that these nectar traits are not aimed at any of the pollinator group. Moreover, our results could suggest that insects operated in the three remaining populations did not prefer nectar sugar composition in nectar offered by *N. ovata*. With the exception of the TUR population, in which fructose participation benefited FRS, in the remaining cases, statistically significant negative correlations were noted. In studies on two *Platanthera* species, we also noted positive selection only on fructose content [17]. To amount the preferences of insects, experiments should be performed. Hexoses, and especially fructose, are preferred by some pollinators due to their lower viscosity, enabling easier absorption [25]. Heil [31] documented that some ants (often observed on *N. ovata* in our studies and by Nilsson [76]) even preferred sucrose-free nectar because they are not able to assimilate this sugar due to lack of invertase. Sucrose-rich nectar may be toxic for some generalists. All the above-mentioned results, at least partially, explain the dominance of hexoses and the high fructose:glucose ratio in the majority of *N. ovata* populations. The concentration of sugars in nectar and the sucrose:hexose ratio also depend on water availability [22]. *N. ovata* populations exist in relatively wet places, and heavy rainfall in 2020 might additionally decrease sugar concentration.

N. ovata nectar is rich in amino acids, we noted 28 distinct AAs (20 proteogenic and eight non-proteogenic), and 26 were common for all populations. In the nectar of specialist orchids from the *Platanthera* genus, 23 AAs were found in total, from nine to 20 in each population [17]. Moreover, the nectar of other orchids was composed of the lower number of AAs—20 in *Gymnadenia conopsea* [28] and 17 in *Limodorum abortivum* and *Epipactis atropurpurea* [29]. In three populations (OPA, LUB, and WIS), no relationship between RS and AAs was found, while in two others (TUR and SKA), only single statistically significant correlations were noted—the percentage of Cys positively influenced FRS in TUR, while in SKA negatively influenced FRS. In the last population, an increased amount of Gln increased FRS. In the remaining three populations (POG, ZAB1, and ZAB2) we noted a larger influence of AAs on RS; it concerned mainly FRS and almost all of the correlations were negative. Pollinia removal depended only on Pro in POG, on GABA and Cit in ZAB1, and on Val and Gly in ZAB2. The most abundant in all *N. ovata* populations were Ala and Glu, but they weakly affected RS, having only a negative influence on FRS (Glu in SKA and POG, and Ala in ZAB1 and POG). Other AAs with a relatively high amount in *N. ovata* nectar were Asp, Cys, Gly, Thr, Asn, Gln, Ile, Phe, and Pro among the group of proteogenic AAs, and Orn and GABA among non-proteogenic ones.

AAs in floral nectar are important for the survival of nectar-feeding animals [14,16,32,33,97], although the role of particular AAs is poorly explained. It is known, for example, that one among the most abundant AAs in *N. ovata* nectar (Ala) influences insects' growth, while the second (Glu) affects pollinators' behavior [28] similarly to Leu and Met [12]. On the other hand, Venjakob, et al. [61] found that Ala and Gly may deter honeybees. In the case of *N. ovata*, the second function of Ala is more probable, as we found a negative correlation of this AA with RS. One of the most common AAs in plant nectar is Pro, which rewards pollinators and acts as a propellant for the lift phase of the flight [98,99]. It triggers the normal insects' salt-receptor neurons, which initiates feeding [19,23,97]. Its accumulation is also interpreted as a plant's answer to stress factors [98]. Pro was present in all populations studied with a quite high amount, but only in POG did it positively correlated with PR, and in ZAB2 with FRS. Two other AAs (Asp and Thr), which belong to the most abundant AAs in *N. ovata* nectar, seem negatively correlated with FRS in ZAB1, and are known as general repellents [14]. Moreover, Glu, Leu, and Met play a potential role in parasitoid rejection [12]. One of the two most abundant non-proteogenic AAs in *N. ovata* nectar, GABA, influences the insect nervous system and muscle activity [11,100]. Its higher amount was connected with a decrease in both PR and FRS in ZAB1. An interesting result was observed in POG, where a higher amount of this AA was negatively correlated with fruiting, while its percentage was positively correlated with FRS. This indicates that not only the amount of a particular AA, but also relationships between them, may be important in shaping plant-pollinator interaction. BABA, although a less common AA in nature than GABA, was present in all *N. ovata* populations with a relatively high amount. It contributes to protecting plants from pathogens [101,102]. One of the important nectar traits is its taste, which attracts or discourages visitors and depends on some AAs [27,28,103]. AA compositions influence pollinator taste perception and pollinating behavior through specific neurological or phago-stimulating pathways [12]. Some authors suggest that the taste function is even more important than the nutritive value [34,35]. We observed a potential influence of the amount of AAs from a particular taste group only in POG and ZAB1. The positive correlation between taste group IV and PR was noted only in POG. In the remaining cases, nectar taste could shape FRS, always in a negative way. This may indicate that insects present in the majority of *N. ovata* populations were not sensitive to nectar taste or did not prefer this taste. Similar results were obtained for other nectariferous orchids [17,27].

We found inter-population variation in flower structure and the amount of particular nectar components, similar to other studies [19,27,41,55,56,58,79]. One of the sources of this variability is differences in soils in which *N. ovata* grows. Soil properties influenced mainly nectar composition. Production of nectar is costly, even to 30% of flower costs [104]; thus, it requires adequate soil resources. Our studies suggest that more important than the participation of particular chemical elements in the soil is their proportionality. The most important in shaping nectar character were the C:N and N:P ratios. In POG, where C:N was the highest, the sum of sugars was 1.5–2 times higher, and the amount of AAs was three to even almost eight times higher than in other populations. The increase of the C:N ratio in soil caused a higher sum of sugars, glucose, and fructose as well as some AAs (Asp, Glu, Asn, Ser, Trp, and His). Simultaneously, the increase in the same soil characteristic had a negative effect on the percentage of Gly, Thr, Tau, Met, and GABA. A higher N:P ratio negatively influenced the total amount of sugars and fructose and the percentage of Glu and Tyr, while it increased the percentage of Cys and Ile. The importance of soil traits for nectar traits or plant condition was noted by other authors [27,28,34,39,69].

It should be noted that the other factors, such as weather conditions, may shape plant properties and pollinators' assemblages and their activity [12,81,96]. The weather condition in our studies differed between seasons: 2020 was rainier than 2019. It could cause that higher values characterized some plant parameters in the second year of the study. In 19 out of 63 cases, we noted statistically significant changes between years; in 15 cases, we observed an increase of these values (3 cases of floral display and 11 of flower traits). On the other hand, in 2020, in most cases, the level of fruiting was lower. The explanation of year-to-year changes of plant traits due to weather should be undertaken with caution because only some traits and only in some populations differed between seasons. Moreover, in neighboring populations in the Biebrza National Park, the same traits often changed in opposite directions from year to year. It can indicate the greater role of other factors than the weather in these changes.

In our study, we tried to answer the question: In which way is *N. ovata* adapted to a wide range of pollinators? As a generalist with reference to pollinators, *N. ovata* depends on an exceptionally high number of insects in the pollination process—almost 300 species of visitors, among which at least 50 species attached pollinia [76]. How can the demands of such a wide range of insects, differing in sizes, mouth apparatus, nutritional needs, and behaviors, be met? The answer is simple: The plants' offer should also be wide. In the case of *N. ovata*, this rich offer includes the wide range of nectar components (e.g., a large number of AAs) and differentiation of their amounts as well as the variability of flower structures. This indicates that this species did not evolve flower traits, which filter flower visitors; thus, they are not dedicated to a certain group of pollinators. Generally, our results fit the generalistic character of *N. ovata*, but the level of generalization at the species level seems to be higher than at the population level. The lack of, or poorly matched, interactions between flower structure and nectar chemistry and the levels of RS in *N. ovata* populations from northeast Poland confirm this statement. This finding is in agreement with the results of studies on other generalists [90]. Jacquemyn and Brys [105] found that a large variation in flower traits in *Orchis purpurea* populations is maintained by the lack of strong selection pressures on these traits. Differentiation of flower traits enables pollination by whatever flower visitors have a suitable size and appropriate behavior. The probability that whatever species among almost 300 *N. ovata* visitors will serve as an effective pollinator is quite high. However, the variation of FRS among populations suggests that despite the high number of potential pollinators of this species, their abundance in particular populations was extremely differentiated. A low level of fruiting in some populations and a high ratio of PR:FRS (especially in SKA) indicates pollinators' deficiency. In SKA, the problem with pollinators is deeper due to anthropogenic impact. It exists in disturbed and fragmented habitats in a restricted area, less abundant in plants, being hosts of *N. ovata* visitors. Significantly lower PR in such populations may reflect unsuitable conditions for insects. The high levels of fruiting in populations from Biebrza National Park resulted from the relatively unchanged environment in this area. Natural habitats are suitable for many plant species connected to insects pollinating *N. ovata* flowers. The higher RS in populations from BNP could also be a result of their larger sizes in comparison to others. The minimum population size is often required to attract sufficient pollinators. This assumption is supported by the results of Brys, et al. [79] study, which found a significant relationship between RS and population size in *N. ovata*.

Our results contribute to the knowledge about the reproductive strategy of *N. ovata* and fit into studies that explain the causes and consequences of generalization in plants. However, in the course of this study, new questions arose, which required further analysis. For example, why does this orchid invest so many resources into nectar production if it is not an effective allurement of insects, as in other nectariferous species, and which nectar components are the most important for its fitness? The yellow-green color of *N. ovata* flowers does not attract pollinators because it does not contrast with the surrounding vegetation. The color purity is typical of many generalist insect-pollinated plants [71]. In such cases, other flower traits (odor and nectar) play a key role. Floral nectar (its concentration and composition) is rarely detectable by a pollinator at a distance [11]. The fragrance is a key floral attractant for most wasps and beetles—the insect groups that are pollinators *of N. ovata*—and also other generalistic plants accompanying this orchid; also, they often possess the same odor compounds [76,95,96,106]. A probable scenario is that at the first step insects are attracted by the fragrance emitted by *N. ovata* flowers, which can explain the high level of PR; however, after probing nectar, which seems tasteless to the visitors, they do not further penetrate the flowers, thus causing a decrease in FRS. If so, why does *N. ovata* produce such ineffectual rewards? The answer partially explains Johnson and Hobbhahn [71] hypothesis that generalist pollination in orchids comes with high reproductive costs. According to the authors, these higher costs also include pollinia losses and inefficiency of pollination, characteristic of most orchids.

Supposing that, in the evolution process, all acts are intentional/on purpose, these high costs also contribute to *N. ovata* fitness. Each of the flower traits developed in this orchid is equally important in shaping RS, even those that seem to be negligible in our studies. They may operate side by side as "comprehensive consumer infinity" for pollinators. It seems that the wide flower variability and complexity of their action is the advantage of this species, which enables the maintenance of populations under different environmental conditions. The results of our studies also have conservation implications; protection of this orchid requires the protection of its wide spectrum of insect partners and their hosts and, thus, the entire habitats in which *N. ovata* exists.

4. Materials and Methods

4.1. Study Species

Neottia ovata is a long-lived, shade-tolerant forest herb with a wide geographical range covering Western Europe to Eastern Siberia [107]. It usually grows on moderately dry to wet soils with a wide range of pH (pH = 5.5–7.5) [108,109]. Yellow-green flowers (15–30) develop on a flexible raceme. Flowers open and age sequentially and remain receptive for 2–3 days. Each flower possesses two pollinia attached to each other, being removed as a pair. *N. ovata* is self-compatible but has a mechanism (well-developed rostellum) to prevent self-fertilization [76,78]. Flowers emit a distinct and somewhat sweet scent and secrete nectar on the labellum [107]. Nectar is produced in the shallow cavity at the wide lip base and in a lesser quantity in a central longitudinal groove along the elongated part of the two-lobbed lip [77,84]. *N. ovata* attracts many insect species. Nilsson [76] observed 283 visitors in Swedish populations, mainly unspecialized anthophilous insects such as ichneumonids, sawflies, and beetles, among them at least 50 species (belonging to the Hymenoptera, Coleoptera, and Diptera) with attached pollinia. After landing, a visiting insect licks the nectar secreted in a groove. Following the nectar trail, the insect is guided to the lip base and the gynostemium [76,77]. Fruits become ripe at the end of June. In one capsule, 218–1774 seeds develop [78].

4.2. Study Area

This study was performed in eight populations of *N. ovata* in northeast Poland. Five of them were localized on mineral islands among pit bogs in the Biebrza National Park (OPA, LUB, POG, ZAB1, and ZAB2), two in the Suwałki Landscape Park (TUR and WIS), and one in Knyszyńska Forest (SKA). Studies were conducted during two years (2019 and 2020), excluding WIS, which was observed only in 2020. Populations differed in size and existed under different environmental conditions (Table 8).

Table 8. Habitat characteristics for *Neottia ovata* populations in northeast Poland.

Region	Population	Habitat Characteristics
Biebrza National Park (BNP)	OPA	Mineral elevation with domination of *Betula pendula* in tree layer, and in undergrowth layer species characteristic for broadleaved forests
	LUB	At the border of mineral island covered by broadleaved forests
	POG	The border of alder forest and peat bogs, partly in open area with domination of grasses and sedges, and partly under shrubs and trees canopy
	ZAB1	Mineral island dominated by open space, covered mainly by grasses and sedges, with patches of shrubs and trees at the border
	ZAB2	Mineral, island dominated by open space, covered mainly by grasses and sedges, with patches of shrubs and single trees at the border
Suwalki Landscape Park (NLP)	TUR	Under canopy of fragment of alder forest with loose undergrowth layer
	WIS	Shallow lowland springs rich in mosses and *Equisetum telmateia*
Knyszynska Forest (KF)	SKA	The small patch of birch forest with domination of sedges, at the foot of the railway embankment

4.3. Fieldwork and Floral Trait Measurements

Because *N. ovata* populations in northeast Poland are small and only 10–20% of the population flowers each season [80], we have started observations on 20–22 individuals (whenever available) from each population. The final sample size was in many cases lower because some shoots were damaged during flowering or before fruiting. We have quantified three floral display traits directly in the field during the peak of flowering: height of shoots, length of inflorescence, and number of flowers. Next, we collected the five lowest flowers from each inflorescence. All five were used for the evaluation of nectar composition, while two out of those were drawn randomly to measure morphological variables such as flower width, labellum length and width, length of longitudinal groove with nectar on the labellum, and length and width of the cavity with nectar at the base of the labellum. The size of the groove and cavity were considered measures of nectar quantity. Samples from all populations were collected during three days under sunny weather. The measures were taken using an opto-digital microscope DSX110 (Olympus Life Science, Waltham, MA, USA) in the Laboratory of Insect Evolutionary Biology and Ecology, Faculty of Biology, University of Bialystok.

To assess the level of reproductive success (RS), we marked shoots and counted the number of flowers per inflorescence in full blooming. During the maturation of capsules, FRS and PR were quantified. FRS was evaluated as the proportion of developed fruits to the number of flowers on the inflorescence and was given in percent. PR was determined in the percent (PR to the total number of pollinaria for each inflorescence). We also evaluated the efficiency of pollination as the ratio of PR:FRS; the higher the index, the lower the pollination efficiency within a population.

4.4. Soil Analysis

Three soil samples were taken from each population at a depth of 5–10 cm. Samples were dried at room temperature, ground, and sieved (1 mm). Two types of pH were measured with a Hach-Lange pH meter (Hach Company, Loveland, CO, USA) in a 1:2.5 soil water mixture and 1:2 soil KCl solution (1 M) mixture [110]. About 25–50 mg of soil was used for total soil organic carbon analysis by dry combustion at 900 °C using the TOC-A Shimadzu analyzer with SSM-5000A combustion module (Shimadzu Corporation, Kyoto, Japan). About 0.5–1 g of dry soil samples were treated with 10 mL of 10% HCl and connected to the gas-tight Scheibler apparatus according to the CO_2 volumetric carbonates analysis method [111]. Total nitrogen content was measured with the Spectroquant nitrogen cell test (Merck KGaA, Darmstadt, Germany) according to Koroleff's method [112]. Soil samples were treated with an oxidizing agent in a thermoreactor, then acidified with sulphuric and phosphoric acid. Nitrogen was measured photometrically with 2,6-dimethylphenol (DMP). Total phosphorus content was measured by perchloric acid digestion followed by the molybdate photometrical test. The absorbance was measured with a spectrofluorometer SpectraMax M2 (Molecular Devices, Sunnyvale, CA, USA).

4.5. Nectar Analysis

4.5.1. Nectar Isolation

Nectar chemistry was studied in 2020. Five flowers per individual were used for nectar analyses. Our preliminary analyses showed that the nectar amount from the lower number of flowers was not enough to correct the detection of nectar components. The flower nectar isolation was performed using a water washing method [113]. Five flowers per sample were placed into the 2 mL Eppendorf tube containing 1 mL of distilled water and shaken in a laboratory thermomixer (120 rpm, 21 °C, 45 min; Eppendorf Corporate, Hamburg, Germany) for the nectar efflux. Then, the flowers were removed from the tubes, and the mixture of water with nectar was evaporated to dryness by centrifugal vacuum concentrator (45 °C, Eppendorf Concentrator Plus, Eppendorf Corporate, Hamburg, Germany). The obtained pellet was dissolved in 20 µL of distilled water, then transferred into the centrifuge tube with a filter and centrifuged to remove impurities ($9000\times g$, 5 min; MPW-55, MPW Med. Instruments, Gliwice, Poland). The purged extract was collected in a glass vial with a 250 µL insert with polymer feet.

4.5.2. Sugar and Amino Acid Determination

Determination and quantification of sugars and AAs were performed using the high-performance liquid chromatography (HPLC) method. An Agilent 1260 Infinity Series HPLC apparat (Agilent Technologies, Inc., Santa Clara, CA, USA) with quaternary pump with an in-line vacuum degasser, thermostatted column, and refrigerated autosampler with autoinjector sample loop was used.

For sugar analysis, a ZORBAX Carbohydrate Analysis Column (4.6 mm × 250 mm, 5 μm) (Agilent Technologies, Inc., Santa Clara, CA, USA) at a temperature of 30 °C and a refractive index detector (RID) was applied. The mobile phase was a solution of acetonitrile/water (70:30, v/v) at a flow rate of 1.4 mL/min. The injection volume was 10 μL. The total time of analysis was 15 min [17].

Meanwhile, for AA detection, an automatic program of derivatization was set. Thus, the o-phthalaldehyde (OPA) and 9-fluorenylmethyl chloroformate (FMOC) reagents were used for the derivatization of primary and secondary AAs [17]. The Agilent Zorbax Eclipse Plus C18 (4.6 × 150 mm, 5 μm) column (Agilent Technologies, Inc., Santa Clara, CA, USA) at a temperature of 40 °C was used to separate individual AA. Detection of primary AAs was performed by a photodiode array detector (DAD) at 388 nm, while detection of secondary AAs was performed by a fluorescence detector (FMOC) with an excitation wavelength of 266 nm and an emission wavelength of 305 nm. The injection volume was 5 μL; the flow rate was 1 mL/min. Eluent A of the mobile phase was 40 mM NaH_2PO_4 (pH 7.8, adjusted by 10 M NaOH solution), while eluent B was a mixture including acetonitrile/methanol/water (45:45:10, $v/v/v$). The gradient was the following: 0–5 min, 100–90% A; 5–25 min, 90–59.5% A; 25–30 min, 59.5–37% A; 30–35 min, 37–18% A; 35–37 min, 18–0% A; 37–40 min, 0% A; and 40–43 min, 100% A.

The analytical data were integrated using the Agilent OpenLab CDS ChemStation software (Agilent Technologies, Inc., Santa Clara, CA, USA) for liquid chromatography systems. Identification of sugars and AAs was performed by comparing retention times of individual sugars and AAs in the reference vs. test solution. The concentration of these compounds was assayed based on comparisons of peak areas obtained for the samples investigated with those of the reference solutions.

4.6. Statistical Analysis

The R programming language/statistical environment was used to perform all statistical computations and analyses, as well as to prepare graphics and transform data for tabular representation [114,115]. The dataset of AAs and sugars were checked for equal variances and normal distribution in each of the populations with the Shapiro-Wilk test and Levene's test [114,116], respectively (both failed for all or some of the groups/variables). Interestingly, in the dataset of the sum of sugars (glucose + fructose + sucrose) for each population were normally distributed and had homogenous variances. The dataset of floral display and flower structure in N. ovata populations in northeast Poland in 2019–2020 were also tested using Shapiro-Wilk and Levene's tests with the result that data were normally distributed and had homogenous variances.

Differences among populations in floral display and flower traits were tested using one-way ANOVA ("stats" package). The influence (monotonous relation) of analyzed parameters on reproductive success (PR and FRS) was checked separately with Spearman's correlation coefficient (r_s) for each population. The same test was used to evaluate the influence of soil parameters on flower display, floral traits, and nectar chemistry, but in this case, correlations were made at the population level between soil characters and average values of analyzed traits.

Dataset of the sum of sugars was also subjected to one-way ANOVA followed by Tukey's post-hoc test. Sugar and AA datasets were supplied to the Kruskal–Wallis test (to perform a non-parametric alternative to the one-way ANOVA test) followed by a pairwise Wilcoxon Rank Sum test with Benjamini-Hochberg adjustment that compared the median values of different parameters between populations [117–119]. Composition of sugars and AAs was tested between populations using Permutational Multivariate Analyses of Variance (PermANOVA) in "vegan" package [120]. Furthermore, a set of descriptive statistics (n, mean, standard error, quartiles) was calculated for AAs and sugars (Figure S1). For all tests, the significance level was α = 0.05. To analyze the effect of AAs on insect chemoreceptors, all identified and determined AAs were grouped into four classes [12]: I. Asn, Gln, Ala, Cys, Gly, Ser, Thr, and Tyr (no effect on the chemoreceptors of fly); II. Arg, Asp, Glu, His, and Lys (inhibition of fly chemoreceptors); III. Pro and Hyp (stimulate the salt cell); and IV. Ile, Leu, Met, Phe, Trp, and Val (ability to stimulate the sugar cell) and presented as a ternary plot [121].

Principal component analysis (PCA) was used to simplify the exploration of AAs. To build the PCA model, the "FactoMineR" package was used [122]. Data (except for β-Ala, which was present in only a few samples) were transformed using Tukey's Ladder of Power [123] with λ that maximizes the Shapiro-Wilk's W statistic using the "rcompanion" package [119]. Starting λ was set to -10, and ending to 10, while the interval between λ was to 0.005. Ala was scaled using $y = -x^{-0.06}$; all other AAs were scaled using $y = x^{\lambda}$ (Table S1). Two tests that indicate the suitability of the AA dataset for structure detection and reduction were performed: Bartlett's test of sphericity [124] and the Kaiser-Mayer-Olkin test of factorial adequacy (KMO) ("psych" package [125]). The p-value from Bartlett's test of sphericity was approximately equal to 0, while the calculated overall measure of sampling adequacy (MSA) from the KMO test was equal to 0.92. MSA for individual AAs ranged from 0.53 to 0.95 (Table S2). Thus, according to Kaiser [126], the MSA value is high enough to perform PCA. Unit variance scaling of the data (scale.unit = TRUE) was applied; thus, PCA was performed on a correlation matrix, rather than on a covariance matrix. Different PCA models, i.e., without and with different data transformation techniques, as well as supplementary variables, were also created and investigated. Finally, six AAs did not participate in the creation of the final PCA model. Instead, they were used as supplementary variables to help interpret the dimensions of variability. According to Cattell's rule, two components should be selected [127], while Kaiser's rule indicated that three components should be retained [128]. Studying the $\cos^2$ plot (Figure S2) led to the selection of the first two components that explain about 73.5% of the variance (Figure S3). All biplots were created using the "factoextra" package [129]. Furthermore, uniform manifold approximation and projection (UMAP) were performed on a raw AA dataset with the exclusion of β-Ala to provide an additional source for detecting sample and population similarity (Figure S4) ("umap" package [130]).

Supplementary Materials: The following are available online at https://www.mdpi.com/1422-0067/22/4/2214/s1, Figure S1: Boxplots of amino acids concentration for *Neottia ovata* populations. Colored dots are individual samples. The crossed square shows the mean. The lower and upper hinges correspond to the lower (Q1) and upper (Q3) quartiles. Thus box length shows the interquartile range (IQR). The thicker line inside boxes corresponds to the median. The lower whisker extends from the hinge to the smallest value at most Q1 $- 1.5 \times$ IQR of the hinge. The upper whisker extends from the hinge to the largest value no further than Q3 $+ 1.5 \times$ IQR. Data beyond the end of the whiskers, indicated with an asterisk symbol, are outliers, Figure S2: Scree plot showing the proportion of explained variance by the principal components, Figure S3: $\cos^2$ for the amino acids selected as active variables in the principal component analysis model, representing the quality of representation for variables on the factor map (Dim1-3), Figure S4: Uniform manifold approximation and projection of all amino acids in *Neottia ovata* populations, except for β-Ala. Individuals (populations) are color-coded and labeled with a number corresponding to Id used in Table S3, Table S1: Amino acids transformation using Tukey's Ladder of Power, Table S2: Kaiser-Meyer-Olkin test results sorted in descending order by the measure of sampling adequacy (MSA) (overall MSA = 0.92), Table S3: Amino acids dataset used in PCA and UMAP analyses.

Author Contributions: Conceptualization, E.B.; methodology, E.B., A.B. and M.C.; software, A.B.; validation, E.B., M.C., J.B., E.J., P.M. and P.Z.; formal analysis, E.B. and A.B.; investigation, M.C., J.B., P.M. and P.Z.; resources, M.C., J.B. and P.Z; data curation, E.B. and A.B.; writing—original draft preparation, E.B.; writing—review and editing, E.B. and A.B.; visualization, A.B.; supervision, E.B.; project administration, E.B.; funding acquisition, E.B., A.B. and P.Z. All authors have read and agreed to the published version of the manuscript.

Funding: This work was funded by the Ministry of Science and Higher Education as part of subsidies for maintaining research potential awarded to the Faculty of Biology of the University of Bialystok (SWB-1, SWB-3, and SWB-6).

Institutional Review Board Statement: Not applicable.

Informed Consent Statement: Not applicable.

Data Availability Statement: Data is contained within the current article and supplementary material.

Acknowledgments: We will like to thank Rector of University in Bialystok and Dean of Department of Biology of University in Bialystok for financial support. A.B. thanks Adam Bajguz for statistical and programming consultations. P.Z. would like to thank J. Chołostiakow-Gromek MSc. for assistance in the preparation of soil samples and laboratory analysis.

Conflicts of Interest: The authors declare no conflict of interest.

References

1. Fenster, C.B.; Armbruster, W.S.; Wilson, P.; Dudash, M.R.; Thomson, J.D. Pollination syndromes and floral specialization. *Annu. Rev. Ecol. Evol. Syst.* **2004**, *35*, 375–403. [CrossRef]
2. Tremblay, R.L.; Ackerman, J.D.; Zimmerman, J.K.; Calvo, R.N. Variation in sexual reproduction in orchids and its evolutionary consequences: A spasmodic journey to diversification. *Biol. J. Linn. Soc.* **2005**, *84*, 1–54. [CrossRef]
3. Ackerman, J. Mechanisms and evolution of food-deceptive pollination systems in orchids. *Lindleyana* **1986**, *1*, 108–113.
4. Cozzolino, S.; Widmer, A. Orchid diversity: An evolutionary consequence of deception? *Trends Ecol. Evol.* **2005**, *20*, 487–494. [CrossRef]
5. Dressler, R. *The Orchids Natural History and Classification*; Harvard University Press: Cambridge, MA, USA, 1981; p. 356. [CrossRef]
6. Whitten, W.M.; Blanco, M.A.; Williams, N.H.; Koehler, S.; Carnevali, G.; Singer, R.B.; Endara, L.; Neubig, K.M. Molecular phylogenetics of *Maxillaria* and related genera (Orchidaceae: Cymbidieae) based on combined molecular data sets. *Am. J. Bot.* **2007**, *94*, 1860–1889. [CrossRef] [PubMed]
7. Ackerman, J.D.; Rodriguez-Robles, J.A.; Melendez, E.J. A meager nectar offering by an epiphytic orchid is better than nothing. *Biotropica* **1994**, *26*, 44–49. [CrossRef]
8. Neiland, M.R.M.; Wilcock, C.C. Fruit set, nectar reward, and rarity in the Orchidaceae. *Am. J. Bot.* **1998**, *85*, 1657–1671. [CrossRef] [PubMed]
9. Baker, H.G.; Baker, I. Floral nectar sugar constituents in relation to pollinator type. In *Handbook of Experimental Pollination Biology*; Jones, C.E., Little, R.J., Eds.; Van Nostrand Reinhold Company Inc.: New York, NY, USA, 1983; pp. 131–141.
10. Baker, H.G.; Baker, I. The predictive value of nectar chemistry to the recognition of pollinator types. *Isr. J. Bot.* **1990**, *39*, 157–166. [CrossRef]
11. Nepi, M. Beyond nectar sweetness: The hidden ecological role of non-protein amino acids in nectar. *J. Ecol.* **2014**, *102*, 108–115. [CrossRef]
12. Nicolson, S.W.; Thornburg, R.W. Nectar chemistry. In *Nectaries and Nectar*; Nicolson, S.W., Nepi, M., Pacini, E., Eds.; Springer: Dordrecht, The Netherlands, 2007; pp. 215–264. [CrossRef]
13. Parachnowitsch, A.L.; Manson, J.S.; Sletvold, N. Evolutionary ecology of nectar. *Ann. Bot.* **2019**, *123*, 247–261. [CrossRef] [PubMed]
14. Petanidou, T.; Van Laere, A.; Ellis, W.N.; Smets, E. What shapes amino acid and sugar composition in Mediterranean floral nectars? *Oikos* **2006**, *115*, 155–169. [CrossRef]
15. Roy, R.; Schmitt, A.J.; Thomas, J.B.; Carter, C.J. Review: Nectar biology: From molecules to ecosystems. *Plant Sci.* **2017**, *262*, 148–164. [CrossRef]
16. Fowler, R.E.; Rotheray, E.L.; Goulson, D. Floral abundance and resource quality influence pollinator choice. *Insect Conserv. Divers.* **2016**, *9*, 481–494. [CrossRef]
17. Brzosko, E.; Bajguz, A. Nectar composition in moth-pollinated *Platanthera bifolia* and *P. chlorantha* and its importance for reproductive success. *Planta* **2019**, *250*, 263–279. [CrossRef]
18. Galetto, L.; Bernardello, G.; Rivera, G.L. Nectar, nectaries, flower visitors, and breeding system in five terrestrial Orchidaceae from central Argentina. *J. Plant Res.* **1997**, *110*, 393–403. [CrossRef]
19. Nocentini, D.; Pacini, E.; Guarnieri, M.; Martelli, D.; Nepi, M. Intrapopulation heterogeneity in floral nectar attributes and foraging insects of an ecotonal Mediterranean species. *Plant Ecol.* **2013**, *214*, 799–809. [CrossRef]

20. Nilsson, L.A. Deep flowers for long tongues: Reply from L.A. Nilsson. *Trends Ecol. Evol.* **1998**, *13*, 509. [CrossRef]
21. Nilsson, L.A. Deep flowers for long tongues. *Trends Ecol. Evol.* **1998**, *13*, 259–260. [CrossRef]
22. Petanidou, T. Sugars in Mediterranean floral nectars: An ecological and evolutionary approach. *J. Chem. Ecol.* **2005**, *31*, 1065–1088. [CrossRef]
23. Willmer, P. Pollination by butterflies and moths. In *Pollination and Floral Ecology*; Willmer, P., Ed.; Princeton University Press: Princeton, NJ, USA, 2011; pp. 322–336. [CrossRef]
24. Baker, H.G.; Baker, I. Starchy and starchless pollen in the Onagraceae. *Ann. Mo. Bot. Gard.* **1982**, *69*, 748. [CrossRef]
25. Heyneman, A.J. Optimal sugar concentrations of floral nectars—Dependence on sugar intake efficiency and foraging costs. *Oecologia* **1983**, *60*, 198–213. [CrossRef]
26. Pyke, G.H.; Waser, N.M. The production of dilute nectars by hummingbird and honeyeater flowers. *Biotropica* **1981**, *13*, 260–270. [CrossRef]
27. Gijbels, P.; Ceulemans, T.; Van den Ende, W.; Honnay, O. Experimental fertilization increases amino acid content in floral nectar, fruit set and degree of selfing in the orchid *Gymnadenia conopsea*. *Oecologia* **2015**, *179*, 785–795. [CrossRef] [PubMed]
28. Gijbels, P.; Van den Ende, W.; Honnay, O. Phenotypic selection on nectar amino acid composition in the Lepidoptera pollinated orchid species *Gymnadenia conopsea*. *Oikos* **2014**, *124*, 421–427. [CrossRef]
29. Pais, M.; Neves, H.; Maria, P.; Vasconcelos, A. Amino acid and sugar content of the nectar exudate from *Limodorum abortivum* (Orchidaceae). Comparison with *Epipactis atropurpurea* nectar composition. *Apidologie* **1986**, *17*, 125–136. [CrossRef]
30. Stpiczyńska, M.; Pielecki, J. Sekrecja, resorbcja i skład chemiczny nektaru podkolana zielonawego *Plantanthera chlorantha* (Custer) Rchb. (Orchidaceae). *Ann. UMCS. Sect. EEE Hortic.* **2002**, *10*, 173–179.
31. Heil, M. Postsecretory hydrolysis of nectar sucrose and specialization in ant/plant mutualism. *Science* **2005**, *308*, 560–563. [CrossRef]
32. Levin, E.; McCue, M.D.; Davidowitz, G. More than just sugar: Allocation of nectar amino acids and fatty acids in a Lepidopteran. *Proc. Biol. Sci.* **2017**, *284*, 20162126. [CrossRef]
33. Mevi-Schütz, J.; Erhardt, A. Amino acids in nectar enhance butterfly fecundity: A long-awaited link. *Am. Nat.* **2005**, *165*, 411–419. [CrossRef]
34. Gardener, M.C.; Gillman, M.P. The taste of nectar—A neglected area of pollination ecology. *Oikos* **2002**, *98*, 552–557. [CrossRef]
35. Zhang, T.-F.; Duan, Y.-W.; Liu, J.-Q. Pollination ecology of *Aconitum gymnandrum* (Ranunculaceae) at two sites with different altitudes. *Acta Phytotaxon. Sin.* **2006**, *44*, 362–370. [CrossRef]
36. Goldberg, L. Patterns of nectar production and composition, and morphology of floral nectaries in *Helicteres guazumifolia* and *Helicteres baruensis* (Sterculiaceae): Two sympatric species from the Costa Rican tropical dry forest. *Rev. Biol. Trop.* **2009**, *57*, 161–177.
37. Gottsberger, G.; Arnold, T.; Linskens, H.F. Intraspecific variation in the amino acid content of floral nectar. *Bot. Acta* **1989**, *102*, 141–144. [CrossRef]
38. Pacini, E.; Nepi, M.; Vesprini, J.L. Nectar biodiversity: A short review. *Plant Syst. Evol.* **2003**, *238*, 7–21. [CrossRef]
39. Gottsberger, G.; Schrauwen, J.; Linskens, H.F. Amino acids and sugars in nectar, and their putative evolutionary significance. *Plant Syst. Evol.* **1984**, *145*, 55–77. [CrossRef]
40. Alexandersson, R.; Johnson, S.D. Pollinator-mediated selection on flower-tube length in a hawkmoth-pollinated *Gladiolus* (Iridaceae). *Proc. Biol. Sci.* **2002**, *269*, 631–636. [CrossRef] [PubMed]
41. Boberg, E.; Ägren, J. Despite their apparent integration, spur length but not perianth size affects reproductive success in the moth-pollinated orchid *Platanthera bifolia*. *Funct. Ecol.* **2009**, *23*, 1022–1028. [CrossRef]
42. Little, K.J.; Dieringer, G.; Romano, M. Pollination ecology, genetic diversity and selection on nectar spur length in *Platanthera lacera* (Orchidaceae). *Plant Spec. Biol.* **2005**, *20*, 183–190. [CrossRef]
43. Maad, J.; Nilsson, L.A. On the mechanism of floral shifts in speciation: Gained pollination efficiency from tongue- to eye-attachment of pollinia in *Platanthera* (Orchidaceae). *Biol. J. Linn. Soc.* **2004**, *83*, 481–495. [CrossRef]
44. Moré, M.; Amorim, F.W.; Benitez-Vieyra, S.; Medina, A.M.; Sazima, M.; Cocucci, A.A. Armament imbalances: Match and mismatch in plant-pollinator traits of highly specialized long-spurred orchids. *PLoS ONE* **2012**, *7*, e41878. [CrossRef]
45. Ollerton, J.; Winfree, R.; Tarrant, S. How many flowering plants are pollinated by animals? *Oikos* **2011**, *120*, 321–326. [CrossRef]
46. Van der Niet, T.; Peakall, R.; Johnson, S.D. Pollinator-driven ecological speciation in plants: New evidence and future perspectives. *Ann. Bot.* **2014**, *113*, 199–211. [CrossRef] [PubMed]
47. Vereecken, N.J.; Dafni, A.; Cozzolino, S. Pollination syndromes in Mediterranean orchids—implications for speciation, taxonomy and conservation. *Bot. Rev.* **2010**, *76*, 220–240. [CrossRef]
48. Grindeland, J.M.; Sletvold, N.; Ims, R.A. Effects of floral display size and plant density on pollinator visitation rate in a natural population of *Digitalis purpurea*. *Funct. Ecol.* **2005**, *19*, 383–390. [CrossRef]
49. Hodges, S.A. The influence of nectar production on hawkmoth behavior, self pollination, and seed production in *Mirabilis multiflora* (Nyctaginaceae). *Am. J. Bot.* **1995**, *82*, 197–204. [CrossRef]
50. Kindlmann, P.; Jersakova, J. Effect of floral display on reproductive success in terrestrial orchids. *Folia Geobot.* **2006**, *41*, 47–60. [CrossRef]
51. Maad, J. Phenotypic selection in hawkmoth-pollinated *Platanthera bifolia*: Targets and fitness surfaces. *Evolution* **2000**, *54*, 112–123. [CrossRef]

52. Vallius, E.; Arminen, S.; Salonen, V. Are There Fitness Advantages Associated with a Large Inflorescence in *Gymnadenia conopsea* ssp. *conopsea*? Available online: http://www.r-b-o.eu/rbo_public?Vallius_et_al_2006.html (accessed on 29 November 2020).
53. Calvo, R.N. Inflorescence size and fruit distribution among individuals in three orchid species. *Am. J. Bot.* **1990**, *77*, 1378–1381. [CrossRef]
54. Pellegrino, G.; Bellusci, F.; Musacchio, A. The effects of inflorescence size and flower position on female reproductive success in three deceptive orchids. *Bot. Stud.* **2010**, *51*, 351–356.
55. Pacini, E.; Nepi, M. Nectar production and presentation. In *Nectaries and Nectar*; Nicolson, S.W., Nepi, M., Pacini, E., Eds.; Springer: Dordrecht, The Netherlands, 2007; pp. 167–214. [CrossRef]
56. Sun, M.; Gross, K.; Schiestl, F.P. Floral adaptation to local pollinator guilds in a terrestrial orchid. *Ann. Bot.* **2014**, *113*, 289–300. [CrossRef]
57. Martén-Rodríguez, S.; John Kress, W.; Temeles, E.J.; Meléndez-Ackerman, E. Plant–pollinator interactions and floral convergence in two species of *Heliconia* from the Caribbean Islands. *Oecologia* **2011**, *167*, 1075–1083. [CrossRef] [PubMed]
58. Trunschke, J.; Sletvold, N.; Ågren, J. Interaction intensity and pollinator-mediated selection. *New Phytol.* **2017**, *214*, 1381–1389. [CrossRef] [PubMed]
59. Olesen, J.M.; Jordano, P. Geographic patterns in plant-pollinator mutualistic networks. *Ecology* **2002**, *83*, 2416. [CrossRef]
60. Ebeling, A.; Klein, A.-M.; Schumacher, J.; Weisser, W.W.; Tscharntke, T. How does plant richness affect pollinator richness and temporal stability of flower visits? *Oikos* **2008**, *117*, 1808–1815. [CrossRef]
61. Venjakob, C.; Leonhardt, S.; Klein, A.-M. Inter-individual nectar chemistry changes of field Scabious, *Knautia arvensis*. *Insects* **2020**, *11*, 75. [CrossRef]
62. Ghazoul, J. Floral diversity and the facilitation of pollination. *J. Ecol.* **2006**, *94*, 295–304. [CrossRef]
63. Juillet, N.; Gonzalez, M.A.; Page, P.A.; Gigord, L.D.B. Pollination of the European food-deceptive *Traunsteinera globosa* (Orchidaceae): The importance of nectar-producing neighbouring plants. *Plant Syst. Evol.* **2007**, *265*, 123–129. [CrossRef]
64. Duffy, K.J.; Stout, J.C. The effects of plant density and nectar reward on bee visitation to the endangered orchid *Spiranthes romanzoffiana*. *Acta Oecol. Int. J. Ecol.* **2008**, *34*, 131–138. [CrossRef]
65. Lachmuth, S.; Henrichmann, C.; Horn, J.; Pagel, J.; Schurr, F.M. Neighbourhood effects on plant reproduction: An experimental-analytical framework and its application to the invasive *Senecio inaequidens*. *J. Ecol.* **2017**, *106*, 761–773. [CrossRef]
66. Pauw, A. Can pollination niches facilitate plant coexistence? *Trends Ecol. Evol.* **2013**, *28*, 30–37. [CrossRef]
67. David, T.I.; Storkey, J.; Stevens, C.J. Understanding how changing soil nitrogen affects plant–pollinator interactions. *Arthropod Plant Interact.* **2019**, *13*, 671–684. [CrossRef]
68. Biesmeijer, J.C.; Roberts, S.P.; Reemer, M.; Ohlemuller, R.; Edwards, M.; Peeters, T.; Schaffers, A.P.; Potts, S.G.; Kleukers, R.; Thomas, C.D.; et al. Parallel declines in pollinators and insect-pollinated plants in Britain and the Netherlands. *Science* **2006**, *313*, 351–354. [CrossRef] [PubMed]
69. Hejcman, M.; Schellberg, J.; Pavlu, V. *Dactylorhiza maculata*, *Platanthera bifolia* and *Listera ovata* survive N application under P limitation. *Acta Oecol.* **2010**, *36*, 684–688. [CrossRef]
70. Tremblay, R.L. Trends in the pollination ecology of the Orchidaceae: Evolution and systematics. *Can. J. Bot.* **1992**, *70*, 642–650. [CrossRef]
71. Johnson, S.D.; Hobbhahn, N. Generalized pollination, floral scent chemistry, and a possible case of hybridization in the African orchid *Disa fragrans*. *South Afr. J. Bot.* **2010**, *76*, 739–748. [CrossRef]
72. Nilsson, L.A.; Jonsson, L.; Rason, L.; Randrianjohany, E. Monophily and pollination mechanisms in *Angraecum arachnites* Schltr. (Orchidaceae) in a guild of long-tongued hawk-moths (Sphingidae) in Madagascar. *Biol. J. Linn. Soc.* **1985**, *26*, 1–19. [CrossRef]
73. Pemberton, R.W. Biotic resource needs of specialist orchid pollinators. *Bot. Rev.* **2010**, *76*, 275–292. [CrossRef]
74. Nilsson, L.A. The evolution of flowers with deep corolla tubes. *Nature* **1988**, *334*, 147–149. [CrossRef]
75. Nilsson, L.A. Pollination ecology of *Epipactis palustris* (L.) Crantz (Orchidaceae). *Bot. Not.* **1978**, *131*, 355–368.
76. Nilsson, L.A. The pollination ecology of *Listera ovata* (Orchidaceae). *Nord. J. Bot.* **1981**, *1*, 461–480. [CrossRef]
77. Claessens, J.; Kleynen, J. *The Flower of the European Orchid. Form and Function*; Claessens & Kleynen (Privately Published): Guelle, Spain, 2011; p. 440.
78. Talalaj, I.; Ostrowiecka, B.; Wlostowska, E.; Rutkowska, A.; Brzosko, E. The ability of spontaneous autogamy in four orchid species: *Cephalanthera rubra*, *Neottia ovata*, *Gymnadenia conopsea*, and *Platanthera bifolia*. *Acta Biol. Cracov. Bot.* **2017**, *59*, 51–61. [CrossRef]
79. Brys, R.; Jacquemyn, H.; Hermy, M. Pollination efficiency and reproductive patterns in relation to local plant density, population size, and floral display in the rewarding *Listera ovata* (Orchidaceae). *Bot. J. Linn. Soc.* **2008**, *157*, 713–721. [CrossRef]
80. Brzosko, E. The dynamics of *Listera ovata* populations on mineral islands in the Biebrza National Park. *Acta Soc. Bot. Pol.* **2002**, *71*, 243–251. [CrossRef]
81. Tamm, C.O. Survival and flowering of some perennial herbs. II. The behaviour of some orchids on permanent plots. *Oikos* **1972**, *23*, 23. [CrossRef]
82. Brzosko, E.; Wróblewska, A. Low allozymic variation in two island populations of *Listera ovata* (Orchidaceae) from NE Poland. *Ann. Bot. Fenn.* **2003**, *40*, 309–315.
83. Brzosko, E.; Wróblewska, A. How genetically variable are *Neottia ovata* (Orchidaceae) populations in northeast Poland? *Bot. J. Linn. Soc.* **2012**, *170*, 40–49. [CrossRef]

84. Kowalkowska, A.K.; Krawczyńska, A.T. Anatomical features related with pollination of *Neottia ovata* (L.) Bluff & Fingerh. (Orchidaceae). *Flora* **2019**, *255*, 24–33. [CrossRef]
85. Amorim, F.W.; Wyatt, G.E.; Sazima, M. Low abundance of long-tongued pollinators leads to pollen limitation in four specialized hawkmoth-pollinated plants in the Atlantic Rain forest, Brazil. *Naturwissenschaften* **2014**, *101*, 893–905. [CrossRef]
86. Amorim, F.W.; de Ávila, R.S., Jr.; de Camargo, A.J.A.; Vieira, A.L.; Oliveira, P.E. A hawkmoth crossroads? Species richness, seasonality and biogeographical affinities of Sphingidae in a Brazilian Cerrado. *J. Biogeogr.* **2009**, *36*, 662–674. [CrossRef]
87. Boberg, E.; Alexandersson, R.; Jonsson, M.; Maad, J.; Ägren, J.; Nilsson, L.A. Pollinator shifts and the evolution of spur length in the moth-pollinated orchid *Platanthera bifolia*. *Ann. Bot.* **2014**, *113*, 267–275. [CrossRef] [PubMed]
88. Sublett, C.A.; Cook, J.L.; Janovec, J.P. Species richness and community composition of sphingid moths (Lepidoptera: Sphingidae) along an elevational gradient in southeast Peru. *Zoologia* **2019**, *36*, 1–11. [CrossRef]
89. Ollerton, J.; Killick, A.; Lamborn, E.; Watts, S.; Whiston, M. Multiple meanings and modes: On the many ways to be a generalist flower. *Taxon* **2007**, *56*, 717–728. [CrossRef]
90. Peralta, G.; Vázquez, D.P.; Chacoff, N.P.; Lomáscolo, S.B.; Perry, G.L.W.; Tylianakis, J.M. Trait matching and phenological overlap increase the spatio-temporal stability and functionality of plant-pollinator interactions. *Ecol. Lett.* **2020**, *23*, 1107–1116. [CrossRef]
91. Nicolson, S.W.; Fleming, P.A. Nectar as food for birds: The physiological consequences of drinking dilute sugar solutions. *Plant Syst. Evol.* **2003**, *238*, 139–153. [CrossRef]
92. Martins, D.J.; Johnson, S.D. Hawkmoth pollination of aerangoid orchids in Kenya, with special reference to nectar sugar concentration gradients in the floral spurs. *Am. J. Bot.* **2007**, *94*, 650–659. [CrossRef]
93. Vandelook, F.; Janssens, S.B.; Gijbels, P.; Fischer, E.; Van den Ende, W.; Honnay, O.; Abrahamczyk, S. Nectar traits differ between pollination syndromes in Balsaminaceae. *Ann. Bot.* **2019**, *124*, 269–279. [CrossRef] [PubMed]
94. Johnson, S.D.; Nicolson, S.W. Evolutionary associations between nectar properties and specificity in bird pollination systems. *Biol. Lett.* **2008**, *4*, 49–52. [CrossRef]
95. Stpiczyńska, M.; Nepi, M.; Zych, M. Nectaries and male-biased nectar production in protandrous flowers of a perennial umbellifer *Angelica sylvestris* L. (Apiaceae). *Plant Syst. Evol.* **2015**, *301*, 1099–1113. [CrossRef]
96. Zych, M.; Junker, R.R.; Nepi, M.; Stpiczynska, M.; Stolarska, B.; Roguz, K. Spatiotemporal variation in the pollination systems of a supergeneralist plant: Is *Angelica sylvestris* (Apiaceae) locally adapted to its most effective pollinators? *Ann. Bot.* **2019**, *123*, 415–428. [CrossRef]
97. Heil, M. Nectar: Generation, regulation and ecological functions. *Trends Plant Sci.* **2011**, *16*, 191–200. [CrossRef]
98. Carter, C.; Shafir, S.; Yehonatan, L.; Palmer, R.G.; Thornburg, R. A novel role for proline in plant floral nectars. *Naturwissenschaften* **2006**, *93*, 72–79. [CrossRef]
99. Nepi, M.; Soligo, C.; Nocentini, D.; Abate, M.; Guarnieri, M.; Cai, G.; Bini, L.; Puglia, M.; Bianchi, L.; Pacini, E. Amino acids and protein profile in floral nectar: Much more than a simple reward. *Flora* **2012**, *207*, 475–481. [CrossRef]
100. Felicioli, A.; Sagona, S.; Galloni, M.; Bortolotti, L.; Bogo, G.; Guarnieri, M.; Nepi, M. Effects of nonprotein amino acids on survival and locomotion of *Osmia bicornis*. *Insect Mol. Biol.* **2018**, *27*, 556–563. [CrossRef]
101. Nepi, M.; Bini, L.; Bianchi, L.; Puglia, M.; Abate, M.; Cai, G. Xylan-degrading enzymes in male and female flower nectar of *Cucurbita pepo*. *Ann. Bot.* **2011**, *108*, 521–527. [CrossRef] [PubMed]
102. Park, S.; Thornburg, R.W. Biochemistry of nectar proteins. *J. Plant Biol.* **2009**, *52*, 27–34. [CrossRef]
103. Baker, H.G.; Baker, I. The occurrence and significance of amino acids in floral nectar. *Plant Syst. Evol.* **1986**, *151*, 175–186. [CrossRef]
104. Southwick, E.E. Photosynthate allocation to floral nectar: A neglected energy investment. *Ecology* **1984**, *65*, 1775–1779. [CrossRef]
105. Jacquemyn, H.; Brys, R. Lack of strong selection pressures maintains wide variation in floral traits in a food-deceptive orchid. *Ann. Bot.* **2020**, *126*, 445–453. [CrossRef]
106. Bergström, G.; Groth, I.; Pellmyr, O.; Endress, P.K.; Thien, L.B.; Hübener, A.; Francke, W. Chemical basis of a highly specific mutualism: Chiral esters attract pollinating beetles in Eupomatiaceae. *Phytochemistry* **1991**, *30*, 3221–3225. [CrossRef]
107. Kotilínek, M.; Těšitelová, T.; Jersáková, J. Biological flora of the British Isles: *Neottia ovata*. *J. Ecol.* **2015**, *103*, 1354–1366. [CrossRef]
108. Vakhrameeva, M.G.; Tatarenko, I.V.; Varlygina, T.I.; Torosyan, G.K.; Zagulski, M.N. *Orchids of Russia and Adjacent Countries (within the Borders of the Former USSR)*; A.R.G. Gantner Verlag K.G.: Ruggell, Liechtenstein, 2008; p. 690.
109. Tsiftsis, S.; Tsiripidis, I.; Karagiannakidou, V.; Alifragis, D. Niche analysis and conservation of the orchids of east Macedonia (NE Greece). *Acta Oecol. Int. J. Ecol.* **2008**, *33*, 27–35. [CrossRef]
110. Jackson, M.L. *Soil Chemical Analysis*; Scientific Publishers: Jodhpur, India, 2014.
111. Allison, L.E.; Moodie, C.D. Carbonate. In *Methods of Soil Analysis. Part 2. Chemical and Microbiological Properties*; Black, C.A., Evans, D.P., White, J.L., Ensminger, L.E., Clark, F.E., Eds.; American Society of Agronomy: Madison, WI, USA, 1965; pp. 1379–1396. [CrossRef]
112. Koroleff, F. Direct spectrophotometric determination of ammonia in precipitation. *Tellus* **1966**, *18*, 562–565. [CrossRef]
113. Morrant, D.S.; Schumann, R.; Petit, S. Field methods for sampling and storing nectar from flowers with low nectar volumes. *Ann. Bot.* **2009**, *103*, 533–542. [CrossRef]
114. R Core Team R: A Language and Environment for Statistical Computing (R version 4.0.2, Taking Off Again). R Foundation for Statistical Computing. Available online: https://www.R-project.org/ (accessed on 22 June 2020).

115. Wickham, H.; Averick, M.; Bryan, J.; Chang, W.; McGowan, L.; François, R.; Grolemund, G.; Hayes, A.; Henry, L.; Hester, J.; et al. Welcome to the Tidyverse. *J. Open Source Softw.* **2019**, *4*, 1686. [CrossRef]
116. Fox, J.; Weisberg, S. *An R Companion to Applied Regression*, 3rd ed.; SAGE Publications, Inc.: Thousand Oaks, CA, USA, 2019.
117. Burda, M. Paircompviz: Multiple Comparison Test Visualization (R Package Version 1.28.0). Available online: https://bioconductor.org/packages/release/bioc/html/paircompviz.html (accessed on 29 November 2020).
118. Graves, S.; Piepho, H.-P.; Selzer, L.; Dorai-Raj, S. MultcompView: Visualizations of Paired Comparisons (R Package Version 0.1-8). Available online: https://cran.r-project.org/web/packages/multcompView/index.html (accessed on 29 November 2020).
119. Mangiafico, S. Rcompanion: Functions to Support Extension Education Program Evaluation (R Package Version 2.3.26). Available online: https://cran.r-project.org/web/packages/rcompanion/index.html (accessed on 29 November 2020).
120. Oksanen, J.; Blanchet, F.G.; Friendly, M.; Kindt, R.; Legendre, P.; McGlinn, D.; Minchin, P.R.; O'Hara, R.B.; Simpson, G.L.; Solymos, P.; et al. Vegan: Community Ecology Package (R Package Version 2.5-7). Available online: https://CRAN.R-project.org/package=vegan (accessed on 30 January 2020).
121. Hamilton, N.E.; Ferry, M. ggtern: Ternary diagrams using ggplot2. *J. Stat. Softw.* **2018**, *87*, 1–17. [CrossRef]
122. Lê, S.; Josse, J.; Husson, F. FactoMineR: An R package for multivariate analysis. *J. Stat. Softw.* **2008**, *25*, 1–18. [CrossRef]
123. Tukey, J.W. *Exploratory Data Analysis*; Addison-Wesley Pub. Co.: Reading, MA, USA, 1977; p. 688.
124. Bartlett, M.S. Tests of significance in factor analysis. *Br. J. Stat. Psychol.* **1950**, *3*, 77–85. [CrossRef]
125. Revelle, W. *Psych: Procedures for Personality and Psychological Research*; (R Package Version 1.8.10); Northwestern University: Evanston, IL, USA, 2017; Available online: https://CRAN.R-project.org/package=psych (accessed on 31 October 2018).
126. Kaiser, H.F. An index of factorial simplicity. *Psychometrika* **1974**, *39*, 31–36. [CrossRef]
127. Cattell, R.B. The scree test for the number of factors. *Multivar. Behav. Res.* **1966**, *1*, 245–276. [CrossRef] [PubMed]
128. Kaiser, H.F. The application of electronic computers to factor analysis. *Educ. Psychol. Meas.* **1960**, *20*, 141–151. [CrossRef]
129. Kassambara, A.; Mundt, F. Factoextra: Extract and Visualize the Results of Multivariate Data Analyses (R Package Version 1.0.6). Available online: https://CRAN.R-project.org/package=factoextra (accessed on 16 January 2020).
130. Konopka, T. umap: Uniform Manifold Approximation and Projection (R Package Version 0.2.7.0). Available online: https://CRAN.R-project.org/package=umap (accessed on 29 November 2020).

Article

Functional Characterization of a *Dendrobium officinale* Geraniol Synthase DoGES1 Involved in Floral Scent Formation

Conghui Zhao [1,2,†], Zhenming Yu [1,3,†], Jaime A. Teixeira da Silva [4], Chunmei He [1], Haobin Wang [1], Can Si [1], Mingze Zhang [1], Danqi Zeng [1] and Jun Duan [1,3,*]

1. Key Laboratory of South China Agricultural Plant Molecular Analysis and Genetic Improvement & Guangdong Provincial Key Laboratory of Applied Botany, South China Botanical Garden, Chinese Academy of Sciences, Guangzhou 510650, China; zhaoconghui@scbg.ac.cn (C.Z.); zhenming311@scbg.ac.cn (Z.Y.); hechunmei2012@scbg.ac.cn (C.H.); wanghaobin17@scbg.ac.cn (H.W.); cans2013@163.com (C.S.); zhangmingze@scbg.ac.cn (M.Z.); zengdanqi20@scbg.ac.cn (D.Z.)
2. College of Life Sciences, University of Chinese Academy of Sciences, No. 19A Yuquan Road, Shijingshan District, Beijing 100049, China
3. Center of Economic Botany, Core Botanical Gardens, Chinese Academy of Sciences, Guangzhou 510650, China
4. Independent Researcher, P.O. Box 7, Miki-Cho Post Office, Ikenobe 3011-2, Kagawa-ken 761-0799, Japan; jaimetex@yahoo.com
* Correspondence: duanj@scib.ac.cn; Tel.: +86-020-37252978
† These authors contributed equally to this work.

Received: 1 September 2020; Accepted: 21 September 2020; Published: 23 September 2020

Abstract: Floral scent is a key ornamental trait that determines the quality and commercial value of orchids. Geraniol, an important volatile monoterpene in orchids that attracts pollinators, is also involved in responses to stresses but the geraniol synthase (GES) responsible for its synthesis in the medicinal orchid *Dendrobium officinale* has not yet been identified. In this study, three potential geraniol synthases were mined from the *D. officinale* genome. DoGES1, which was localized in chloroplasts, was characterized as a geraniol synthase. *DoGES1* was highly expressed in flowers, especially in petals. *DoGES1* transcript levels were high in the budding stage of *D. officinale* flowers at 11:00 a.m. DoGES1 catalyzed geraniol in vitro, and transient expression of *DoGES1* in *Nicotiana benthamiana* leaves resulted in the accumulation of geraniol in vivo. These findings on DoGES1 advance our understanding of geraniol biosynthesis in orchids, and lay the basis for genetic modification of floral scent in *D. officinale* or in other ornamental orchids.

Keywords: floral volatiles; geraniol; MEP pathway; orchids; terpene synthase

1. Introduction

Plants emit an astonishing number of volatile metabolites during growth and development, and these have various roles, some with biological effects, that are considered beneficial to plants and humans [1]. For ornamental plants, floral volatiles have a dual function, to attract pollinators, and in defense against pests, herbivores, and pathogens [2–4]. Orchids, economically important floricultural crops, possess an abundance of floral volatile terpenes. Among them, monoterpenes, especially geraniol, linalool, and their oxygenated derivatives, are predominant components of floral scents [4,5]. Geraniol is an acyclic monoterpene alcohol released from several ornamental plants, such as citronella, geranium, herbs, roses, and orchids (*Phalaenopsis bellina* and *Dendrobium officinale*) [5–8], and is extensively used in fragrance and cosmetics industries because of its pleasant rose-like scent.

Geraniol is synthesized from geranyl pyrophosphate (GPP), the universal five-carbon precursor for the biosynthesis of all monoterpenes, and is catalyzed by a terpene synthase (TPS), which has been designated as geraniol synthase (GES, EC 3.1.7.11) [6,9]. GPP, as an immediate precursor of monoterpenes, is proceeded by a condensation reaction of two C5-isoprene building units, namely isopentenyl diphosphate (IPP) and dimethylallyl diphosphate (DMAPP) [9]. Recent studies have thoroughly characterized two well-established pathways, the cytosolic mevalonic acid (MVA) pathway and the plastidic methylerythritol phosphate (MEP) pathway, that generate IPP and DMAPP, [7–9]. Several enzymes, including 1-deoxy-D-xylulose 5-phosphate synthase (DXS), 1-deoxy-D-xylulose 5-phosphate reductoisomerase (DXR), 4-hydroxy-3-methylbut-2-en-1-yl diphosphate synthase (HDS), and GPP synthase (GPPS), contribute to GPP biosynthesis [10], providing the GPP substrate for GES to generate geraniol. Taken together, GES is a mono-TPS that specifically catalyzes the formation of geraniol from GPP in the MEP pathway.

In plants, two kinds of enzymatic reactions can produce geraniol from GPP, either a TPS-based canonical pathway, which is catalyzed by GES in chloroplasts/plastids [6], or a phosphatase-based non-canonical pathway, which is catalyzed by nudix hydrolase (NUDX) in the cytoplasm (Figure 1) [9,11]. Thus far, the *GES* gene has already been identified and functionally characterized in multiple horticultural plants, including *CitTPS16* in *Citrus sinensis* [12], *LoTPS3* in *Lathyrus odoratus* [13], *GES* in *Ocimum basilicum* [6], and *PbGDPS* in *P. bellina* [14], all of which can produce geraniol from GPP in vitro. However, no TPS with GES activity has been identified in *Rosa rugosa* to date. Only one *NUDX* gene, *RhNUDX1*, converts GPP into geranyl monophosphate (GP), which is then hydrolyzed to geraniol by a petal-derived phosphatase [11]. In orchids, *PbGDPS*, which encodes GPP synthase, may play a key role in regulating the biosynthesis of monoterpenes (geraniol and linalool) in *P. bellina* [14]. In addition, the transcript levels of two TPS genes (*PbTPS5* and *PbTPS10*) are consistent with the production of geraniol and linalool in *P. bellina* [15], although their functionality has not yet been verified. Although geraniol is an important floral volatile compound in *D. officinale*, a medicinal orchid [7], the *GES* gene responsible for geraniol biosynthesis in *D. officinale* has not yet been characterized.

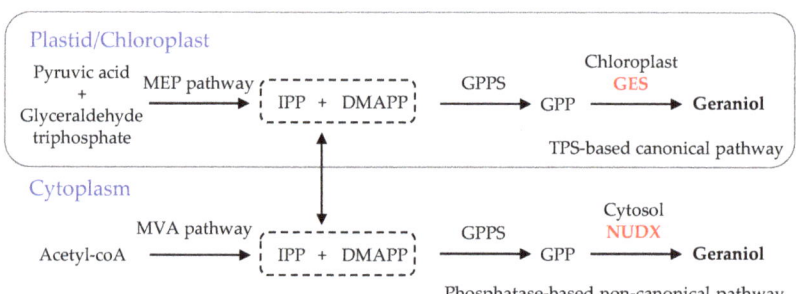

Figure 1. The pathway of the *GES/NUDX* genes responsible for the formation of geraniol in planta [9–11]. C5 precursors DMAPP and IPP are generated by the cytosol mevalonic acid (MVA) and the plastid methylerythritol phosphate (MEP) pathways. DMAPP, dimethylallyl pyrophosphate; GES, geraniol synthase; GPP, geranyl pyrophosphate; GPPS, GPP synthase; IPP, isopentenyl pyrophosphate; NUDX, nudix hydrolase; TPS, terpene synthase.

Herein, using the *D. officinale* genome database [16,17], and according to phylogenetic analysis and sequence homology, three *GES* genes (named *DoGES1–3*), with putative roles in the production of geraniol, were screened. The transcriptional regulatory functions of *DoGES1*, a member of the TPS family, in response to the accumulation of geraniol in *D. officinale* was investigated in different plant tissues (roots, stems, leaves, and flowers), harvest times (8:00, 11:00, 14:00, and 17:00), flower organs (petals, sepals, and gynostemium), and flowering periods (budding, semi-open flowers, fully open flowers). An in vitro assay of recombinant protein in *Escherichia coli* BL21 star (DE3) as well as in vivo transient overexpression in *Nicotiana benthamiana* indicated that *DoGES1* was responsible for geraniol biosynthesis, advancing our understanding of geraniol biosynthesis in *D. officinale*.

2. Results

2.1. Identification of Candidate GES Genes from the D. officinale Genome

From *D. officinale* genomic annotation data, three candidate GES sequences with best matches to known GES proteins [6,12,13,15] were retrieved by BLASTN, and named DoGES1, DoGES2, and DoGES3 (Table S1). Multiple sequence alignment demonstrated that three DoGES proteins had highly conserved aspartate-rich motifs (DDxxD) and NSE/DTE motifs at the C-terminal, and an RRX_8W domain at the N-terminal (Figure 2), suggesting that DoGES1-3 were all TPSs. Among them, DDxxD and NSE/DTE were essential for the cofactor Mg^{2+} or Mn^{2+} to catalyze the synthesis of monoterpenes [18,19], and the RRX_8W domain was also involved in the cyclization of monoterpene synthase [20].

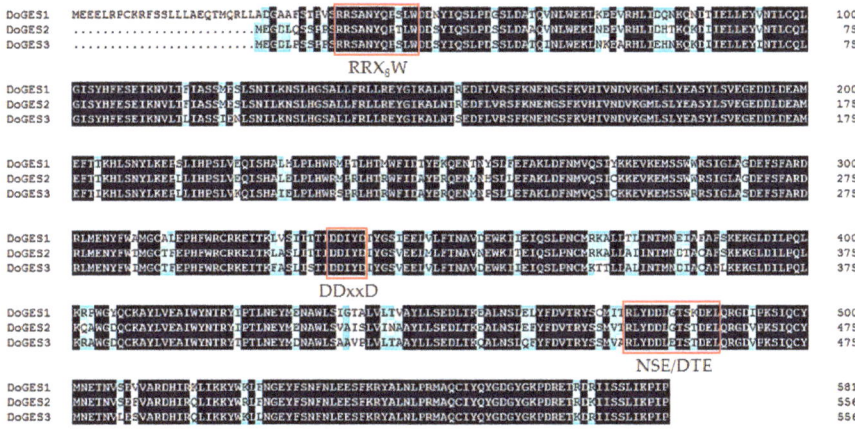

Figure 2. Comparison of deduced amino acid sequences of DoGES proteins in *Dendrobium officinale*. The Asp-rich domain $DD_{XX}D$, the RRX_8W motif, and the NSE/DTE motif, which are highly conserved in plant TPS proteins and required for TPS activity, are indicated. Completely conserved sequences are shaded in black, identical sequences in dark grey, and similar sequences in light grey.

2.2. Phylogenetic Analysis of DoGES Proteins in the D. officinale Genome

To investigate the evolutionary relationship of DoGES proteins with other reported GES proteins, a phylogenetic tree was generated by the neighbor-joining method (Figure 3; Table S2). All three DoGES proteins clustered in the TPS-b subfamily, which is specific to angiosperms and is responsible for encoding monoterpene synthases [21].

Based on the transcription levels in different tissues (roots, stems, leaves, and flowers), *DoGES1* exhibited high expression in flowers, while *DoGES2* and *DoGES3* were mainly expressed in roots and leaves, respectively (Figure S1). Consequently, DoGES1 was selected for our candidate study gene related to floral scent formation.

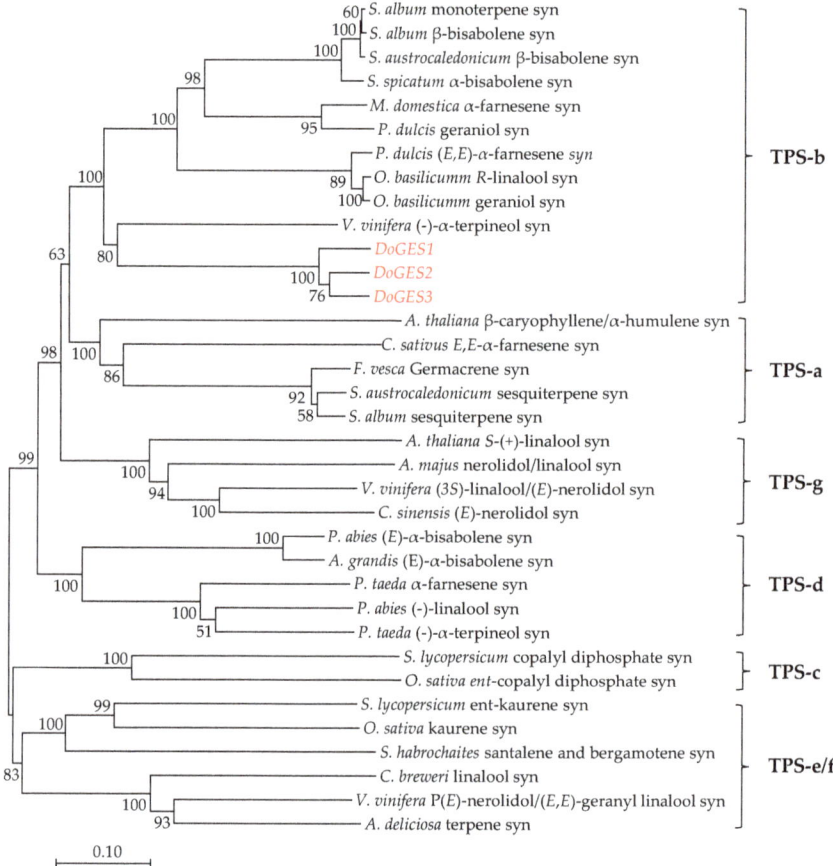

Figure 3. Phylogenetic positioning of GES proteins within representative samples of known plant TPS proteins. The neighbor-joining tree was generated using MEGA 7.0 software after the alignment of full-length DoGES proteins in *D. officinale* with other plant TPS proteins. The seven subfamilies TPS-a-g are delimited based on the taxonomic distribution of the TPS families [19]. All sequences that were used can be retrieved from Supplementary Table S2. syn, synthase.

2.3. Molecular Cloning and Analysis of DoGES1 from D. officinale Flowers

RNA isolated from *D. officinale* flowers during the blossoming period were used as template and amplified via nested PCR. Full-length cDNA sequences of *DoGES1* have a 1749-bp long open reading frame (ORF) that encodes 582 amino acids with a theoretical isoelectric point of 5.34 and a molecular weight of 67.99 kDa (Figure S2). The DoGES1 sequence was submitted to GenBank Data Libraries under accession number MT875214.

The DoGES1 secondary structure, which was determined using the SOPMA program (http://npsa-pbil.ibcp.fr/), shows that it harbors 69.24% α-helixes, 23.37% random coils, 3.78% β-turns, and 3.61% extended strands. The Chlorop 1.1 tool predicted that DoGES1 contains a 34 amino acid long N-terminal chloroplast transit peptide. To determine the subcellular localization of DoGES1, three subcellular localization tools (AtSubP [22], Plant-mPLoc [23], and pLoc-mPlant [24]) were used. All of them demonstrated that DoGES1 was located in chloroplasts, and was thus likely a mono-TPS in the MEP-pathway, but not in the cytosolic MVA pathway.

2.4. Subcellular Localization of DoGES1

To confirm the intracellular localization of DoGES1, pSAT6-EYFP-DoGES1 was transformed to the mesophyll protoplasts of 4-week-old *Arabidopsis thaliana* leaves. Yellow fluorescent signals were visualized by confocal laser scanning microscopy. The images indicate that DoGES1 was located in chloroplasts (Figure 4), similar to *LiTPS2*, which encodes a mono-TPS in lily (*Lilium longiflorum* 'Siberia') [25], indicating that DoGES1 may be responsible for monoterpene synthesis.

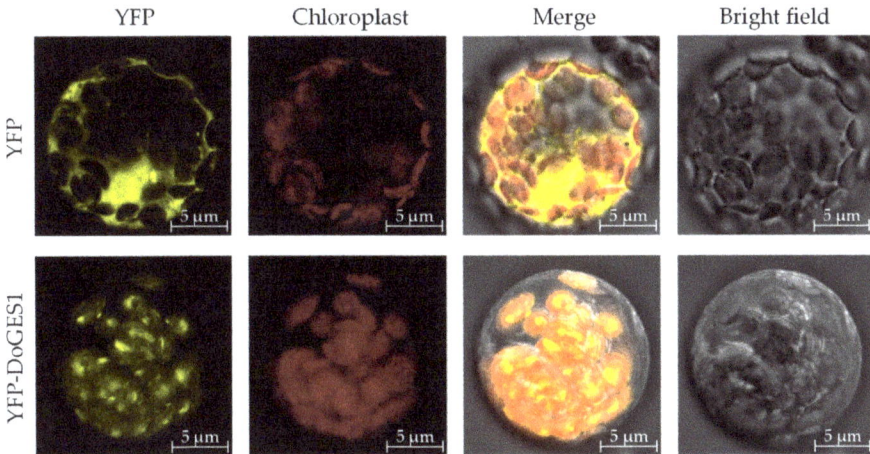

Figure 4. Subcellular localization of DoGES1 in *Dendrobium officinale*. Yellow fluorescence indicates the DoGES1-YFP fusion protein signal. Red fluorescence is chloroplast autofluorescence. The merged images indicate a combination of chloroplast autofluorescence and YFP fluorescence.

2.5. Functional Characterization of Enzyme Encoded by DoGES1 in Escherichia coli

To investigate whether DoGES1 encodes an enzyme that can produce monoterpenes, the full-length ORF sequence of DoGES1 was subcloned into prokaryotic expression vector pET-32a, expressed in *E. coli* BL21 (DE3), and induced by 1 mM isopropyl-β-D-thiogalactopyranoside (IPTG). Recombinant pET-32a-DoGES1 proteins were purified using affinity chromatography on a Ni-NTA agarose column, and were identified and isolated with a single band matching the expected size of DoGES1 in an SDS-PAGE gel (Figure S3). After incubation with GPP as substrate for the synthesis of monoterpenes, recombinant DoGES1 proteins successfully yielded geraniol with two characteristic fragment *m/z* 69 and *m/z* 41 in mass spectra produced by GC–MS (Figure 5), which showed the same mass spectral features of the authentic standard, geraniol. In addition, geraniol was not detected among the protein extracts from an empty vector cell mixture (Figure 5). These results suggest that DoGES1 from *D. officinale* had the capacity to specifically catalyze the formation of geraniol. Consequently, DoGES1 was classified as a geraniol synthase.

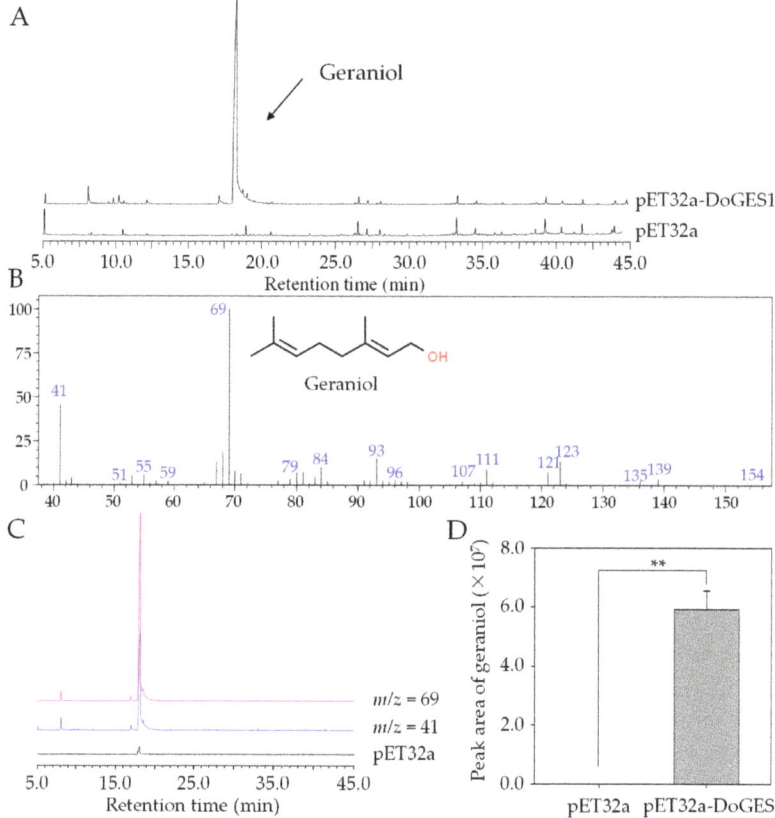

Figure 5. In vitro enzymatic assays of recombinant DoGES1 using GPP as the substrate. (**A**) Total ion chromatogram of the products formed by incubating extracts of empty vector pET32a and pET32a-DoGES1 with GPP. (**B**) Mass spectrum of products generated by the pET32a-DoGES1 enzyme. It is almost identical to the mass spectrum of geraniol, the standard. (**C,D**) Gas chromatograms of products yielded by DoGES1 using GPP as substrate. m/z, mass-to-charge ratio. In (**D**), different letters above error bars (standard deviation) ($n = 3$) indicate significant differences (** indicates $p < 0.01$, Student's t-test) between pET32a and pET32a-DoGES1.

2.6. Ectopic Expression of DoGES1 in N. benthamiana

We ectopically expressed *DoGES1* under the control of the cauliflower mosaic virus (CaMV) 35S promoter in *N. benthamiana* leaves. Positive transgenic leaves were screened by PCR for the presence of the *DoGES1* gene. *DoGES1* was not detected in 6-week-old *N. benthamiana* leaves transformed with the empty vector pCAMBIA3300, while the positively transformed *N. benthamiana* leaves were selected for subsequent analysis as a result of the high transcription levels of *DoGES1*. As expected, a large amount of geraniol was produced in *N. benthamiana* leaves overexpressing *DoGES1* 3 days after treatment, whereas geraniol was not observed in the control group (Figure 6). Therefore, *DoGES1* seemed to be a single-product enzyme that contributed to the biosynthesis of geraniol.

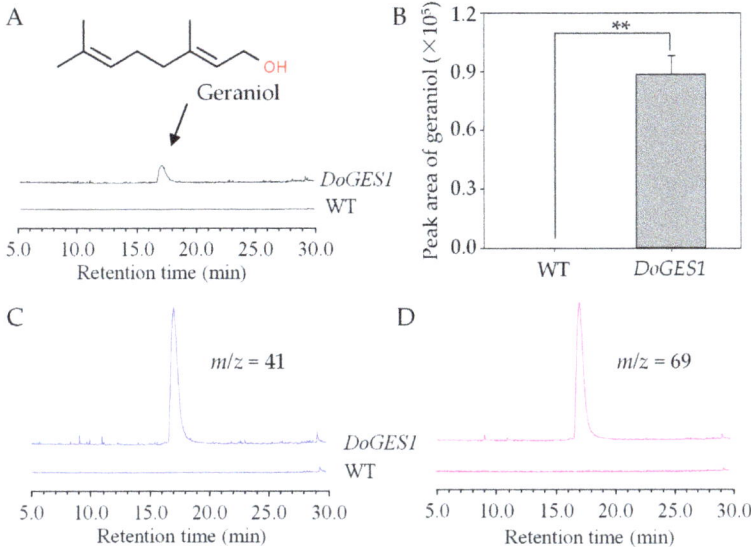

Figure 6. Ectopic expression of *DoGES1* in *Nicotiana benthamiana* leaves. (**A**,**B**) GC-MS analysis of monoterpenes from *N. benthamiana* leaves overexpressing *DoGES1*. The *N. benthamiana* leaves transformed with pCAMBIA3300 served as the control group (WT). (**C**,**D**) Mass spectrum of products generated in *N. benthamiana* leaves overexpressing *DoGES1*. In (**B**), error bars (standard deviation) (n = 3) indicate significant differences between WT and treatments (** indicates $p < 0.01$, Student's *t*-test). *m/z*, mass-to-charge ratio. WT, wild type.

2.7. Temporal-Spatial Expression Patterns Analysis of DoGES1 in D. officinale

To clarify the expression patterns of *DoGES1* in *D. officinale*, different tissues (roots, stems, leaves, and flowers), developmental stages of flowers (budding, semi-open flowers, fully open flowers), flower organs (petal, gynostemium, and labellum), and flower harvest times (8:00, 11:00, 14:00, and 17:00) were measured using real-time quantitative PCR (RT-qRCR). The results indicate that *DoGES1* exhibited the highest transcription levels in flowers, followed by stems, while the roots and the leaves displayed a relatively low level (Figure 7A). As flowers developed, the expression levels of *DoGES1* mRNA also varied. *DoGES1* was slightly expressed in the bud stage, expression increased enormously in semi-open flowers where it peaked, and dropped notably in the fully open flowers (Figure 7B). Among different flower organs, *DoGES1* expression was 4.77- and 12.29-fold higher in petals than in the gynostemium or labellum, respectively (Figure 7C). A comparison of *DoGES1* expression at specific times of the day showed that *DoGES1* showed higher expression at 11:00 than at 8:00, 14:00, and 17:00 (Figure 7D). Furthermore, geraniol content fluctuated among the harvest time points (from 8:00 to 17:00), with a trend of first increasing and then decreasing; the highest amount was observed at 14:00 (Figure S4). Thus, *DoGES1* was upregulated at first, then downregulated and possessed the highest value at 11:00, corresponding to the increase and then decrease in geraniol content at the same harvest time points (from 8:00 to 17:00). These results suggest that floral monoterpenes, such as geraniol encoded by *DoGES1*, might originate from petals and are emitted at 14:00, with a considerably high release during the budding stage in *D. officinale* flowers.

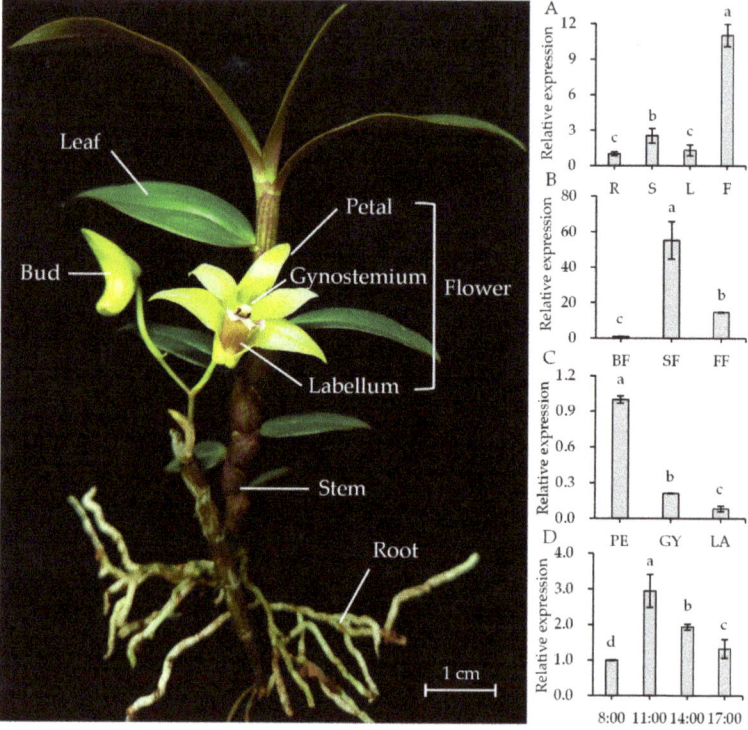

Figure 7. Expression levels of *DoGES1* in different *Dendrobium officinale* tissues. (**A**) Transcript levels of *DoGES1* in roots, stems, leaves and flowers. (**B**) Transcript levels of *DoGES1* in three developmental stages of *D. officinale* flowers. (**C**) Transcript levels of *DoGES1* in three *D. officinale* flower organs. (**D**) Transcript levels of *DoGES1* among the harvest time points. R, roots; S, stems; L, leaves; F, flowers; BF, budding flowers; SF, semi-open flowers; FF, fully open flowers; PE, petal; GY, gynostemium; LA, labellum. Three developmental stages of *D. officinale* flowers include budding, semi-open flower, and fully open flower are shown in Figure S6. Four time points (8:00, 11:00, 14:00, and 17:00) indicate the time of day when petals were sampled. Different letters above error bars (standard deviation) ($n = 10$) indicate significant differences among different treatments ($p < 0.05$, Duncan's multiple range test).

2.8. Activation of DoGES1 Gene Expression in Response to Methyl Jasmonate

Methyl jasmonate (MeJA), which is involved in responses to various stresses, regulates terpene metabolism [26]. To explore the response of the *DoGES1* gene after the application of MeJA, *DoGES1* transcript levels were quantified by qRT-PCR. Compared to the non-treated control, *DoGES1* was significantly upregulated between 2.41- and 21.49-fold, with the highest expression at 3 h after MeJA treatment (Figure 8A). Furthermore, a significant increase ($p < 0.05$) in geraniol was induced by MeJA, increasing by 243.47% (Figure 8B). This finding suggests the involvement of *DoGES1* in the MeJA-dependent biosynthesis of geraniol.

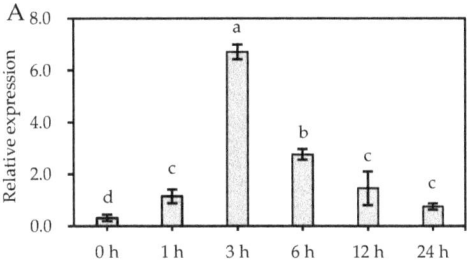 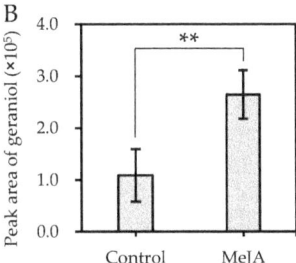

Figure 8. Transcript accumulation of the *DoGES1* gene (**A**), and the accumulation of geraniol (**B**) in response to methyl jasmonate (MeJA). The semi-open flowers of *Dendrobium officinale* were sprayed with 100 μM MeJA for 24 h. In (**A**), different letters above error bars (standard deviation) ($n = 10$) indicate significant differences under MeJA treatment for 24 h ($p < 0.05$, Duncan's multiple range test). In (**B**), error bars (standard deviation) ($n = 10$) indicate significant differences between the control and MeJA treatment (** indicates $p < 0.01$, Student's *t*-test).

3. Discussion

Geraniol, an important acyclic monoterpene with a distinctive rose-like scent, is widely used in the flavor and fragrance industries [27]. In plants, there are two biosynthetic pathways of geraniol (Figure 1). One depends on GES, and is a rate-limiting enzyme located in chloroplasts or plastids. The universal five-carbon precursors IPP and DMAPP produce GPP by plastidic GPPS in the MEP pathway, then GPP as substrate is catalyzed by mono-TPS via a common ionization-dependent reaction [28,29]. The *GES* gene, which was first cloned and functionally identified in sweet basil (*O. basilicum*), was shown to have the function of geraniol catalysis [6]. Subsequently, *GES* has been fairly extensively studied in many plant species, such as *C. sinensis* [12], *L. odoratus* [13], *Gardenia jasminoides* [30], *N. tabacum* [31], and *P. bellina* [15]. The second pathway for the production of geraniol is a phosphatase-based non-canonical pathway for monoterpene biosynthesis that involves NUDX [9,11]. After comparing the aroma components and the different gene expression profiles in two rose (*R. hybrida*) varieties, an important cytosolic enzyme, RhNUDX1 [11], was identified. When RhNUDX1 was incubated with GPP, this recombinant protein showed diphosphohydrolase activity and overexpression of *RhNUDX1* resulted in the accumulation of geraniol in *N. benthamiana*. Moreover, *RhNUDX1*-RNAi rose lines exhibited a relatively lower level of geraniol. Interestingly, *A. thaliana* NUDX1 efficiently hydrolyzed GPP to GP, but was not responsible for the production of geraniol while GES oversaw geraniol production [32]. Therefore, the proteins in the NUDX family might all be able to bind and act upon hydrolysis of GPP to GP, but only rose RhNUDX1 can promote the generation of geraniol [11], which is attributed to differential GES/NUDX enzyme localization [9,32,33]. Mono-TPS, located in chloroplasts/plastids, uses GPP as substrate, where GPP is also generated. In this study, a 34-aa chloroplast transit peptide was found in the N-terminal of DoGES1, so DoGES1 was targeted to the chloroplast (Figure 4). Therefore, chloroplast DoGES1 may function as the chloroplast/plastid-localized GES, but not as the cytoplasmic NUDX, resulting in the formation of geraniol via the canonical pathway. Similarly, two GES proteins were also located in plastids. Transient expression of *Valeriana officinalis* *VoGES* and *Lippia dulcis LdGES* in *N. benthamiana* leaves showed that GFP signal was located in the plastid [34], and HcTPS7 from *Hedychium coronarium* contained an 80 aa peptide in the N-terminal, targeting the protein to plastids [35].

The TPS family is typically divided into seven subfamilies, namely TPS-a, -b, -c, -d, -e/f, -g, and -h based on amino acid sequences and phylogeny [20]. TPS-a mainly codes for sesquiterpene synthase or diterpene synthase in monocotyledonous and dicotyledonous plants, respectively [21]. TPS-b and -g are both primarily involved in monoterpene synthesis, and TPS-b harbors an N-terminal RRX_8W motif that is a key metal binding domain for divalent metal ions, while TPS-g lacks the RRX_8W motif [29]. In the present study, an important TPS gene *DoGES1*, encoding geraniol synthase, was isolated from *D.*

officinale flowers. It encodes 582 amino acids containing DDxxD and RRX$_8$W motifs (Figure 2), and was clustered into the TPS-b subfamily (Figure 3), sharing 80% homology with mono-TPS of *L. longiflorum* 'Siberia' LiTPS2 [25]. DoGES1 recombinant protein accepted GPP as a substrate and singly generated geraniol in *E. coli* (Figure 5), which is associated with *O. basilicum* GES that utilizes GPP to uniquely generate geraniol in vitro [6]. Furthermore, *DoGES1* was ectopically expressed in *N. benthamiana* leaves, resulting in significantly increased geraniol content in transgenic leaves, compared to the control group (Figure 6). Although some TPS proteins have been shown to catalyze the formation of multiple products [25,30,31], GES is highly specific and produces only geraniol, similar to linalool synthase, which makes a single acyclic monoterpene alcohol, linalool [6,7,19,20]. Therefore, DoGES1 mainly acts as a geraniol synthase in the synthesis of floral volatiles.

In higher plants, the emission of volatile terpenes is often temporal and spatially specific [3,4,36]. For example, *PbGDPS*, which participates in the generation of geraniol and linalool in *P. bellina*, was specifically expressed in flowers and highly expressed on the fifth day after flower initiation [4,5,14]. The released of a high content of linalool in *Osmanthus fragrans* 'Dangui' petals is caused by the overexpression of mono-TPS (*LIS1* encoding linalool synthase) [37]. Ten species in *Maxillariinae* (Orchidaceae), such as *Maxillaria picta*, *M. cerifera*, and *M. marginata*, release volatile monoterpenes, mostly from the sepals and at the start of flowering [38]. In *D. officinale*, *DoGES1* was highly expressed in flowers, especially in petals, but had lower expression levels in leaves, stems, and roots (Figure 7). Generally, biosynthesis and emission of volatile compounds are developmentally regulated, usually enriching expression in an initial stage of development such as young leaves, unfertilized flowers, and unripe fruits [9,33]. With the continuous development of flowers, *DoGES1* substantially increased from the budding stage and peaked at the semi-open flower stage, decreasing at the fully open flower stage (Figure 6). This implies that GPP also accumulated extensively in semi-open flowers and generated an abundant amount of geraniol that was released in flowers, then, in fully open flowers, geraniol was heavily reduced, attributed to a reduction in the expression of *DoGES1*.

Environmental factors such as light or temperature normally influence the emission of volatile aroma scent [2]. The emission of *P. bellina* flower scent was regulated by light and the circadian clock [15]. In constant light (500 µmol·m^{-2} s^{-1}), *P. violacea* emitted mostly monoterpenes whose levels decreased in constant darkness. PbNAC1, which regulates the synthesis of monoterpenes, interacts with long hypocotyl 5 (HY5), which is a positive transcription factor involved in the responsiveness of plants to light [5,39]. *DoGES1* demonstrated fluctuations throughout the day, increasing at 8:00, then decreasing from 11:00 to 17:00, peaking at 11:00 (Figure 7), corresponding to the increase and then decrease in geraniol content at the same harvest time points, with the highest value observed at 14:00, suggesting that the emission of monoterpenes may have a diurnal rhythm in *D. officinale* petals. In addition, the MeJA-induced biosynthesis of terpenes (for example, geraniol) was observed (Figure 8) in *D. officinale* semi-open flowers, mainly due to the upregulation of *DoGES1* expression. The same finding that MeJA treatment resulted in the enhanced expression of 24/36 *CsTPS* genes and an increase in the amount of monoterpene volatiles such as linalool, geraniol, and their derivatives, was reported in *Camellia sinensis* leaves [40]. The CGTCA-motif and three MYC motifs in the *DoGES1* promoter (Figure S5) can interact with the MYC2 transcription factor of the JA signaling pathway [41]. Thus, *cis*-elements (CGTCA-motif and MYC) in the MeJA-induced *DoGES1* gene may be able to activate the JA signaling pathway, thereby regulating the enhancement of monoterpene (geraniol), although this possibility needs to be further explored.

4. Materials and Methods

4.1. Plant Materials and MeJA Treatment

D. officinale 'Zhongke 5' plants were cultivated in a greenhouse under controlled environment conditions as mentioned previously [7] in the South China Botanical Garden, Chinese Academy of Sciences (Guangzhou, China). The roots, stems, leaves, and flowers (including petal, gynostemium, and labellum) from 2-year-old adult *D. officinale* plants (Figure 7), as well as three different flowering stages (budding, semi-open flowers, and fully open flowers, Figure S6), were sampled and stored at −80 °C. Additionally, the petals of semi-open flowers at different harvest times (8:00, 11:00, 14:00, and 17:00) were sampled and stored at −80 °C. Four-week-old *A. thaliana* Col-0 for subcellular localization and *N. benthamiana* plants for ectopic expression experiments were both grown in a growth room at 22 °C with a 16 h photoperiod. For the MeJA treatment, *D. officinale* flowers during the semi-open flower stage were sprayed with 100 μM MeJA for 0, 1, 3, 6, 12, and 24 h; 0 h was used as the control group. All samples were harvested, frozen in nitrogen liquid, and stored at −80 °C.

4.2. Molecular Cloning of the DoGES1 Gene and Bioinformatics Analysis

Based on the annotated *GES* gene sequence from the *D. officinale* genome database, one gene (*DoGES1*) was screened and cloned. Total RNA from 0.1 g of *D. officinale* flowers for gene cloning were isolated by the Quick RNA Isolation Kit (Huayueyang, Beijing, China) following the manufacturer's protocol. The first strand cDNA of *DoGES1* was synthesized using the Reverse Transcription System (Promega Co., Madison, WI, USA) according to the manufacturer's instruction. Full-length ORF sequences of *DoGES1* were cloned using the KOD-plus Mutagenesis Kit (Toyobo, Osaka, Japan). The PCR program was as follows: 98 °C for 2 min, 35 cycles of 98 °C for 10 s, 60 °C for 20 s and 72 °C for 30 s, then constant 72 °C for 5 min. PCR products were recovered using the HiPure Gel Pure Micro Kit (Magen, Guangzhou, China). Specific primers are listed in Table S3.

The DoGES1 amino acid sequences were submitted to Clustal X 2.0 to conduct multiple sequence alignment [42]. A phylogenetic tree was constructed by MEGA 7.0 software [43] based on the neighbor-joining computational method [44]. The secondary structure of the DoGES1 protein was determined using the SOPMA program (http://npsa-pbil.ibcp.fr/). In order to study the subcellular localization of the DoGES1 protein, AtSubP [22], Plant-mPLoc [23], and pLoc-mPlant [24], all possessing good accuracy (>70%), were used on their corresponding websites.

4.3. Prokaryotic Expression and Purification of DoGES1 Protein

To express the DoGES1 protein in *E. coli*, the ORF of *DoGES1* was ligated into the pET32a vector using the InFusion® HD Cloning Kit (Takara, Dalian, China). The recombinant plasmid pET32a-DoGES1 was transformed into *E. coli* BL21 (DE3) competent cells (TsingKe Bio., Guangzhou, China). The identified positive clones were incubated at 37 °C with shaking at 180 rpm for 8 h in Luria–Bertani liquid medium supplemented with 50 μg mL^{-1} kanamycin. Culture medium was diluted to OD$_{600nm}$ between 0.5 and 0.6 and placed at 37 °C. Then, 1 mM isopropyl-β-D-thiogalactopyranoside (IPTG) was added to cultures, which were incubated for 8 h at 18 °C with shaking at 160 rpm. Finally, cell cultures were harvested by centrifugation at 12,000× *g* for 10 min. The precipitate was resuspended in fresh lysis buffer (10 mM imidazole, 300 mM NaCl, 50 mM NaH$_2$PO$_4$, pH 8.0) and lysed by sonication (60 Hz, repeated cycles of 5 s sonication and 5 s suspension) for 40 min. The clear lysate was collected by centrifugation at 10,000× *g* for 15 min.

Protein was purified with Ni-NTA Agarose (Qiagen, Hilden, Germany), eluted with elution buffer (250 mM imidazole, 300 mM NaCl, 50 mM NaH$_2$PO$_4$, pH 8.0), then desalinated in PD-10 desalting columns (GE Healthcare, Chicago, IL, USA) as previously described [7]. Purified DoGES1 protein was examined by sodium dodecyl sulfate polyacrylamide gel electrophoresis (SDS-PAGE).

4.4. In Vitro Enzyme Assay of DoGES1

The reaction mixture consisted of 100 μg of DoGES1 protein and 1 mL of MOPSO buffer (10 mM, pH 7.0 containing 5 mM dithiothreitol, 10 mM MgCl$_2$, and 10 mM GPP as substrate). The mixtures were overlaid with hexane in a total volume of 200 μL and incubated at 30 °C for 1 h. Reaction products were extracted by mixing vigorously for 5 min to obtain the enzymatic products, and 1 μL of the dehydrated extract was collected for gas chromatography-mass spectrometry (GC-MS) analysis.

4.5. Transient Expression of DoGES1 in N. benthamiana

The complete ORF sequences of DoGES1 (excluding the termination codon) were digested with *Bam*HI (Takara) and *Sac*I (Takara), and subcloned into the plant binary expression vector pCAMBIA3300 (CAMBIA, Canberra, Australia). The DoGES1-pCAMBIA3300 recombinant plasmid was transformed into *Agrobacterium tumefaciens* strain GV3101 cells via the freeze-thaw method as previously described [45]. Six-week-old leaves of *N. benthamiana* were injected with *A. tumefaciens* GV3101 cultures, and then maintained at 22 °C in darkness for 3 d. Infected 6-week-old vegetative leaves of *N. benthamiana* were collected and stored at −80 °C.

4.6. Quantification of Volatile Monoterpenes Using GC-MS

Volatile monoterpenes were analyzed by GC-MS equipped with a 30-m Supelcowax-10 column (0.25 mm diameter × 0.25 μm film thickness). About 500 mg of infected tobacco leaves or *D. officinale* flowers were blended with 3 mL dichloromethane containing 5 nmol ethyl decanoate (Sigma-Aldrich, St. Louis, MO, USA; CAS number 110-38-3, 98% purity) as the internal standard, and incubated at 25 °C while shaking at 100 rpm for 8 h. The extraction was passed through anhydrous Na_2SO_4 to remove remaining water, and filtered through a 0.22 μm PVDF membrane filter (Anpel Laboratory Technologies Inc., Shanghai, China), then concentrated to 500 μL under a nitrogen flow, and subjected to GC-MS (QP2010 SE, Shimadzu Co., Kyoto, Japan) analysis. The reaction program was carried out following our previously published protocol [7]. Products were identified by comparing mass spectra and retention times against the NIST 2008 mass spectra library (https://chemdata.nist.gov/) and the mass spectrum of the standard, geraniol (Sigma-Aldrich; CAS number 106-24-1, 98% purity).

4.7. Subcellular Location of DoGES1 Protein

The ORF sequence of DoGES1 was cloned into pSAT6-EYFP-N1 at the *Nco*I site, which was driven by the CaMV 35S promoter. The recombinant DoGES1-YFP plasmid was transformed into 4-week-old *A. thaliana* protoplasts that were isolated from rosette leaves by PEG-mediated transformation as described previously [46]. After incubation at 22 °C for 14 h in darkness, YFP fluorescence signals were excited at 514 nm and with an emission wavelength of 527 nm using a Leica TCS SP8 STED 3× microscope (Wetzlar, Hesse, Germany).

4.8. Real-Time Quantitative PCR Analysis

To analyze *DoGES1* gene expression patterns, qRT-PCR was performed. Total RNA from three flowering periods (budding, semi-open flowers, and fully open flowers), different tissues (roots, stems, leaves, flowers, petals, gynostemium, and labellum), and the flowers from different sampling times (8:00, 11:00, 14:00, and 17:00) were isolated and reverse transcribed, as described above. The total reaction volume was 10 μL containing 0.4 μL of each primer, 1 μL of template cDNA, 5 μL of SYBR Green PCR Master Mix (Novogene, Beijing, China), and 3.2 μL of ddH_2O. The PCR reaction was carried out with the LightCycler® 480 Instrument (Roche Diagnostics, Mannheim, Germany) as described previously [7]. The relative abundance of DoEF-1α (GenBank accession no.: JF825419) was used as an internal standard and calculated using the $2^{-\Delta\Delta CT}$ method [47]. Specific primers are shown in Table S3.

5. Conclusions

In the present study, we identified the *DoGES1* gene from *D. officinale*. It contributed to the regulation and production of geraniol biosynthesis. DoGES1 was located in chloroplast, and could utilize GPP to singly produce geraniol in vitro. Separately, *N. benthamiana* leaves overexpressing *DoGES1* considerably accumulated geraniol in vivo, which was consistent with the main monoterpene geraniol in *D. officinale* flowers. Our work also demonstrated that *DoGES1* was highly expressed in the petals during the semi-open flower stage at 11:00, and was activated by exogenous MeJA treatment.

These results indicate that DoGES1 could effectively control the biosynthesis of geraniol in *D. officinale*, laying the foundation for biotechnological modification of floral scent profiles in orchids.

Supplementary Materials: The following are available online at http://www.mdpi.com/1422-0067/21/19/7005/s1. Table S1. Three candidate GES sequences from the *D. officinale* genome. Table S2. The reported TPS proteins from other plant species used in the phylogenetic analysis. Table S3. Gene-specific primers used in the experiments. Figure S1. Transcription levels of *DoGES1*, *DoGES2*, and *DoGES3* in different tissues. Figure S2. Agarose gel electrophoresis of cDNA amplification of the *DoGES1* gene. Figure S3. SDS-PAGE analysis of DoGES1 recombinant protein expressed in *Escherichia coli* BL21. Figure S4. Content of geraniol in semi-open *D. officinale* flowers at 8:00, 11:00, 14:00, and 17:00. Figure S5. Putative regulatory *cis*-elements in the *DoGES1* promoter. Figure S6. Different developmental stages of *D. officinale* 'Zhongke 5' flowers.

Author Contributions: Conceptualization, J.D.; methodology, C.Z. and Z.Y.; software, Z.Y.; validation, C.Z., C.H., H.W., C.S., M.Z., and D.Z.; formal analysis, C.Z., Z.Y. and J.A.T.d.S.; investigation, C.Z.; resources, J.D.; data curation, C.Z. and Z.Y.; writing—original draft preparation, C.Z., Z.Y. and J.A.T.d.S.; writing—review and editing, J.A.T.d.S. and Z.Y.; visualization, C.Z. and Z.Y.; supervision, J.D.; project administration, J.D.; funding acquisition, J.D. All authors have read and agreed to the published version of the manuscript.

Funding: This research was funded by the National Key Research and Development Program of China, grant number 2018YFD1000400, Key Area R&D Project of Guangdong Province, grant number 2020B020221001, and National Natural Science Foundation of China, grant number 31871547.

Acknowledgments: The authors are grateful to Yongxia Jia for her assistance with GC-MS analysis, and to Rufang Deng for her assistance with subcellular localization.

Conflicts of Interest: The authors declare no conflict of interest.

Abbreviations

DMAPP	Dimethylallyl diphosphate
DXR	1-Deoxy-D-xylulose 5-phosphate reductoisomerase
DXS	1-Deoxy-D-xylulose 5-phosphate synthase
GC–MS	Gas chromatography–mass spectrometry
GES	Geraniol synthase
GPP	Geranyl pyrophosphate
GPPS	GPP synthase
HDS	4-Hydroxy-3-methylbut-2-en-1-yl diphosphate synthase
IPP	Isopentenyl diphosphate
IPTG	Isopropyl-β-D-thiogalactopyranoside
MeJA	Methyl jasmonate
MEP	Methylerythritol phosphate
MVA	Mevalonic acid
SDS-PAGE	Sodium dodecyl sulfate polyacrylamide gel electrophoresis
YFP	Yellow fluorescent protein

References

1. Goff, S.A.; Klee, H.J. Plant volatile compounds: Sensory cues for health and nutritional value? *Science* **2006**, *311*, 815–819. [CrossRef]
2. Holopainen, J.K.; Gershenzon, J. Multiple stress factors and the emission of plant VOCs. *Trends Plant Sci.* **2010**, *15*, 176–184. [CrossRef] [PubMed]
3. Bouwmeester, H.; Schuurink, R.C.; Bleeker, P.M.; Schiestl, F. The role of volatiles in plant communication. *Plant J.* **2019**, *100*, 892–907. [CrossRef] [PubMed]
4. Ramya, M.; Jang, S.; An, H.R.; Lee, S.Y.; Park, P.M.; Park, P.H. Volatile organic compounds from orchids: From synthesis and function to gene regulation. *Int. J. Mol. Sci.* **2020**, *21*, 1160. [CrossRef]
5. Hsiao, Y.Y.; Pan, Z.J.; Hsu, C.C.; Yang, Y.P.; Hsu, Y.C.; Chuang, Y.C.; Shih, H.H.; Chen, W.H.; Tsai, W.C.; Chen, H.H. Research on orchid biology and biotechnology. *Plant Cell Physiol.* **2011**, *52*, 1467–1486. [CrossRef] [PubMed]
6. Iijima, Y.; Gang, D.R.; Fridman, E.; Lewinsohn, E.; Pichersky, E. Characterization of geraniol synthase from the peltate glands of sweet basil. *Plant Physiol.* **2004**, *134*, 370–379. [CrossRef]

7. Yu, Z.; Zhao, C.; Zhang, G.; Teixeira da Silva, J.A.; Duan, J. Genome-wide identification and expression profile of TPS gene family in *Dendrobium officinale* and the role of *DoTPS10* in linalool biosynthesis. *Int. J. Mol. Sci.* **2020**, *21*, 5419. [CrossRef] [PubMed]
8. Chacón, M.G.; Marriott, A.; Kendrick, E.G.; Styles, M.Q.; Leak, D.J. Esterification of geraniol as a strategy for increasing product titre and specificity in engineered *Escherichia coli*. *Microb. Cell Factories* **2019**, *18*, 105. [CrossRef]
9. Sun, P.; Schuurink, R.C.; Caissard, J.C.; Hugueney, P.; Baudino, S. My way: Noncanonical biosynthesis pathway for plant volatiles. *Trends Plant Sci.* **2016**, *21*, 884–894. [CrossRef]
10. Dudareva, N.; Klempien, A.; Muhlemann, J.K.; Kaplan, I. Biosynthesis, function and metabolic engineering of plant volatile organic compounds. *New Phytol.* **2013**, *198*, 16–32. [CrossRef]
11. Magnard, J.L.; Roccia, A.; Caissard, J.C.; Vergne, P.; Sun, P.; Hecquet, R.; Dubois, A.; Hibrand-Saint Oyant, L.; Jullien, F.; Nicolè, F.; et al. Plant volatiles. Biosynthesis of monoterpene scent compounds in roses. *Science* **2015**, *349*, 81–83. [CrossRef] [PubMed]
12. Li, X.; Xu, Y.; Shen, S.; Yin, X.; Klee, H.; Zhang, B.; Chen, K.; Hancock, R. Transcription factor CitERF71 activates the terpene synthase gene *CitTPS16* involved in the synthesis of E-geraniol in sweet orange fruit. *J. Exp. Bot.* **2017**, *68*, 4929–4938. [CrossRef] [PubMed]
13. Bao, T.; Shadrack, K.; Yang, S.; Xue, X.; Li, S.; Wang, N.; Wang, Q.; Wang, L.; Gao, X.; Cronk, Q. Functional characterization of terpene synthases accounting for the volatilized-terpene heterogeneity in *Lathyrus odoratus* cultivar flowers. *Plant Cell Physiol.* **2020**, *61*, pcaa100. [CrossRef] [PubMed]
14. Hsiao, Y.Y.; Jeng, M.F.; Tsai, W.C.; Chuang, Y.C.; Li, C.Y.; Wu, T.S.; Kuoh, C.S.; Chen, W.H.; Chen, H.H. A novel homodimeric geranyl diphosphate synthase from the orchid *Phalaenopsis bellina* lacking a DD$(X)_{2-4}$D motif. *Plant J.* **2008**, *55*, 719–733. [CrossRef] [PubMed]
15. Chuang, Y.C.; Hung, Y.C.; Tsai, W.C.; Chen, W.H.; Chen, H.H. PbbHLH4 regulates floral monoterpene biosynthesis in *Phalaenopsis* orchids. *J. Exp. Bot.* **2018**, *69*, 4363–4377. [CrossRef]
16. Yan, L.; Wang, X.; Liu, H.; Tian, Y.; Lian, J.; Yang, R.; Hao, S.; Wang, X.; Yang, S.; Li, Q.; et al. The genome of *Dendrobium officinale* illuminates the biology of the important traditional Chinese orchid herb. *Mol. Plant* **2015**, *8*, 922–934. [CrossRef]
17. Zhang, G.; Xu, Q.; Bian, C.; Tsai, W.C.; Yeh, C.M.; Liu, K.; Yoshida, K.; Zhang, L.; Chang, S.; Chen, F.; et al. The *Dendrobium catenatum* Lindl genome sequence provides insights into polysaccharide synthase, floral development and adaptive evolution. *Sci. Rep.* **2016**, *6*, 19029. [CrossRef]
18. Jiang, S.Y.; Jin, J.; Sarojam, R.; Ramachandran, S.A. Comprehensive survey on the terpene synthase gene family provides new insight into its evolutionary patterns. *Genome Biol. Evol.* **2019**, *11*, 2078–2098. [CrossRef]
19. Chen, F.; Tholl, D.; Bohlmann, J.; Pichersky, E. The family of terpene synthases in plants: A mid-size family of genes for specialized metabolism that is highly diversified throughout the kingdom. *Plant J.* **2011**, *66*, 212–229. [CrossRef]
20. Bohlmann, J.; Meyer-Gauen, G.; Croteau, R. Plant terpenoid synthases: Molecular biology and phylogenetic analysis. *Proc. Natl. Acad. Sci. USA* **1998**, *95*, 4126–4133. [CrossRef]
21. Gao, F.; Liu, B.; Li, M.; Gao, X.; Fang, Q.; Liu, C.; Ding, H.; Wang, L.; Gao, X. Identification and characterization of terpene synthase genes accounting for volatile terpene emissions in flowers of *Freesia × hybrida*. *J. Exp. Bot.* **2018**, *69*, 4249–4265. [CrossRef]
22. Kaundal, R.; Saini, R.; Zhao, P.X. Combining machine learning and homology-based approaches to accurately predict subcellular localization in *Arabidopsis*. *Plant Physiol.* **2010**, *154*, 36–54. [CrossRef] [PubMed]
23. Chou, K.C.; Shen, H.B. Plant-mPLoc: A top-down strategy to augment the power for predicting plant protein subcellular localization. *PLoS ONE* **2010**, *5*, e11335. [CrossRef] [PubMed]
24. Cheng, X.; Xiao, X.; Chou, K.C. PLoc-mPlant: Predict subcellular localization of multi-location plant proteins by incorporating the optimal GO information into general PseAAC. *Mol. Biosyst.* **2017**, *13*, 1722–1727. [CrossRef] [PubMed]
25. Zhang, T.; Guo, Y.; Shi, X.; Yang, Y.; Chen, J.; Zhang, Q.; Sun, M. Overexpression of *LiTPS2* from a cultivar of lily (*Lilium* 'Siberia') enhances the monoterpenoids content in tobacco flowers. *Plant Physiol. Biochem.* **2020**, *151*, 391–399. [CrossRef] [PubMed]
26. Martin, D.M.; Gershenzon, J.; Bohlmann, J. Induction of volatile terpene biosynthesis and diurnal emission by methyl jasmonate in foliage of Norway spruce. *Plant Physiol.* **2003**, *132*, 1586–1599. [CrossRef] [PubMed]
27. Chen, W.; Viljoen, A.M. Geraniol—A review of a commercially important fragrance material. *S. Afr. J. Bot.* **2010**, *76*, 643–651. [CrossRef]

28. Vranová, E.; Coman, D.; Gruissem, W. Network analysis of the MVA and MEP pathways for isoprenoid synthesis. *Annu. Rev. Plant Biol.* **2013**, *64*, 665–700. [CrossRef]
29. Tholl, D. Terpene synthases and the regulation, diversity and biological roles of terpene metabolism. *Curr. Opin. Plant Biol.* **2006**, *9*, 297–304. [CrossRef]
30. Ye, P.; Liang, S.; Wang, X.; Duan, L.; Jiang-Yan, F.; Yang, J.; Zhan, R.; Ma, D. Transcriptome analysis and targeted metabolic profiling for pathway elucidation and identification of a geraniol synthase involved in iridoid biosynthesis from *Gardenia jasminoides*. *Ind. Crop Prod.* **2019**, *132*, 48–58. [CrossRef]
31. Hamachi, A.; Nisihara, M.; Saito, S.; Rim, H.; Takahashi, H.; Islam, M.; Uemura, T.; Ohnishi, T.; Ozawa, R.; Maffei, M.E.; et al. Overexpression of geraniol synthase induces heat stress susceptibility in *Nicotiana tabacum*. *Planta* **2019**, *249*, 235–249. [CrossRef]
32. Liu, J.; Guan, Z.; Liu, H.; Qi, L.; Zhang, D.; Zou, T.; Yin, P. Structural insights into the substrate recognition mechanism of *Arabidopsis* GPP-bound NUDX1 for noncanonical monoterpene biosynthesis. *Mol. Plant* **2018**, *11*, 218–221. [CrossRef] [PubMed]
33. Zhou, F.; Pichersky, E. More is better: The diversity of terpene metabolism in plants. *Curr. Opin. Plant Biol.* **2020**, *55*, 1–10. [CrossRef] [PubMed]
34. Dong, L.; Miettinen, K.; Goedbloed, M.; Verstappen, F.W.; Voster, A.; Jongsma, M.A.; Memelink, J.; van der Krol, S.; Bouwmeester, H.J. Characterization of two geraniol synthases from *Valeriana officinalis* and *Lippia dulcis*: Similar activity but difference in subcellular localization. *Metab. Eng.* **2013**, *20*, 198–211. [CrossRef]
35. Yue, Y.; Yu, R.; Fan, Y. Characterization of two monoterpene synthases involved in floral scent formation in *Hedychium coronarium*. *Planta* **2014**, *240*, 745–762. [CrossRef] [PubMed]
36. Nagegowda, D.A.; Gutensohn, M.; Wilkerson, C.G.; Dudareva, N. Two nearly identical terpene synthases catalyze the formation of nerolidol and linalool in snapdragon flowers. *Plant J.* **2008**, *55*, 224–239. [CrossRef] [PubMed]
37. Han, Y.; Wang, H.; Wang, X.; Li, K.; Dong, M.; Li, Y.; Zhu, Q.; Shang, F. Mechanism of floral scent production in *Osmanthus fragrans* and the production and regulation of its key floral constituents, β-ionone and linalool. *Hortic. Res.* **2019**, *6*, 106. [CrossRef]
38. Flach, A.; Dondon, R.C.; Singer, R.B.; Koehler, S.; Amaral Mdo, C.; Marsaioli, A.J. The chemistry of pollination in selected Brazilian *Maxillariinae* orchids: Floral rewards and fragrance. *J. Chem. Ecol.* **2004**, *30*, 1045–1056. [CrossRef]
39. Chuang, Y.C.; Lee, M.C.; Chang, Y.L.; Chen, W.H.; Chen, H.H. Diurnal regulation of the floral scent emission by light and circadian rhythm in the *Phalaenopsis* orchids. *Bot. Stud.* **2017**, *58*, 50–58. [CrossRef]
40. Zhou, H.; Shamala, L.F.; Yi, X.; Yan, Z.; Wei, S. Analysis of terpene synthase family genes in *Camellia sinensis* with an emphasis on abiotic stress conditions. *Sci. Rep.* **2020**, *10*, 933. [CrossRef]
41. Hong, G.J.; Xue, X.Y.; Mao, Y.B.; Wang, L.J.; Chen, X.Y. *Arabidopsis* MYC2 interacts with DELLA proteins in regulating sesquiterpene synthase gene expression. *Plant Cell* **2012**, *24*, 2635–2648. [CrossRef] [PubMed]
42. Larkin, M.A.; Blackshields, G.; Brown, N.P.; Chenna, R.; McGettigan, P.A.; McWilliam, H.; Valentin, F.; Wallace, I.M.; Wilm, A.; Lopez, R.; et al. Clustal W and Clustal X version 2.0. *Bioinformatics* **2007**, *23*, 2947–2948. [CrossRef] [PubMed]
43. Kumar, S.; Stecher, G.; Tamura, K. MEGA7: Molecular evolutionary genetics analysis version 7.0 for bigger datasets. *Mol. Biol. Evol.* **2016**, *33*, 1870–1874. [CrossRef] [PubMed]
44. Saitou, N.; Nei, M. The neighbor-joining method: A new method for reconstructing phylogenetic trees. *Mol. Biol. Evol.* **1987**, *4*, 406–425.
45. Yu, Z.; Zhang, G.; Teixeira da Silva, J.A.; Yang, Z.; Duan, J. The β-1,3-galactosetransferase gene *DoGALT2* is essential for stigmatic mucilage production in *Dendrobium officinale*. *Plant Sci.* **2019**, *287*, 110179. [CrossRef]
46. Yu, Z.; He, C.; Teixeira da Silva, J.A.; Luo, J.; Yang, Z.; Duan, J. The GDP-mannose transporter gene (*DoGMT*) from *Dendrobium officinale* is critical for mannan biosynthesis in plant growth and development. *Plant Sci.* **2018**, *277*, 43–54. [CrossRef]
47. Livak, K.J.; Schmittgen, T.D. Analysis of relative gene expression data using real-time quantitative PCR and the $2^{-\Delta\Delta CT}$ method. *Methods* **2001**, *25*, 402–408. [CrossRef]

© 2020 by the authors. Licensee MDPI, Basel, Switzerland. This article is an open access article distributed under the terms and conditions of the Creative Commons Attribution (CC BY) license (http://creativecommons.org/licenses/by/4.0/).

Article

DoRWA3 from *Dendrobium officinale* Plays an Essential Role in Acetylation of Polysaccharides

Can Si [1,2], Jaime A. Teixeira da Silva [3], Chunmei He [1], Zhenming Yu [1], Conghui Zhao [1], Haobin Wang [1], Mingze Zhang [1] and Jun Duan [1,*]

1. Key Laboratory of South China Agricultural Plant Molecular Analysis and Genetic Improvement, South China Botanical Garden, Chinese Academy of Sciences, Guangzhou 510650, China; cans2013@163.com (C.S.); hechunmei26@163.com (C.H.); zhenming311@scbg.ac.cn (Z.Y.); zhaoconghui@scbg.ac.cn (C.Z.); wanghaobin17@scbg.ac.cn (H.W.); zhangmingze@scbg.ac.cn (M.Z.)
2. College of Life Sciences, University of Chinese Academy of Sciences, Beijing 100049, China
3. Independent researcher, P.O. Box 7, Miki-cho Post Office, Ikenobe 3011-2, Miki-cho, Kita-gun, Kagawa-ken 761-0799, Japan; jaimetex@yahoo.com
* Correspondence: duanj@scib.ac.cn; Tel.: +86-020-37252978

Received: 7 August 2020; Accepted: 24 August 2020; Published: 28 August 2020

Abstract: The acetylation or deacetylation of polysaccharides can influence their physical properties and biological activities. One main constituent of the edible medicinal orchid, *Dendrobium officinale*, is water-soluble polysaccharides (WSPs) with substituted *O*-acetyl groups. Both *O*-acetyl groups and WSPs show a similar trend in different organs, but the genes coding for enzymes that transfer acetyl groups to WSPs have not been identified. In this study, we report that REDUCED WALL ACETYLATION (RWA) proteins may act as acetyltransferases. Three *DoRWA* genes were identified, cloned, and sequenced. They were sensitive to abscisic acid (ABA), but there were no differences in germination rate and root length between wild type and 35S::DoRWA3 transgenic lines under ABA stress. Three DoRWA proteins were localized in the endoplasmic reticulum. *DoRWA3* had relatively stronger transcript levels in organs where acetyl groups accumulated than *DoRWA1* and *DoRWA2*, was co-expressed with polysaccharides synthetic genes, so it was considered as a candidate acetyltransferase gene. The level of acetylation of polysaccharides increased significantly in the seeds, leaves and stems of three 35S::DoRWA3 transgenic lines compared to wild type plants. These results indicate that DoRWA3 can transfer acetyl groups to polysaccharides and is a candidate protein to improve the biological activity of other edible and medicinal plants.

Keywords: acetyl groups; *Dendrobium officinale*; REDUCED WALL ACETYLATION; endoplasmic reticulum

---

1. Introduction

Polysaccharides, which are extracted from many edible and medicinal plants, have been widely used in food, cosmetics and pharmaceutical industries due to their therapeutic properties and low toxicity [1,2]. The functional properties of polysaccharides depend on several structural parameters, particularly the composition of monosaccharides, molecular weight and functional groups [3]. Acetyl groups, which are substituted at the backbone or sidechain of polysaccharides, can expose more hydroxyl groups in water, thus influence the solubility, gelation, surface structure and other physical properties of polysaccharides [4–6]. Furthermore, in plants, the deacetylation or acetylation of polysaccharides can affect their molecular weight, structure and conformation, and thus influence their biological activity, conferring various activities (antibacterial, antibiofilm, antioxidant, anticoagulant and immunoregulatory) [7,8].

Thus far, three different protein families have been shown to be involved in the O-acetylation of polysaccharides, REDUCED WALL ACETYLATION (RWA), ALTERED XYLOGLUCAN9 (AXY9), and TRICHOME BIREFRINGENCE LIKE (TBL), with 4, 1 and 46 members in *Arabidopsis thaliana* [9–13]. The single mutant *rwa2* showed an indistinguishable phenotype and had a 17% lower degree of acetylation (DA) compared with the wild type (WT) [10] while the quadruple mutant *rwa1rwa2rwa3rwa4* displayed a severely dwarfed phenotype and 63% lower DA in rosette leaves [14]. Similarly, *RWA* genes in hybrid aspen downregulated the acetylation of wood, including xylan and xyloglucan, by 15–20%, but this did not affect the height or stem diameter of plants significantly [15]. A single mutant *axy9.2* in *A. thaliana* had smaller leaves and 35% less DA in rosette leaves [11]. The single knockout mutant *tbl27* showed 14% lower DA of xyloglucan, while the double mutant *rwa2tbl27* showed as much as 24% lower DA in rosette leaves [12]. Although the biosynthetic pathway for O-acetylation of polysaccharides is fairly clear, very little is known about the mechanism of O-acetylation in edible and medicinal plants.

Dendrobium officinale Kimura et Migo could be a good model species to address these research challenges. First, as a traditional Chinese medicine (TCM), the in vivo and in vitro biological activities of water-soluble polysaccharides (WSPs), which are the major medicinal ingredients of *D. officinale*, have antioxidant, antitumor, antidiabetic, anti-inflammation, and immunomodulating activities [16]. Secondly, WSPs contain mannose, glucose and acetyl groups substituted at the O-2 or O-3 site of mannosyl residues [17], and the primary structure, such as mannose, β-(1→4)-Man linkage and acetyl groups, mainly contribute to the bioactivity of WSPs [18]. Finally, the O-acetyl content accounts for as much as 2.9% (w/w, dry weight) of the polysaccharides [19]. Despite this, until now, the mechanism of O-acetylation in *D. officinale* had not yet been reported.

This article focuses on *D. officinale RWA* genes. Bioinformatics tools were used to obtain basic information about *DoRWA* genes such as gene structure, cis-elements and conserved domains. The expression levels of three *DoRWA* genes in different organs and developmental stages, and in response to abiotic stresses, were also assessed. Three DoRWA proteins were transformed into the protoplasts of *A. thaliana* with a localization marker to assess the localization of these proteins. Most importantly, *35S::DoRWA3* transgenic lines were constructed to verify the biological functions of *DoRWA3*. The exploration of *RWA* genes in this orchid would facilitate the targeting of the genes coding for acetyltransferase.

2. Results

2.1. Isolation and Sequence Analysis of the DoRWA Genes

Three *DoRWA* genes, named *DoRWA1*, *DoRWA2* and *DoRWA3*, were identified in the *D. officinale* genome [20]. Their open reading frames (ORFs) were 1638, 1638 and 1617 bp long, encoding 545, 545 and 538 aa. The molecular weights (MWs) of the three genes were 63.825, 63.780 and 63.634 Da, and their isoelectric points (pIs) were 8.84, 9.01 and 8.94. The ORF sequences of three genes were submitted to NCBI with the accession numbers MT199223, MT199224 and MT199225. BlastP results revealed that DoRWA1 had 99% similarity with RWA1 of *Dendrobium catenatum* (XP_020674439.1) and 92% similarity with RWA1 of *Phalaenopsis equestris* (XP_020584729.1); DoRWA2 had 99% similarity with RWA1 of *D. catenatum* (XP_028548609.1) and 92% similarity with RWA3L of *P. equestris* (XP_020577546.1); DoRWA3 had 99% similarity with RWA4 of *D. catenatum* (XP_020684246.1) and 84% similarity with RWA4L of *P. equestris* (XP_020573535.1).

2.2. Bioinformatics of the DoRWA Genes

The RWA protein sequences from three plants were aligned (Figure S1). The highest similarity of these proteins was 71.59%, indicating that RWA was considerably conserved in these plants. These proteins had the same domain Cas1-AcylT (411–459 aa), accounting for 73.77–78.60% of the full length (538–584 aa) (Figure 1). The exon–intron structures and length of the three genes varied. *DoRWA1* (37 kb) and *DoRWA2* (14 kb) had a similar gene structure and contained 16 exons and

15 introns, while *DoRWA3* (6 kb) had 15 exons and 14 introns (Figure S2). The protein sequences of the three plants were also used to construct a phylogenetic tree using the Neighbor-Joining (N-J) method. DoRWA1 and DoRWA2 were clustered as one branch, and DoRWA3 was clustered with AtRWA2, PtRWA-C and PtRWA-D (Figure 1). Using the homologous *A. thaliana* protein, RWA2 was used to conduct a protein–protein interaction network analysis in which RWA2 was correlated with other TBL proteins [12,13] (Figure 2), such as pectin *O*-acetyltransferase TBR [21], xylan *O*-aceryltransferase TBL3 and TBL31 [22], and xyloglucan *O*-acetyltransferase TBL27 [23], indicating that DoRWA2 and DoTBL proteins were active in the same metabolic pathway.

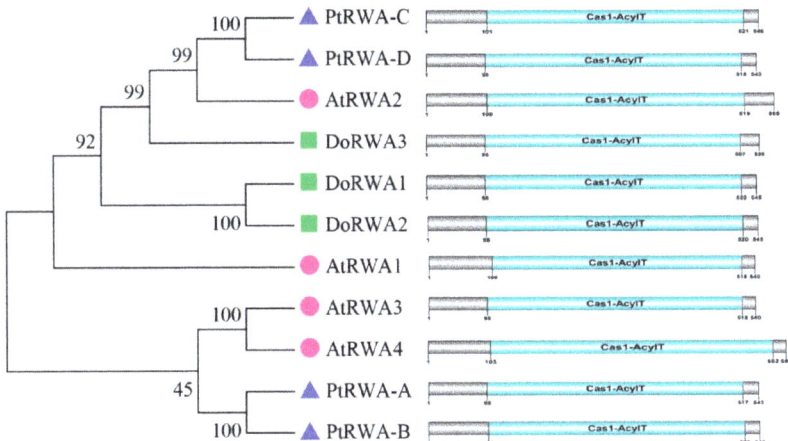

Figure 1. Phylogenetic tree and conserved domain analysis of REDUCED WALL ACETYLATION (RWA) proteins in *Dendrobium officinale*, *Arabidopsis thaliana* and *Populus trichocarpa*. The phylogenetic tree was constructed by the neighbor-joining (N-J) method in MEGA 7.0 software. Domains were drawn by DOG2.0 software. The gene IDs of RWA proteins are: DoRWA1 (MT199223), DoRWA2 (MT199224), DoRWA3 (MT199225), AtRWA1 (At5g46340), AtRWA2 (At3g06550), AtRWA3 (At2g34410), AtRWA4 (At1g29890), PtRWA-A (Potri.001g352300), PtRWA-B (Potri.011g079400), PtRWA-C (Potri.010g148500), and PtRWA-D (Potri.008g102300). The blue area indicates the Cas1-AcylT conserved domain.

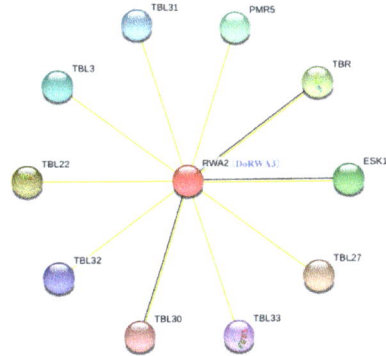

Figure 2. Protein–protein association networks of DoRWA3 using RWA2, a homologous *Arabidopsis thaliana* protein. Yellow lines represent "textmining" and black lines represent "co-expression".

2.3. Analysis of cis-Elements, and Expression Patterns of DoRWA Genes under Cold and ABA Treatments

The *cis*-elements of the three genes contained hormone-responsive elements (methyl jasmonate, abscisic acid (ABA), auxin, gibberellin) and abiotic stress-responsive elements (low temperature)

(Figure 3A). All three *DoRWA* genes had the ABA-responsive element while *DoRWA1* and *DoRWA2* (but not *DoRWA3*) had a low temperature-responsive element. To verify these predictions, the expression patterns of the three *DoRWA* genes in response to cold and ABA were assessed. In the cold treatment, the relative expression level (fragments per kilobase per million, FPKM) of *DoRWA1* was upregulated, *DoRWA2* was downregulated, but *DoRWA3* showed no difference (Figure 3B). In the ABA treatment, the transcript levels of the three genes were upregulated at first, then peaked, but were finally downregulated in three organs (roots, stems, leaves), with peak expression at 6 h in roots and stems, and at 3 h in leaves (Figure 3C).

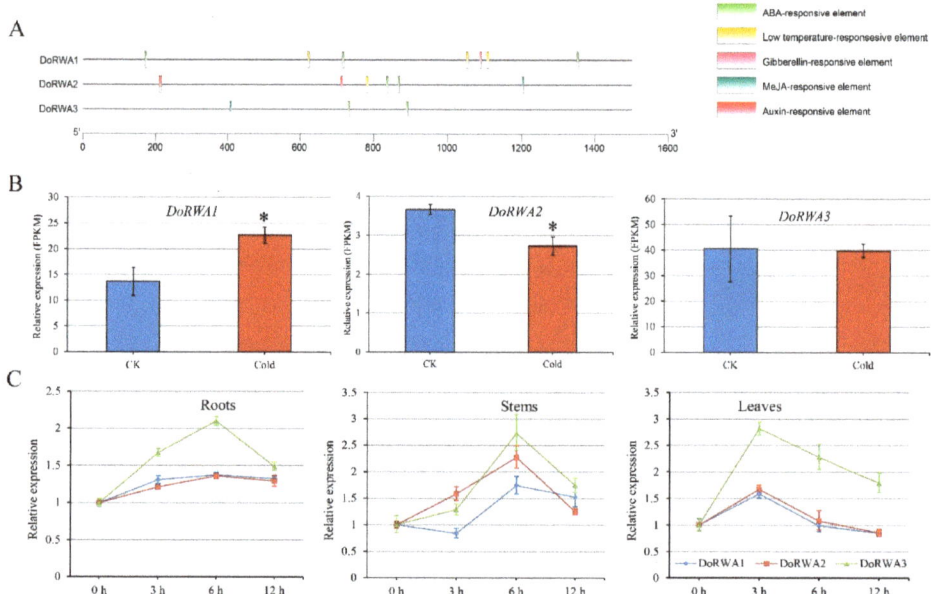

Figure 3. Analysis of *cis*-elements and relative expression levels of three *DoRWA* genes under cold stress (4 °C) and abscisic acid (ABA), (100 µM) treatment. (**A**) Analysis of *cis*-elements in the promoter region of three *DoRWA* genes. Different colors represent different *cis*-elements. (**B**) Relative expression level of three *DoRWA* genes under CK (20 °C) and cold stress (4 °C) using the FPKM value. (**C**) Relative expression level of three *DoRWA* genes in the ABA treatment. The transcript level of *DoRWA* genes at 0 h was set as 1. Each data bar represents mean ± SD (n = 3). * indicates $p < 0.05$ between the expression level of three *DoRWA* genes in CK and cold treatment according to Duncan's multiple range test (DMRT).

2.4. Cellular Localization of DoRWA Proteins

The three DoRWA proteins (DoRWA1, DoRWA2, DoRWA3) had 10, 10 and 11 transmembrane helices, which indicated they may be localized in a membranous organelle (Figure S3A,B,D). The fluorescent signals showed that all three DoRWA proteins were co-localized with the ER-rk (Figure 4), which is also localized in the endoplasmic reticulum (ER) [24]. Furthermore, the YFP fluorescence of DoRWA3 was not localized in the Golgi apparatus (GA) (Figure S3D). These results indicate that the ER plays an important role in the acetylation of polysaccharides.

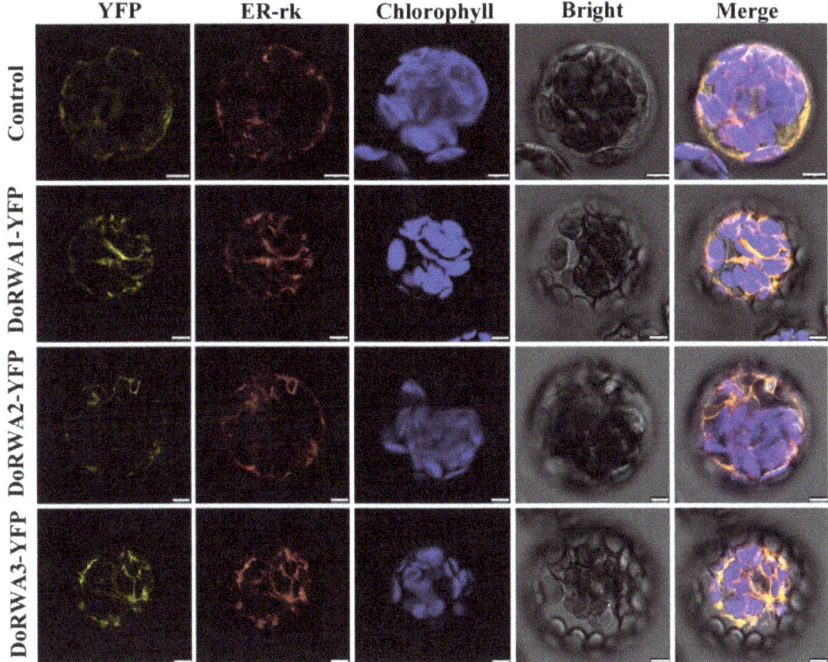

Figure 4. Subcellular localization of YFP, DoRWA1-YFP, DoRWA2-YFP and DoRWA3-YFP with the endoplasmic reticulum localization marker ER-rk in the *Arabidopsis thaliana* mesophyll protoplasts. Bar = 5 μm.

2.5. WSPs and O-Acetyl Groups Mainly Accumulated in the Stems of D. officinale

WSPs and O-acetyl groups showed a similar trend, accumulating the most in stems, followed by flowers and leaves, and the least in roots of seedlings and adult plants (Figure 5). The content of WSPs in stems was 65.55 mg/g in seedlings and 267.48 mg/g in adult plants. The content of O-acetyl groups of WSPs in stems was 23.22 mg/g in seedlings and 70.96 mg/g in adult plants.

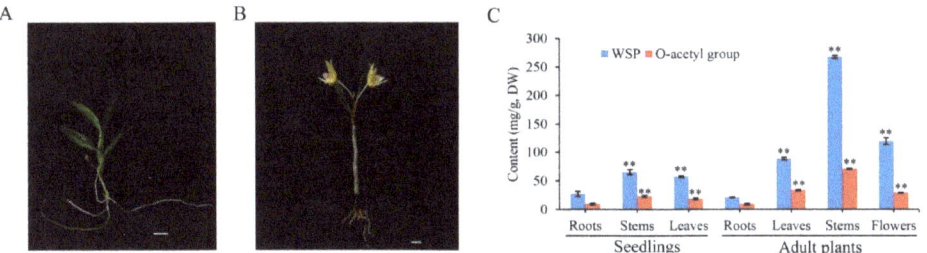

Figure 5. The metabolic accumulation of water-soluble polysaccharides (WSPs) and O-acetyl groups in different organs of *Dendrobium officinale*. (**A**) Ten-month-old seedling. Bar = 1 cm. (**B**) Adult plant. Bar = 1 cm. (**C**) The content of WSPs and related O-acetyl groups in different organs of seedlings and adult plants. Each data bar represents the mean ± SD ($n = 3$). ** indicates $p < 0.01$ between the content of WSPs and O-acetyl groups in the roots of seedlings and in other organs of seedlings and adult plants according to DMRT.

2.6. Expression Patterns of DoRWA Genes in Different Organs of Seedlings and Adult Plants

In *D. officinale* seedlings, the expression profiles of *DoRWA1* and *DoRWA2* were similar, with higher transcript levels in roots than in stems and leaves, while the relative expression of *DoRWA3* was higher in stems than in roots and leaves (Figure 6A). In adult plants, the expression levels of *DoRWA1* and *DoRWA2* showed fewer differences in several organs, but the expression of *DoRWA3* was relatively higher in stems, leaves and flowers where acetyl groups accumulated (Figure 6B). The mRNA ratio of *DoRWA3* was 5.80- and 14.41-fold higher than *DoRWA1* and *DoRWA2* in stems, and 4.42- and 25.94-fold higher than in leaves of seedlings (Figure 6A). Similar results were also found in adult plants (Figure 6B).

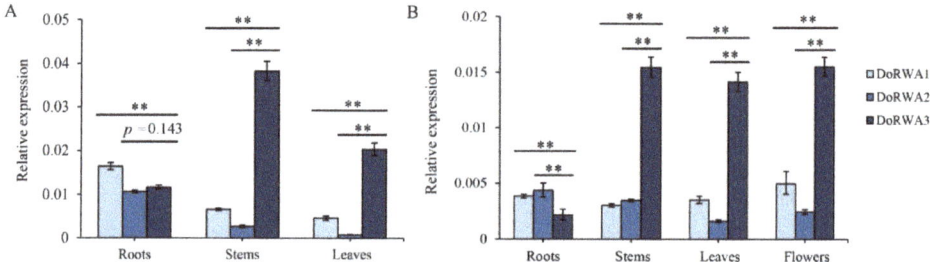

Figure 6. Expression profiles of *DoRWA* genes in different organs. Relative expression levels of three *DoRWA* genes in different organs of (**A**) seedlings and (**B**) adult plants. In A and B, *EF-1α* served as the control. Each data bar represents the mean ± SD ($n = 3$). ** indicates $p < 0.01$ between the expression level of *DoRWA3* and *DoRWA1* (*DoRWA2*) according to DMRT.

2.7. Co-Expression of DoRWA3 with Synthetic Genes of Polysaccharides

In the four developmental stages, the content of WSPs increased from S1 to S3, decreased from S3 to S4, and peaked at S3 [25]. The transcript levels of key genes related to polysaccharides in *D. officinale*, such as *cellulose synthesis-like* (*CSL*) [25], *GDP-mannose pyrophosphorylation* (*GMP*) [26], *UDP glucose 4-epimerase* (*UGE*) [27] and *GDP-mannose transporter* (*GMT*) [28] peaked at S2 or S3, corresponding to trends in the content of WSPs (Figure 5C). The expression profile of *DoRWA3* peaked at S2, and showed a close association with the key genes described above (Figure 7), This indicates that *DoRWA3* may be a candidate gene responsible for coding the enzyme that transfers acetyl groups to WSPs.

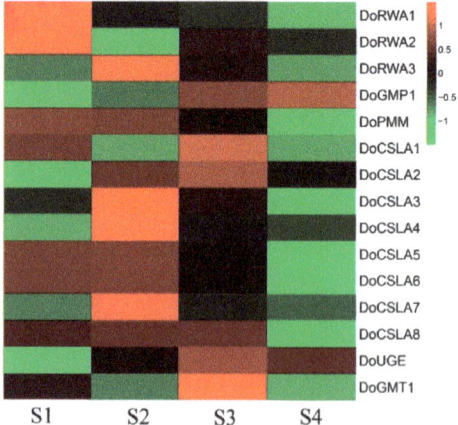

Figure 7. Expression profiles of three *DoRWA* genes and other polysaccharide-related genes in four developmental stages. RPKM values of different genes were log-transformed (log of mean RPKM).

2.8. DoRWA3 Overexpression Increased the Acetylation Level of Polysaccharides in A. thaliana

The semi-quantitative PCR and qRT-PCR results (Figure 8B,D) indicated that *DoRWA3* was successfully inserted into the *A. thaliana* genome and could be transcribed normally. There were no differences in the phenotype (color, size, flowering time, etc.) between WT and the three overexpression (OE) transgenic lines (Figure 8C). Simultaneously, the transcript level of four *AtRWA* genes in WT and three OE transgenic lines were also tested. There were no differences in the relative expression levels of *AtRWA1*, *AtRWA3*, and *AtRWA4* between WT and transgenic lines, but the expression level of *AtRWA2* was lower in the three transgenic lines compared with WT (Figure 9).

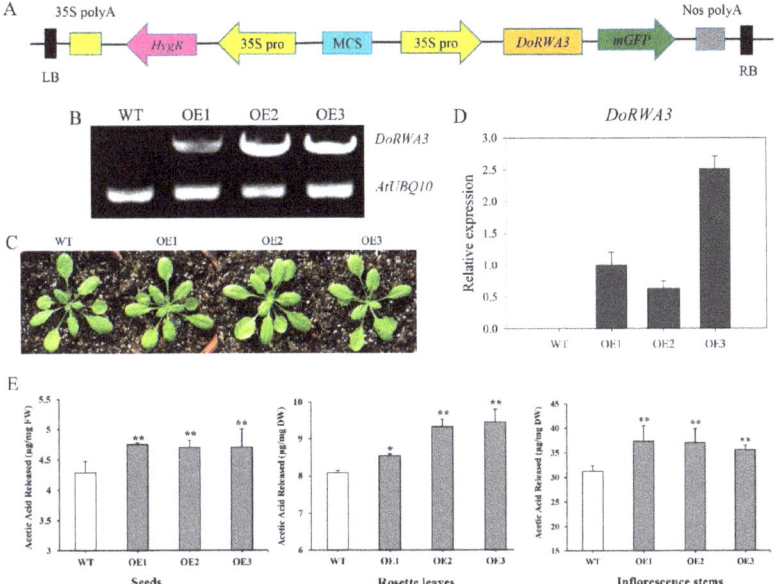

Figure 8. Overexpression of *DoRWA3* increased the acetylation level of polysaccharides in *A. thaliana*. (**A**) The pCAMBIA1302 vector used for *A. thaliana* transformation. (**B**) Semi–quantitative PCR of the *DoRWA3* gene in wild type (WT) and three overexpression (OE) transgenic (35S::DoRWA3) lines: OE1, OE2 and OE3. (**C**) Phenotype of the WT and three OE transgenic lines. (**D**) qRT-PCR analysis of *DoRWA3* in the WT and three OE transgenic lines. (**E**) The content of released acetic acid in different organs of the WT and three OE transgenic lines. The transcript level of *DoRWA3* in OE1 was set as 1. FW, fresh weight; DW, dry weight. Data bars represent the mean ± SD ($n = 3$). * and ** indicate $p < 0.05$ and $p < 0.01$ between the WT and OE transgenic lines according to DMRT.

In seeds, the content of released acetic acid was about 1.11-, 1.10- and 1.10-fold higher in the three transgenic OE lines than in WT (Figure 8E). The corresponding values were 1.06-, 1.15- and 1.17-fold higher in rosette leaves (Figure 8E) and 1.20-, 1.19- and 1.14-fold higher in inflorescence stems (Figure 8E). The exogenously inserted *DoRWA3* gene increased the level of acetylation of polysaccharides in seeds, leaves and stems of transgenic *A. thaliana* by 10–11%, 6–17%, and 14–20%, respectively. The released acetic acid of stems (31–38 mg/g) was higher than leaves (8–9.5 mg/g) and seeds (4–5 mg/g), indicating that stems accumulated more *O*-acetyl groups in polysaccharides of *A. thaliana*, relative to *D. officinale*.

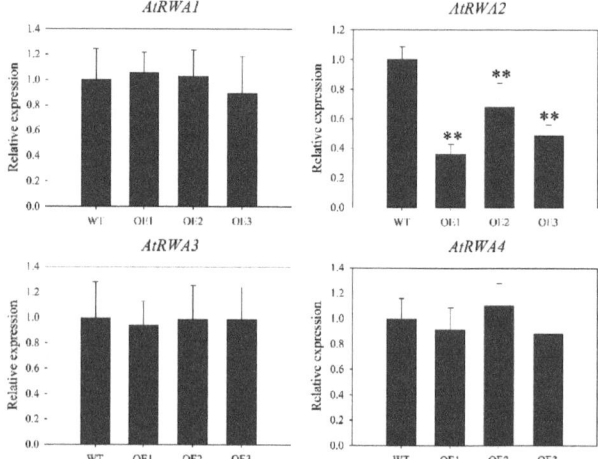

Figure 9. Relative expression levels of four *AtRWA* genes in the wild type (WT) and three overexpression (OE) *35S::DoRWA3* transgenic lines (OE1, OE2, OE3). The transcript level of *AtRWA* genes in the WT was set as 1. Each data bar represents mean ± SD ($n = 3$). ** indicate $p < 0.01$ between the transcript levels of three *DoRWA* genes WT and OE transgenic lines according to DMRT.

2.9. ABA Sensitivity Was Not Affected by Constitutive Expression of DoRWA3

The germination rate of WT and transgenic lines was almost 100% (Figure S4A,C). ABA treatment reduced root length of all plants: in WT, root length decreased from 5.95 cm (control) to 5.11 cm (2 µM ABA), but there were no significant ($p > 0.05$) differences between the WT and transgenic plants (Figure S4D). These findings indicate that ABA sensitivity may not be affected by the exogenous *DoRWA3*.

3. Discussion

Acetyl groups affect the biological activities of polysaccharides in many edible and medicinal plants, such as *Cyclocarya paliurus* [29], *Dendrobium huoshanense* [30], *D. officinale* [31] and *Plantago asiatica* [32], *Amorphophallus konjac* [33] and *Aloe vera* [34]. Konjac glucomannan (KGM), which is extracted from the corm of *A. konjac* [35], consisting of β-1,4-linked mannose and glucose residues (molar ratio: 1.6:1) with substituted acetyl groups. Deacetylated KGM has less health benefits than KGM [36]. Acemannan extracted from *A. vera* leaves contains the β-1,4-linked mannose and glucose (molar ratio: 3:1) with substituted acetyl groups [37]. After treating with alkaline to remove acetyl groups, the solubility, hydrophilicity and bioactivities of acemannan are reduced [5].

Cell wall acetylation has been shown to play broad roles in plant abiotic and biotic responses. *rwa2-1* and *rwa2-3* mutants showed enhanced resistance to the fungus *Botrytis cinerea* [10]. Some genes related to hormones and oxidative stress, such as auxin, ABA, jasmonic acid (JA), cytokinin (CK), light, cold and drought are down or upregulated in the transcriptome of mutant *rwa2-3* compared with the WT, but no difference in the phenotype (growth and root length) was observed when the WT and *rwa2-3* were treated with hormones (auxin, ABA, JA and CK) [38]. In this study, many *cis*-elements in the promoter region of *DoRWA* genes were associated with hormones and abiotic stress (Figure 3A). qRT-PCR proved that *DoRWA* genes were responsive to ABA (Figure 3C), but there were no differences in germination rate and root length between *A. thaliana* WT and OE transgenic lines when treated with ABA (Figure S4), indicating that *DoRWA3* might not participate in the ABA-dependent signaling pathway.

The expression levels of four *RWA* genes in *A. thaliana* varied in different organs, but they all had relatively higher expression levels in inflorescence stems than in leaves and flowers, while *AtRWA2* expression was also relatively higher in leaves than in flowers [10]. The expression levels of four

RWA genes in *Populus tremula* varied in different organs (seeds, roots, leaves, buds and flowers) and treatment (drought): *PtRWA-A* was always expressed more than *PtRWA-B*, and *PtRWA-C* more than *PtRWA-D* [15]. In different organs, *DoRWA1* and *DoRWA2* showed the highest mRNA ratio in roots where the content of acetyl groups was lowest compared with other organs (Figure 6A), most likely because they had similar gene structures and conserved domains (Figure 1 and Figure S2); *DoRWA3* had relatively higher expression levels in organs where acetyl groups accumulated, namely stems, leaves and flowers (Figure 6). The mRNA ratio of *DoRWA3* showed significantly higher expression levels than *DoRWA1* and *DoRWA2* in stems, leaves and flowers (Figure 6). In four developmental stages, *DoRWA3*, but not *DoRWA1* and *DoRWA2*, was co-expressed with the synthetic genes of WSPs (Figure 7). Thus, we hypothesize that *DoRWA3* is a key gene coding for an acetyltransferase, while *DoRWA1* and *DoRWA2* may be redundant.

The AtRWA2 protein was localized in the ER or GA in *Nicotiana benthamiana* leaves or carrot protoplasts [9,10], showing species specificity. Since polysaccharides are synthesized in the GA and the related GDP-mannose transporter protein is also localized in the GA [28], it was expected that the DoRWA proteins would also be localized in the GA. However, they were localized in the ER (Figure 4), indicating that ER may be responsible for the upstream acetylation of polysaccharides.

The phenotype (growth and morphology) of a single mutant was not different from the WT [10], while triple and quadruple *rwa* mutants showed a severely dwarfed phenotype [14]. Similarly, the three 35S::DoRWA3 transgenic lines were indistinguishable from the WT (Figure 8C). This suggests that a single *rwa* mutant or *DoRWA3* OE transgenic lines have no effect on the growth of *A. thaliana*.

A minor reduction in the acetylation level was detected in the inflorescence stems of single mutants *rwa1*, *rwa2*, *rwa3* and *rwa4* [9], indicating that the four *AtRWA* genes may influence the level of cell wall acetylation in *A. thaliana*. To verify if these four *AtRWA* genes participate in increasing the acetylation level in cell wall polymers, their transcript levels were detected in WT and three OE transgenic lines. The expression patterns of *AtRWA1*, *AtRWA3* and *AtRWA4* were similar, while the expression profile of *AtRWA2* was lower in the three transgenic lines compared with the WT (Figure 9), indicating that the four *AtRWA* genes did not play a vital role in increasing the acetylation level of polysaccharides in OE transgenic lines.

The level of acetylation of cell wall residues from leaves was significantly reduced in a single mutant *rwa2-1* compared with WT, while only a small difference was found in the stems [10]. The quadruple mutant *rwa1rwa2rwa3rwa4* showed the largest reduction of acetylation level, about a 63% decrease [14]. In our study, the content of acetic acid increased significantly in the three transgenic lines compared with the WT in three organs: 6–17% in leaves, 10–11% in seeds and 14–20% in stems (Figure 8E). These findings indicate that exogenous *DoRWA3* could increase the acetylation level of polysaccharides in *A. thaliana*. *DoRWA3* could be considered as a candidate gene to improve the biological activities of polysaccharides in other edible and medicinal plants.

4. Materials and Methods

4.1. Plant Materials and Hormone Treatment

The young seedlings of *D. officinale* were cultured on half-strength (macro- and micronutrients) Murashige and Skoog (MS) medium [39], containing 0.5% activated carbon, 2% sucrose, and 0.5% agar (pH 5.7). Adult *D. officinale* plants were planted in a ground bark substrate in a greenhouse of South China Botanical Garden (Guangzhou, Guangdong, China) under ambient conditions. The seeds of *A. thaliana* Columbia (Col-0) were placed at 4 °C and continual darkness for 3 d, then transferred to a substrate containing nutritive soil and vermiculite (v/v, 2:1) under a controlled environment (80% humidity, 22 °C, 16-h photoperiod). For the ABA treatment, 10-month-old plantlets of *D. officinale* were treated with 100 µM ABA for 0, 3, 6 and 12 h, 0 h was regarded as the control group. The roots, stems, leaves, and flowers of young seedlings and adult plants were collected, frozen in nitrogen liquid, and stored at −80 °C.

4.2. RNA Extraction, cDNA Synthesis and qRT-PCR Analysis

Total RNA was extracted by an sodium dodecyl sulfate (SDS) method [40]. Briefly, 0.2 g of fresh sample was ground into a powder with liquid nitrogen. Extraction buffer (100 mM Tris-HCl, pH = 8.0; 50 mM EDTA, pH = 8.0; 500 mM NaCl; 1% SDS; 4% β-mercaptoethanol) was added and vortexed. After centrifuging at 12,000 rpm for 5 min, 1/3 (v/v) KAC (pH 4.8, 5 mol/L) was added to the solution to remove polysaccharides. After centrifuging at 12,000 rpm for 10 min, the supernatant was washed with chloroform:isoamyl alcohol (v/v, 24:1) and precipitated by 100% isopropanol. The precipitate was washed twice with 75% ethanol, dissolved in RNAase-free water, and stored at −80 °C. Any contaminating DNA was removed by Recombinant DNAase I (TaKaRa Bio Inc., Dalian, China) following the manufacturer's protocol.

Purified RNA was reverse transcribed to cDNA by the GoScript™ Reverse Transcription System Protocol (Promega, Madison, WI, USA) according to the manufacturer's instructions. cDNA was further used for gene cloning and qRT-PCR analysis. For qRT-PCR analysis, the iTaq™ Univeral SYBR® Green Supermix (Bio-Rad Laboratories Co. Ltd., Hercules, CA, USA) was used as the polymerase in the following reaction system: stage 1 (95 °C for 2 min); stage 2 (40 cycles of 95 °C for 15 s, 60 °C for 1 min); stage 3 (95 °C for 15 s, 60 °C for 1 min, 95 °C for 15 s, 60 °C for 15 s). *Actin* (JX294908) and *EF-1α* gene [41] from *D. officinale*, and *Actin2* (At3g18780), *UBC* (At5g25760) and *PP2AA3* (At1g13320) [10] from *A. thaliana* were used as reference genes. The primers designed for qRT-PCR analysis are listed in Table S1. The $2^{-\Delta\Delta CT}$ method [42] was used to calculate the relative expression levels of different genes. All treatments were sampled as three biological and technical replicates.

4.3. Identification and Cloning of DoRWA Genes

Four *A. thaliana* RWA proteins were downloaded from TAIR (https://www.arabidopsis.org/): AtRWA1 (At5g46340), AtRWA2 (At3g06550), AtRWA3 (At2g34410) and AtRWA4 (At1g29890) [10]. They were used as queries to search for homologous proteins in the *D. officinale* protein database [19] using Bioedit software [43]. All putative *D. officinale* RWA proteins were further identified by BlastP in NCBI (https://blast.ncbi.nlm.nih.gov/Blast.cgi) to discard any repeated proteins or proteins without the conserved domains of the RWA family. The identified *RWA* genes were used to design the specific primers for gene cloning.

RWA genes were cloned from cDNA using KOD FX polymerase (Toyobo Co. Ltd., Osaka, Japan) with the following protocol: stage 1 (94 °C for 3 min); stage 2 (40 cycles of 98 °C for 10 s, 55 °C for 30 s, 72 °C for 2 min); stage 3 (72 °C for 10 min). PCR products were separated on a 1% agarose gel, purified by a DNA Gel Extraction Kit (Dongsheng Co. Ltd., Guangzhou, China), linked to the PMD-18T vector (TaKaRa Bio Inc.), sequenced by Beijing Genome Institute (Shenzhen, Guangdong, China), then submitted to NCBI. The protein sequences of the three genes were submitted to ExPASy (https://web.expasy.org/protparam/) to calculate MWs and theoretical pIs. The primers designed for cloning the three *DoRWA* genes are listed in Table S1.

4.4. Bioinformatics Analysis

The RWA protein sequences from *D. officinale*, *A. thaliana* and *Populus trichocarpa* were initially aligned with DNAMAN 7.0 software (Lynnon Biosoft Crop., San Ramon, CA., USA). To complete a phylogenetic analysis, all RWA proteins from these three plants were further aligned by MUSCLE (https://www.ebi.ac.uk/Tools/msa/muscle/), then were used to construct a Neighbor-Joining (N-J) tree built in MEGA 7.0 software [44] with the following parameters: 1000 bootstrap replications; pair deletion. Protein domain structures were drawn in DOG software [45]. Gene structure, including exons, introns, 5′-UTRs and 3′-UTRs of *DoRWA* genes, were obtained from the *D. officinale* gff database [20], then submitted to GSDS version 2.0 [46] to draw the exon-intron structure. The promoter region (from 0 to −1500 bp) of *DoRWA* genes were obtained from the *D. officinale* scaffold database [20], submitted to PlantCare [47] to discover all *cis*-elements, then the type and number of *cis*-elements were assessed and submitted to TBtools software [48]. STRING Version 11.0 was used to analyze the protein–protein association networks [49].

4.5. Protoplast Isolation and Subcellular Localization of DoRWA Proteins

Protoplasts were isolated following the protocol described by Schapire et al. [50]. At first, the enzyme solution was made as follows: 1.5% (*w/v*) cellulase R-10 (Yakult Pharmaceutical Industry Co. Ltd., Tokyo, Japan), 0.3% (*w/v*) macerozyme R-10 (Yakult Pharmaceutical Industry), 20 mM KCl (Mackline, Shanghai, China), 20 mM MES (pH 5.7; Sigma-Aldrich, St. Louis, MO, USA), and 0.4 M mannitol (Mackline). The enzyme solution was warmed to 55 °C for 10 min, then 10 mM $CaCl_2$ (Sigma-Aldrich) and 0.1% (*w/v*) bovine serum albumin (Sigma-Aldrich) were added after the solution had cooled down. Next, the lower epidermal surface cell layer was peeled off young leaves of 4- to 5-week-old *A. thaliana* plants by autoclave tape, then leaves were treated with the enzyme solution at 25 °C for 2 h (50 rpm). The resulting harvested protoplasts were washed twice by W5 solution containing 154.5 mM NaCl (Mackline), 125 mM $CaCl_2$, 5 mM KCl, 2 mM MES (pH 5.7) and 5 mM glucose (Aladdin, Shanghai, China). Finally, protoplasts were gently resuspended in MMG solution that contained 0.4 mM mannitol, 15 mM $MgCl_2$ and 4 mM MES (pH 5.7).

To predict transmembrane helices, the sequences of three DoRWA proteins (DoRWA1, DoRWA2, DoRWA3) sequences were submitted to the TMHMM Server v. 2.0 (http://www.cbs.dtu.dk/services/TMHMM/). The full-length coding sequences of the three *DoRWA* genes (stop codon was removed) were inserted into the pSAT6-EYFP-N1 vector [51] at the *Nco*I site using the In-fusion® HD Cloning Kit (TaKaRa Bio Inc.). Since AtRWA2 was localized in the GA [9] or ER [10], to assess in which organelle DoRWA proteins were localized, recombinant protein combined with GA or ER localization marker [24] were transformed into leaf mesophyll protoplasts of 4- to 5-week-old *A. thaliana* plants using PEG-mediated transformation [52]. After maintaining protoplasts at 22 °C for 16 h in the dark, fluorescence signals were visualized with a Leica TCS SP8 STED 3x microscope (Leica Camera AG, Solms, Germany). The primers designed for pSAT6-EYFP-N1-DoRWAs construction are listed in Table S1.

4.6. Content of Water-Soluble Polysaccharides and O-Acetyl Groups in Different Organs

The content of WSPs in different organs (roots, stems, leaves) was measured according to He et al. [25]. The *O*-acetyl groups of WSPs were detected by a modified colorimetric method, as described by Gudlavalleti et al. [53]. Briefly, 0.1 g of each organ was weighed accurately, added to 25 mL of purified water, then warmed at 80 °C for 2 h. After centrifuging at 8000 rpm for 10 min, 1 mL of supernatant was added to 2 mL of freshly formulated alkaline hydroxylamine (mixture of 2 mol/L of hydroxylamine hydrochloride and 3.5 mol/L sodium hydroxide (*v/v*, 1:1)), then vortexed immediately. After 4 min, 1 mL of 4 mol/L hydrochloric acid and 1 mL of 0.37 mol/L ferrous chloride–hydrochlolic acid were added. Absorbance of the mixture was measured with an ultraviolet spectrophotometer (UV-1800PC; AOE Instruments Co. Ltd., Shanghai, China) at 540 nm. The acetylcholine chloride was used as the standard. In the control, hydrochloric acid was added before the formulated alkaline hydroxylamine.

4.7. RNA-Seq Expression Analysis at Four Developmental Stages and under Cold Stress

To develop the expression profiles of *DoRWA* genes and key genes related to the synthesis of WSPs at four developmental stages (S1–S4: WSP content was upregulated from S1–S3, peaked at S3, and downregulated at S4) [25], four raw reads (SRR1917040, SRR1917041, SRR1917042, SRR1917043) [25] corresponding to the S1, S2, S3 and S4 stage, were mapped to the transcriptome sequence database of *D. officinale* [54]. From the mapped database [24], the reads per kilobase per million reads (RPKM) [55] values of three *DoRWA* genes and WSP synthetic genes at S1–S4 stage were first downloaded then were log-transformed to render data suitable for heatmap analysis. To obtain the expression patterns of *DoRWA* genes under cold stress, six raw reads (SRR3210613, SRR3210621, SRR3210626, SRR3210630,

SRR3210635, SRR3210636) [56] were downloaded from NCBI, then mapped to the *D. officinale* genome sequence database [20]. The FPKM value was used for gene expression analysis.

4.8. Generation of 35S::DoRWA3 Transgenic Lines

The full-length coding sequence of *DoRWA3* (stop codon was removed) was linked with the pCAMBIA1302 vector at the *Nco*I site using the In-fusion® HD Cloning Kit (TaKaRa Bio Inc.). Recombinant plasmid was transformed into *Agrobacterium tumefaciens* EHA105. Inflorescences of 5- to 6-week-old WT plants were transfected by *A. tumefaciens* with the floral dip method [57]. Three homologous OE lines (OE1, OE2, OE3) were screened on half-strength MS (1/2 MS) medium containing 25 μg/mL hygromycin B (Roche Holding AG, Basel, Switzerland). The primers designed for the pCAMBIA1302-DoRWA3 construction are listed in Table S1.

4.9. Semi-Quantitative RT-PCR

Total RNAs from the leaves of 1-month-old WT and three OE transgenic lines were extracted and purified as described above. The semi-quantitative PCR reaction was catalyzed by KOD FX polymerase (Toyobo Co. Ltd.) with the following protocol: stage 1 (94 °C for 3 min); stage 2 (40 cycles of 98 °C for 10 s, 56 °C for 30 s, 72 °C for 2 min); stage 3 (72 °C for 10 min). *UBQ10* (At4g05320) from *A. thaliana* was used as the control. PCR products were visualized on a 1% agrose gel under ultraviolet light. The primers designed for semi-quantitative PCR are listed in Table S1.

4.10. Cell Wall Preparation and Determination of Acetyl Esters

The rosette leaves and inflorescence stems of 6-week-old plants, and seeds of WT and three *35S::DoRWA3* transgenic lines, were collected then dried at 80 °C for 12 h (seeds were naturally air-dried for 2 weeks). Samples were ground into power for cell wall extraction.

The alcohol-insoluble residues (AIR) of samples were extracted according to Harholt et al. [58]. Briefly, 30 mg of sample was weighed accurately, washed with 1 mL of 96% ethanol, then kept at 70 °C for 30 min to deactivate enzymes. After centrifuging the mixture at 10,000 g for 5 min, the supernatant was removed, 1 mL of 70% ethanol was added, and the mixture was vortexed. The pellet was washed with 70% ethanol, then centrifuged at 10,000 g for 5 min, and this was repeated until the solution became colorless. Finally, the precipitate was washed in 1 mL of 100% acetone, vortexed immediately, placed at room temperature (RT) for 10 min, then centrifuged at 10,000 g for 5 min. The supernatant was discarded and the pellet was oven-dried at 50 °C until constant weight.

AIR (4 mg) of different samples were accurately weighed, saponified by 400 μL of 1 mol/L NaOH, then centrifuged at 150 rpm overnight (at 28 °C). The solution was neutralized with 400 μL of 1 mol/L HCl, then centrifuged at 12,000 rpm for 10 min. The released acetic acid content in the supernatant was determined by using the Acetic Acid Assay Kit (Megazyme, Wicklow, Ireland) as described by Gill et al. [59]. Briefly, 40 μL of supernatant of different samples was added into a UV- capable 96-well plate, then diluted with 64 μL of ddH$_2$O. An amount of 42 μL of mixture (Soulution 1 and Solution 2; *v/v*, 2.5:1) was transferred into each sample, mixed and incubated at RT for 3 min. Absorbance of the mixture was read at 340 nm (A_0). An amount of 12 μL of 10-fold diluted solution 3 was added to the wells, mixed and incubated at RT for 4 min, then read at 340 nm (A_1). Finally, 12 μL of 10-fold diluted Solution 4 was added to the plate, mixed thoroughly and incubated at RT for 12 min, then read at 340 nm (A_2). Solution 5 (acetic acid solution) served as the standard. In the control, 40 μL of supernatant was replaced by 40 μL of ddH$_2$O. A_0, A_1 and A_2 values were used to calculate ΔA according to the manufacturer's recommendations.

4.11. ABA Treatment and Phenotype Assay

To study the effect of ABA on germination rates, the seeds (48 seeds per sample) of WT and three OE transgenic lines (OE1, OE2 and OE3) were sown on the 1/2 MS medium containing 0 or 1 μM ABA for 4 d. For the root length assay, seedlings growing on 1/2 MS medium for 3 d were transplanted to 1/2 MS medium supplemented with 1 or 2 μM ABA. After vertical culture for 7 d, root length was assessed using ImageJ software (http://rsbweb.nih.gov/ij/).

4.12. Statistical Analysis

All data were plotted in Excel 2013 (Microsoft Inc., Redmond, WA, USA) and Sigmaplot 12.0 (Systat Software Inc., San Jose, CA, USA). Data were analyzed by one-way analysis of variance (ANOVA) and means separated by Duncan's multiple range test (DMRT) ($p < 0.05$, $p < 0.01$) in SPSS version 22.0 software (IBM Corp., Armonk, NY, USA).

5. Conclusions

Three *DoRWA* genes, named *DoRWA1*, *DoRWA2* and *DoRWA3*, were cloned from the medicinal orchid, *D. officinale*. Phylogenetic analysis revealed that DoRWA3 was clustered with the identified acetyltransferase genes (i.e., AtRWA2, PtRWA-C, PtRWA-D) into one branch. Interestingly, the *cis*-elements of the three *DoRWA* genes had the ABA-responsive element and their expression patterns were sensitive to ABA treatment. The results of subcellular localization showed that the three DoRWA proteins were localized in the ER, and not in the GA. The *O*-acetyl groups shared a similar trend as WSPs in different organs. qRT-PCR and RNA-seq results showed that *DoRWA3* was mainly expressed in the organs where the *O*-acetyl groups accumulated, displaying significantly higher expression than *DoRWA1* and *DoRWA2* in different organs, except for roots. *DoRWA3* was co-expressed with key genes related to the synthesis of WSPs, so it is regarded as a candidate gene that codes for an acetyltransferase. The acetylation level of polysaccharides in seeds, leaves and stems of the three *A. thaliana* OE transgenic lines was significantly higher than in WT, indicating that *DoRWA3* has a similar function as *AtRWA2*.

Supplementary Materials: Supplementary materials can be found at http://www.mdpi.com/1422-0067/21/17/6250/s1. Figure S1: Alignment of amino acid sequences of RWA proteins in *Dendrobium officinale*, *Arabidopsis thaliana* and *Populus trichocarpa*; Figure S2: Gene structure of three *DoRWA* genes. Figure S3: Predictions of subcellular localization of DoRWA1 (A), DoRWA2 (B), DoRWA3 (C) and subcellular localization of DoRWA3-YFP with Golgi apparatus localization marker G-rk (D); Figure S4: Comparison of germination rate and root length between wild type (WT) and overexpression (OE) transgenic plants; Table S1: Primers designed for PCR.

Author Contributions: Conceptualization, J.D., C.S. and J.A.T.d.S.; methodology, C.S.; software, C.S., C.H., Z.Y.; validation, C.Z., H.W. and M.Z.; formal analysis, C.S.; investigation, C.S.; resources, J.D.; data curation, J.D.; writing—original draft preparation, C.S. and J.A.T.d.S.; writing—review and editing, C.S. and J.A.T.d.S.; visualization, C.S. and J.A.T.d.S.; supervision, J.D.; project administration, J.D.; funding acquisition, J.D. All authors have read and agreed to the published version of the manuscript.

Funding: This research was funded by the project "Cultivation of new varieties of *Dendrobium officinale* in Guangdong Province", grant number Y334041001.

Acknowledgments: The authors are grateful to Rufang Deng for assistance with subcellular localization analysis. We also thank the editor and reviewers for providing suggestions that improved the quality of this manuscript.

Conflicts of Interest: The authors declare no conflict of interest.

Abbreviations

ABA	Abscisic acid
AIR	Alcohol-insoluble residue
Axy9	Altered xyloglucan9
CSL	Cellulose synthesis-like
CK	Cytokinin
DA	Degree of acetylation
DMRT	Duncan's multiple range test
ER	Endoplasmic reticulum
FPKM	Fragments per kilobase per million

GMP	GDP-mannose pyrophosphorylation
GMT	GDP-mannose transporter
GA	Golgi apparatus
JA	Jasmonic acid
KGM	Konjac glucomannan
MW	Molecular weight
MS	Murashige and Skoog
N-J	Neighbor-Joining
ORF	Open reading frame
OE	Overexpression
PMM	Phosphomannomutase
pI	Isoelectric Points
RPKM	Reads per kilobase per million
RT	Room temperature
TCM	Traditional Chinese medicine
qRT-PCR	Quantitative real time polymerase chain reaction
RWA	Reduced wall acetylation
TBL	Trichome birefringence-like
UGE	UDP glucose 4-epimerase
WSP	Water-soluble polysaccharide

References

1. Zhang, Z.S.; Wang, X.W.; Yu, S.C.; Yin, L.; Zhao, M.X.; Han, Z.P. Synthesized oversulfated and acetylated derivatives of polysaccharide extracted from *Enteromorpha linza* and their potential antioxidant activity. *Int. J. Biol. Macromol.* **2011**, *49*, 1012–1015. [CrossRef] [PubMed]
2. Song, Y.; Yang, Y.; Zhang, Y.Y.; Duan, L.S.; Zhou, C.L.; Ni, Y.Y.; Liao, X.J.; Li, Q.H.; Hu, X.S. Effect of acetylation on antioxidant and cytoprotective activity of polysaccharides isolated from pumpkin (*Cucurbita pepo*, lady godiva). *Carbohyd. Polym.* **2013**, *98*, 686–691. [CrossRef] [PubMed]
3. Chen, Y.; Zhang, H.; Wang, Y.X.; Nie, S.P.; Li, C.; Xie, M.Y. Acetylation and carboxymethylation of the polysaccharide from *Ganoderma atrum* and their antioxidant and immunomodulating activities. *Food Chem.* **2014**, *156*, 279–288. [CrossRef] [PubMed]
4. Chen, F.; Huang, G.L. Preparation and immunological activity of polysaccharides and their derivatives. *Int. J. Biol. Macromol.* **2018**, *12*, 211–216. [CrossRef] [PubMed]
5. Chokboribal, J.; Tachaboonyakiat, W.; Sangvanich, P.; Ruangpornvisuti, V.; Jettanacheawchankit, S.; Thunyakitposal, P. Deacetylation affects the physical properties and bioactivity of acemannan, an extracted polysaccharide from *Aloe vera*. *Carbohyd. Polym.* **2015**, *133*, 556–566. [CrossRef] [PubMed]
6. Du, X.Z.; Li, J.; Chen, J.; Li, B. Effect of degree of deacetylation on physicochemical and gelation properties of konjac glucomannan. *Food Res. Int.* **2012**, *46*, 270–278. [CrossRef]
7. Salah, F.; Ghoul, Y.E.; Mahdhi, A.; Majdoub, H.; Jarroux, N.; Sakli, F. Effect of the deacetylation degree on the antibacterial and antibiofilm activity of acemannan from *Aloe vera*. *Ind. Crop. Prod.* **2017**, *103*, 13–18. [CrossRef]
8. Huang, S.Y.; Chen, F.; Cheng, H.; Huang, G.L. Modification and application of polysaccharide from traditional Chinese medicine such as *Dendrobium officinale*. *Int. J. Biol. Macromol.* **2020**, *157*, 385–393. [CrossRef]
9. Lee, C.H.; Teng, Q.H.; Zhong, R.Q.; Ye, Z.H. The four Arabidopsis *REDUCED WALL ACETYLATION* genes are expressed in secondary wall-containing cells and required for the acetylation of xylan. *Plant Cell Physiol.* **2011**, *52*, 1289–1301. [CrossRef]
10. Manabe, Y.; Nafisi, M.; Verhertbruggen, Y.; Orfila, C.; Gille, S.; Rautengarten, C.; Cherk, C.; Marcus, S.E.; Somerville, S.; Pauly, M.; et al. Loss-of-function mutation of *REDUCED WALL ACETYLATION2* in Arabidopsis leads to reduced cell wall acetylation and increased resistance to *Botrytis cinerea*. *Plant Physiol.* **2011**, *155*, 1068–1078. [CrossRef]
11. Schultink, A.; Naylor, D.; Dama, M.; Pauly, M. The role of the plant-specific ALTERED XYLOGLUCAN9 protein in Arabidopsis cell wall polysaccharide O-acetylation. *Plant Physiol.* **2015**, *167*, 1271–1283. [CrossRef] [PubMed]

12. Gill, S.; Souza, A.D.; Xiong, G.Y.; Benz, M.; Cheng, K.; Schultink, A.; Reca, I.B.; Pauly, M. O-acetylation of *Arabidopsis* hemicellulose xyloglucan requires AXY4 or AXY4L, proteins with a TBL and DUF231 Domain. *Plant Cell* **2011**, *23*, 4041–4053. [CrossRef] [PubMed]
13. Bischoff, V.; Nita, S.; Neumetzler, L.; Schindelasch, D.; Urbain, A.; Eshed, R.; Persson, S.; Delmer, D.; Scheible, W.R. *TRICHOME BIREFRINGENCE* and its homolog AT5G01360 encode plant-specific DUF231 proteins required for cellulose biosynthesis in *Arabidopsis*. *Plant Physiol.* **2010**, *153*, 590–602. [CrossRef] [PubMed]
14. Manabe, Y.; Verhertbruggen, Y.; Gill, S.; Harholt, J.; Chong, S.L.; Pawar, P.M.A.; Mellerowicz, E.J.; Tenkanen, M.; Cheng, K.; Pauly, M.; et al. Reduced wall acetylation proteins play vital and distinct roles in cell wall O-acetylation in *Arabidopsis*. *Plant Physiol.* **2013**, *163*, 1107–1117. [CrossRef] [PubMed]
15. Pawar, P.M.A.; Ratke, C.; Balasubramanian, V.K.; Chong, S.L.; Gandla, M.L.; Adriasola, M.; Sparrman, T.; Hedenström, M.; Szwaj, K.; Derba-Maceluch, M.; et al. Downregulation of RWA genes in hybrid aspen affects xylan acetylation and wood saccharification. *New Phytol.* **2017**, *214*, 1491–1505. [CrossRef]
16. Teixeira da Silva, J.A.; Ng, T.B. The medicinal and pharmaceutical importance of *Dendrobium* species. *Appl. Microbiol. Biotechnol.* **2017**, *101*, 2227–2239. [CrossRef]
17. Xing, X.H.; Cui, S.W.; Nie, S.P.; Phillips, G.O. Study on *Dendrobium officinale* O-acetyl-glucomannan (Dendronan®): Part II. Fine structure of O-acetylated residues. *Carbohyd. Polym.* **2015**, *117*, 422–433. [CrossRef]
18. Huang, X.J.; Nie, S.P.; Cai, H.L.; Zhang, G.Y.; Cui, S.W.; Xie, M.Y.; Phillips, G.O. Study on *Dendrobium officinale* O-acetyl-glucomannan (Dendronan®): Part IV. Immunomogulatory activity *in vivo*. *J. Funct. Foods* **2015**, *15*, 525–532. [CrossRef]
19. Wei, W.; Feng, L.; Bao, W.R.; Ma, D.L.; Leung, C.H.; Nie, S.P.; Han, Q.B. (2016) Structure characterization and immunomodulating effects of polysaccharides isolated from *Dendrobium officinale*. *J. Agric. Food Chem.* **2016**, *64*, 881–889. [CrossRef]
20. Zhang, G.Q.; Xu, Q.; Bian, C.; Tsai, W.C.; Yeh, C.M.; Liu, K.W.; Youshida, K.; Zhang, L.S.; Chang, S.B.; Chen, F.; et al. The *Dendrobium catenatum* Lindl. genome sequence provides insights into polysaccharide synthase, floral development and adaptive evolution. *Sci. Rep.* **2016**, *6*, 19029. [CrossRef]
21. Sinclair, S.A.; Larue, C.; Bonk, L.; Khan, A.; Michel, H.C.; Stein, R.J.; Grolimund, D.; Begerow, D.; Neumann, U.; Haydon, M.J.; et al. Etiolated seedlings development requires repression of photomorphogenesis by a small cell-wall-derived dark signal. *Curr. Biol.* **2017**, *27*, 1–16. [CrossRef] [PubMed]
22. Yuan, Y.X.; Teng, Q.; Zhong, R.Q.; Ye, Z.H. TBL3 and TBL31, two Arabidopsis DUF231 domain proteins, are required for 3-O-monoacetylation of xylan. *Plant Cell Physiol.* **2016**, *57*, 35–45. [CrossRef] [PubMed]
23. Zhu, X.F.; Sun, Y.; Zhang, B.C.; Mansoori, N.; Wan, J.X.; Liu, Y.; Wang, Z.W.; Shi, Y.Z.; Zhou, Y.H.; Zheng, S.J. *TRICHOME BIREFRINGENCE-LIKE27* affects aluminum sensitivity by modulating the O-acetylation of xyloglucan and aluminum-binding capacity in Arabidopsis. *Plant Physiol.* **2014**, *166*, 181–189. [CrossRef] [PubMed]
24. Nelson, B.K.; Cai, X.; Nebenführ, A. A multicolored set of in vivo organelle markers for co-localization studies in Arabidopsis and other plants. *Plant J.* **2007**, *51*, 1126–1136. [CrossRef]
25. He, C.M.; Zhang, J.X.; Liu, X.C.; Zeng, S.J.; Wu, K.L.; Yu, Z.M.; Wang, X.J.; Teixeira da Silva, J.A.; Lin, Z.J.; Duan, J. Identification of genes involved in biosynthesis of mannan polysaccharides in *Dendrobium officinale* by RNA-seq analysis. *Plant Mol. Biol.* **2015**, *88*, 219–231. [CrossRef]
26. He, C.M.; Yu, Z.M.; Teixeira da Silva, J.A.; Zhang, J.X.; Liu, X.C.; Wang, X.J.; Zhang, X.H.; Zeng, S.J.; Wu, K.L.; Tan, J.W.; et al. DoGMP1 from *Dendrobium officinale* contributes to mannose content of water-soluble polysaccharides and plays a role in salt stress response. *Sci. Rep.* **2017**, *7*, 41040. [CrossRef]
27. Yu, Z.M.; He, C.M.; Teixeira da Silva, J.A.; Zhang, G.H.; Dong, W.; Luo, J.P.; Duan, J. Molecular cloning and functional analysis of DoUGE related to water-soluble polysaccharides from *Dendrobium officinale* with enhanced abiotic stress tolerance. *Plant Cell Tissue Organ Cult.* **2017**, *131*, 579–599. [CrossRef]
28. Yu, Z.M.; He, C.M.; Teixeira da Silva, J.A.; Luo, J.P.; Yang, Z.Y.; Duan, J. The GDP-mannose transporter gene (*DoGMT*) from *Dendrobium officinale* is critical for mannan biosynthesis in plant growth and development. *Plant Sci.* **2018**, *277*, 43–54. [CrossRef]
29. Liu, X.; Xie, J.H.; Jia, S.; Huang, L.X.; Wang, Z.J.; Li, C.; Xie, M.Y. Immunomodulatory effects of an acetylated *Cyclocarya paliurus* polysaccharide on murine macrophages RAW264.7. *Int. J. Biol. Macromol.* **2017**, *98*, 576–581. [CrossRef]

30. Hsieh, Y.Y.; Chien, C.; Liao, S.K.S.; Liao, S.F.; Huang, W.T.; Yang, W.B.; Lin, C.C.; Cheng, T.J.R.; Chang, C.C.; Fang, J.M.; et al. Structure and bioactivity of polysaccharides in medicinal plant *Dendrobium huoshanense*. *Bioorgan. Med. Chem.* **2008**, *16*, 6054–6068. [CrossRef]
31. Tong, W.; Yu, Q.; Li, H.; Cui, W.W.; Nie, S.P. Chemical modification and immunoregulatory activity of polysaccharides from *Dendrobium officinale*. *Food Sci.* **2017**, *38*, 155–160.
32. Jiang, L.M.; Nie, S.P.; Huang, D.F.; Fu, Z.H.; Xie, M.Y. Acetylation modification improves immunoregulatory effect of polysaccharide from seeds of *Plantago asiatica* L. *J. Chem.* **2018**, *2018*, 3082026. [CrossRef]
33. Li, J.; Ye, T.; Wu, X.F.; Chen, J.; Wang, S.S.; Lin, L.F.; Li, B. Preparation and characterization of heterogeneous deacetylated konjac glucomannan. *Food Hydrocoll.* **2014**, *40*, 9–15. [CrossRef]
34. Javed, S.; Rahman, A. *Aloe vera* gel in food, health products and cosmetics industry. In *Studies in Natural Products and Chemistry*; Rahman, A.U., Ed.; Elsevier Science & Technology: London, UK, 2014; Volume 41, pp. 261–285.
35. Nishinari, K.; Williams, P.A.; Phillips, G.O. Review of the physicochemical characteristics and properties of konjac mannan. *Food Hydrocoll.* **1992**, *6*, 199–222. [CrossRef]
36. Li, M.Y.; Feng, G.P.; Wang, H.; Yang, R.L.; Xu, Z.L.; Sun, Y.M. Deacetylated konjac glucomannan is less effective in reducing dietary-induced hyperlipidemia and hepatic steatosis in C57BL/6 mice. *J. Agric. Food Chem.* **2017**, *65*, 1556–1565. [CrossRef] [PubMed]
37. Humman, J.H. Composition and applications of *Aloe vera* leaf gel. *Molecules* **2008**, *13*, 1599–1616. [CrossRef]
38. Nafisi, M.; Stranne, M.; Fimognari, L.; Atwell, S.; Martens, H.J.; Pedas, P.R.; Hansen, S.F.; Nawrath, C.; Scheller, H.V.; Kliebenstein, D.J.; et al. Acetylation of cell wall is required for structural integrity of the leaf surface and exerts a global impact on plant stress responses. *Front. Plant Sci.* **2015**, *6*, 550. [CrossRef]
39. Murashige, T.; Skoog, F. A revised medium for rapid growth and bioassays with tobacco tissue cultures. *Physiol. Plant* **1962**, *15*, 473–497. [CrossRef]
40. Chattopadhyay, N.; Kher, R.; Godbole, M. Inexpensive SDS/phenol method for RNA extraction from tissues. *Biotechniques* **1993**, *15*, 24–26.
41. Livak, K.J.; Schmittgen, T.D. Analysis of relative gene expression data using real-time quantitative PCR and $2^{-\Delta\Delta CT}$ method. *Methods* **2001**, *25*, 402–408. [CrossRef]
42. Zhang, G.; Zhao, M.M.; Zhang, D.W.; Guo, S.X. Reference gene selection for real-time quantitative PCR analysis of *Dendrobium officinale*. *Chin. Pharm. J.* **2013**, *48*, 1664–1668.
43. Altschul, S.F.; Madden, T.L.; Schäffer, A.A.; Zhang, J.H.; Zhang, Z.; Miller, W.; Lipman, D.J. Gapped BLAST and PSI-BLAST: A new generation of protein database search programs. *Nucl. Acids Res.* **1997**, *25*, 3389–3402. [CrossRef] [PubMed]
44. Kumar, S.; Stecher, G.; Tamura, K. MEGA7: Molecular evolutionary genetics analysis version 7.0 for bigger datasets. *Mol. Biol. Evol.* **2016**, *33*, 1870–1874. [CrossRef] [PubMed]
45. Ren, J.; Wen, L.P.; Gao, X.J.; Jin, C.J.; Xue, Y.; Yao, X.B. DOG1.0: Illustrator of protein domain structures. *Cell Res.* **2009**, *19*, 271–273. [CrossRef]
46. Hu, B.; Jin, J.P.; Guo, A.Y.; Zhang, H.; Luo, J.C.; Gao, G. GSDS2.0: An upgraded gene feature visualization server. *Bioinformatics* **2015**, *31*, 1296–1297. [CrossRef]
47. Lescot, M.; Déhais, P.; Thijs, G.; Marchal, K.; Moreau, Y.; Peer, Y.V.D.; Rouzé, P.; Rombauts, S. PlantCARE, a database of plant cis-acting regulatory elements and a portal to tools for in silico analysis of promoter sequences. *Nucl. Acids Res.* **2002**, *30*, 325–327. [CrossRef]
48. Chen, C.J.; Chen, H.; Zhang, Y.; Thomas, H.R.; Frank, M.H.; He, Y.H.; Xia, R. TBtools—An integrative toolkit developed for interactive big biological data. *Mol. Plant* **2020**, *13*, 1194–1202. [CrossRef]
49. Szklarczyk, D.; Gable, A.L.; Lyon, D.; Junge, A.; Wyder, S.; Cepas, J.H.; Simonovic, M.; Doncheva, N.T.; Morris, J.H.; Bork, P.; et al. STRING V11: Protein-protein association networks with increased coverage, supporting functional discovery in genome-wide experimental datasets. *Nucl. Acids Res.* **2019**, *47*, 607–613. [CrossRef]
50. Schapire, A.L.; Lois, L.M. A simplified and rapid method for the isolation and transfection of Arabidopsis leaf mesophyll protoplasts for large-scale applications. In *Methods in Molecular Biology*; Walker, J.M., Clifton, N.J., Eds.; Humana Press: New York, NY, USA, 2016; pp. 79–88.
51. Citovsky, V.; Lee, L.Y.; Vyas, S.; Glick, E.; Chen, M.E.; Vainstein, A.; Gafni, Y.; Gelvin, S.B.; Tzfira, T. Subcellular localization of interacting proteins by biomolecular fluorescence complementation in planta. *J. Mol. Biol.* **2006**, *362*, 1120–1131. [CrossRef]

52. Yoo, S.D.; Cho, Y.H.; Sheen, J. Arabidopsis mesophyll protoplasts: A versatile cell system for transient gene expression analysis. *Nat. Protoc.* **2007**, *2*, 1565–1572. [CrossRef]
53. Gudlavalleti, S.K.; Datta, A.K.; Tzeng, Y.L.; Noble, C.; Carlson, R.W.; Stephens, D.S. The *Neisseria meningitides* serogroup a capsular polysaccharide O-3 and O-4 acetyltransferase. *J. Biol. Chem.* **2004**, *279*, 42765–42773. [CrossRef] [PubMed]
54. Zhang, J.X.; He, C.M.; Wu, K.L.; Teixeira da Silva, J.A.; Zeng, S.J.; Zhang, X.H.; Yu, Z.M.; Xia, H.Q.; Duan, J. Transcriptome analysis of *Dendrobium officinale* and its application to the identification of genes associated with polysaccharide synthesis. *Front. Plant. Sci.* **2016**, *7*, 5. [CrossRef] [PubMed]
55. Mortazavi, A.; Williams, B.A.; McCue, K.; Schaeffer, L.; Wold, B. Mapping and quantifying mammalian transcriptomes by RNA-seq. *Nat. Methods* **2008**, *5*, 621–628. [CrossRef] [PubMed]
56. Wu, Z.G.; Jiang, W.; Chen, S.L.; Mantri, N.; Tao, Z.M.; Jiang, C.X. Insights from the cold transcriptome and metabolome of *Dendrobium officinale*: Global reprogramming of metabolic and gene regulation networks during cold accumulation. *Front. Plant Sci.* **2016**, *7*, 1653. [CrossRef] [PubMed]
57. Clough, S.J.; Bent, A.F. Floral dip: A simplified method for *Agrobacterium*-mediated transformation for *Arabidopsis thaliana*. *Plant. J.* **1998**, *16*, 735–743. [CrossRef] [PubMed]
58. Harholt, J.; Jensen, J.K.; Sørensen, S.O.; Orfila, C.; Pauly, M.; Scheller, H.V. ARABINAN DEFICIENT 1 is a putative arabinosyltransferase involved in biosynthesis of pectic arabinan in Arabidopsis. *Plant Physiol.* **2006**, *140*, 49–58. [CrossRef] [PubMed]
59. Gill, S.; Cheng, K.; Skinner, M.E.; Liepman, A.H.; Wilkerson, C.G.; Pauly, M. Deep sequencing of voodoo lily (*Amorphophallus konjac*): An approach to identify relevant genes involved in the synthesis of the hemicellulose glucomannan. *Planta* **2011**, *234*, 515–526. [CrossRef]

© 2020 by the authors. Licensee MDPI, Basel, Switzerland. This article is an open access article distributed under the terms and conditions of the Creative Commons Attribution (CC BY) license (http://creativecommons.org/licenses/by/4.0/).

Article

Symbiotic and Asymbiotic Germination of *Dendrobium officinale* (Orchidaceae) Respond Differently to Exogenous Gibberellins

Juan Chen *, Bo Yan, Yanjing Tang, Yongmei Xing, Yang Li, Dongyu Zhou and Shunxing Guo *

Institute of Medicinal Plant Development, Chinese Academy of Medical Sciences & Peking Union Medical College, Beijing 100193, China; yanbo1823@gmail.com (B.Y.); yanjingtang1@gmail.com (Y.T.); ymxing@implad.ac.cn (Y.X.); 15114605567ly@gmail.com (Y.L.); zhoudy651@gmail.com (D.Z.)
* Correspondence: chenjuan@implad.ac.cn (J.C.); sxguo@implad.ac.cn (S.G.); Tel./Fax: +86-10-57833231 (S.G.)

Received: 7 August 2020; Accepted: 19 August 2020; Published: 25 August 2020

Abstract: Seeds of almost all orchids depend on mycorrhizal fungi to induce their germination in the wild. The regulation of this symbiotic germination of orchid seeds involves complex crosstalk interactions between mycorrhizal establishment and the germination process. The aim of this study was to investigate the effect of gibberellins (GAs) on the symbiotic germination of *Dendrobium officinale* seeds and its functioning in the mutualistic interaction between orchid species and their mycobionts. To do this, we used liquid chromatograph-mass spectrometer to quantify endogenous hormones across different development stages between symbiotic and asymbiotic germination of *D. officinale*, as well as real-time quantitative PCR to investigate gene expression levels during seed germination under the different treatment concentrations of exogenous gibberellic acids (GA_3). Our results showed that the level of endogenous GA_3 was not significantly different between the asymbiotic and symbiotic germination groups, but the ratio of GA_3 and abscisic acids (ABA) was significantly higher during symbiotic germination than asymbiotic germination. Exogenous GA_3 treatment showed that a high concentration of GA_3 could inhibit fungal colonization in the embryo cell and decrease the seed germination rate, but did not significantly affect asymbiotic germination or the growth of the free-living fungal mycelium. The expression of genes involved in the common symbiotic pathway (e.g., calcium-binding protein and calcium-dependent protein kinase) responded to the changed concentrations of exogenous GA_3. Taken together, our results demonstrate that GA_3 is probably a key signal molecule for crosstalk between the seed germination pathway and mycorrhiza symbiosis during the orchid seed symbiotic germination.

Keywords: orchid mycorrhiza; plant hormone; symbiosis germination; gene expression

1. Introduction

Orchidaceae is among the largest families of flowering plants, one that is fascinating and rich in species diversity, with diverse pollination mechanisms and a unique mycorrhizal symbiotic relationship [1]. Orchid mycorrhizae differ from other major types of mycorrhizae in that, besides mineral nutrients (e.g., P and N), this type of fungus supplies carbohydrates to the plant, especially in the early stages of seed germination and seedling development [2]. Some orchid species are obligate symbiotic partners with fungi during their whole life cycle (e.g., mycotrophic orchid *Gastrodia* spp.), while some epiphytic and terrestrial orchids can alternately depend on fungi in their adult stages [3].

Orchid seeds are numerous, ranging from 1300 to 4,000,000 seeds per capsule, but they are extremely small and dust-like, with an undifferentiated embryo, limited storage reserves and lacking an endosperm [4]. Accordingly, seed germination and the subsequent development of the protocorm of almost all orchids is dependent either on mutualistic symbiosis with a compatible fungus, such as the member of Tulasnelloid, Sebacinaoid, and Ceratobasidiaceae under natural field conditions (symbiotic germination, SG) [5] or the replacement of the fungus by an exogenous nutrient substance in medium under controlled conditions (asymbiotic germination, AG) [6]. Thus, symbiotic germination is acknowledged as being a unique and important topic of orchid seed biology.

Morphological and cytological studies have shown that fungi enter the embryo of orchid seeds through the suspensor end, then form hyphae coil (pelotons) in the cortical cells of the embryo and finally the pelotons are digested by the embryo cell while the embryo undergoes dramatic development: from being swollen, turning light green (stage 1) to a ruptured seed coat (stage 2), then forming green protocorms (stage 3) and, finally, having expanded leaves on a developed young seedling (stage 4 and stage 5) [7,8]. Arrays of microtubules and actin microfilaments are reportedly involved in the infection droplet release and symbiosome development during legume–rhizobia interactions and establishment of arbuscular mycorrhiza, even probably in the peloton's lysis of orchid mycorrhizae [9]. In addition, the reserve substance of the embryo also undergoes extreme changes; for example, the lipid body and protein in the embryo cell is gradually degraded, and starch grains appear at the beginning of fungal inoculation but these are gradually depleted with the symbiotic germination progress of *Dendrobium officinale* seeds [8]. All this recent research has sketched a relatively clear outline from a cell ultrastructure perspective of the symbiotic germination process of orchid seeds. Yet the molecular mechanism establishing the mycorrhizal relationship between fungi and orchid seeds at their early germination stage remains unclear.

As is well-known, plant hormones, especially gibberellins (GAs) and abscisic acids (ABA), play crucial roles not only in seed germination but also in mycorrhizal establishment. The key steps in the signal transduction pathway for GAs' biosynthesis, metabolism, and seed germination regulation have been demonstrated clearly, and the involved genes encoding key regulatory enzymes related to GAs' biosynthesis have been identified, such as gibberellin 20 oxidase (GA20ox) and GA3-oxidase (*GA3ox*) related to GA biosynthesis, and gibberellin 2-oxidase (GA2ox) that catalyzes the degradation of GAs [10]. DELLA proteins are reportedly central players in hormone-mediated crosstalk and they can interact through the N-terminal domain with the GA receptor encoded by GID1, through which GAs promote DELLA degradation. In addition, DELLA proteins are recognized as common components of the mycorrhizal signaling pathway and mutations to them can cause rice to fail to form mycorrhizal relationships [11,12]. Thus, we hypothesized that biosynthesis and signal transduction pathway of GAs contribute to crosstalk with the common symbiotic pathway (CSP)—a putative signal transduction pathway shared by arbuscular mycorrhizas and the rhizobium-legume symbiosis—transducing glomeromycotan or rhizobial signal perception from the plasma membrane into the nucleus during the symbiotic germination of *D. officinale* seeds [13].

To address our hypothesis, we took the *D. officinale* (an epiphytic orchid) inoculated with *Tulasnella* sp. as a model system, because *D. officinale* is among the Chinese traditional medicinal plants whose genomes have been sequenced [14], and the genome of the fungus *Tulasnella calospora* (a orchid mycorrhizal fungus) is sequenced [15]. The aims of this study were twofold. (1) To identify the differentially expressed genes (DEGs) involved in the biosynthesis and signal transduction of plant hormones and profile their expression patterns based on transcriptomic data. (2) To quantify and analyze the endogenous hormones' level in the orchid at different germination stages between AG and SG and analyze the effect of exogenous GA_3 upon AG and SG of *D. officinale* seeds. This study provides a new insight for better understanding orchids' seed biology and their symbiotic mechanism and provides important data for cultivation of *D. officinale* and other medicinal orchid plants via mycorrhizal techniques.

2. Results

2.1. Determination of Endogenous Hormones' Level at Different Germination Stages between SG and AG of D. officinale

In the previous study, we experimentally demonstrated that the seed germination of *D. officinale* on the oatmeal agar (OMA) medium with fungi is faster than seed germination on 1/2 MS medium without fungi [7]. *D. officinale* seeds usually take 10 days to develop up to stage 2 in SG, compared to 16 days in AG (Figure 1). After 2 weeks of sowing seeds, more than 50% of the seeds formed the protocorm structures (stage 3) in SG while the protocorm formation took at least 3 weeks in AG. After about 20 days, seeds in SG can develop seedling stage (stages 4), compared to 30 days in AG. It took approximately 5 weeks to finish the germination process in SG and at least two months in AG.

To understand the dynamic changes of endogenous hormones' content during seed germination of *D. officinale*, using liquid chromatograph-mass spectrometer (LC-MS/MS), we quantified the five kinds of endogenous hormones—GA_3, ABA, indole-3-acetic acid (IAA), *trans*-zeatin (ZT), and jasmonic acid (JA)—on the free-living fungus and the differently developed seeds in SG and AG, respectively (stage 0, no germination; stage 2, early germination; stage 3, protocorm; stage 4, seedlings) (Figure 1). These results showed that ungerminated seeds have the highest ABA content (12.78 ng/g·FW), but ABA content decreased as seed germination progressed (Table 1). The GA_3 content rose at the early germination stage (stage 2) but declined as seeds developed. The ABA and GA_3 contents were similar between AG and SG at the same stage, but the ratio of GA_3/ABA was significantly higher in SG than AG ($p < 0.05$) (Figure 2). Interestingly, IAA was dramatically increased in the protocorm stage of SG (25.91 ng/g·FW) when compared to AG (0.48 ng/g·FW). This result likely explains the faster differentiation rate in the protocorm stage in SG (2 weeks) than AG (3 weeks) during *D. officinale* seed germination. Additionally, minute amounts of ZT (0.0075~0.014 ng/g·FW) were detected in both the free-living mycelium of fungus and ungerminated seeds. For ungerminated seeds, JA could not be detected and the free-living mycelium of *Tulasnella* sp. (S6) featured a low JA content (1.63 ng/g·FW), but JA peaked most in the early germination stage (stage 2) in AG (Table 1). Further, all five kinds of hormones were detected in free-living mycelium of mycorrhizal fungus *Tulasnella* sp., albeit their context ranged almost 10-fold (0.44~4.29 ng/g·FW) (Table 1).

2.2. Effect of Exogenous GA_3 Treatment on Phenotypic Changes in D. officinale Seeds under SG and AG Conditions

Different concentrations of GA_3 (0, 0.05, 0.1, 0.5, 1 µM) were added exogenously in medium to observe its effect on the SG and AG groups, respectively (Figure 3). These results showed that GA_3 affected the establishment of the mycorrhizal relationship between fungus and seeds in a dose-dependent manner. Namely, seed germination did not significantly change in the low GA_3 treatment concentration (0.05 µM), though germination was inhibited slightly by the middle GA_3 concentration (0.1 µM), yet germination was completely inhibited when high concentrations of exogenous GA_3 (0.5 µM, 1 µM) in SG compared to the control (SG without any GA_3 treatment) (Figure 3A–T). During the 4 weeks after sowing seeds, their effective germination rate gradually decreased from 40% to 0%, while more exogenous GA_3 was applied in SG. The resulting morphological characters examined under a light microscope showed the clear presence of pelotons in the embryo cell of SG seeds (Figure 3U–W). Seed germination was achieved to seedling differentiation stage (stage 4) under low GA_3 treatment concentration (0.05 µM) in SG (Figure 3B,G,L). No fungal mycelium colonized the seed embryo in the 0.1 µM GA_3 treatment in SG, indicating that exogenous GA_3 at a high concentration probably inhibited the signal recognition that normally occurs between the fungus and the seed, leading to failed fungal colonization (Figure 3S–Y). Neither seed germination in 1/2 MS medium (without fungus) nor the mycelium growth of the fungus on PDA were inhibited or displayed conspicuous morphological changes at any exogenous GA_3 concentration (Figure 3F–J,U1–Y1,A1–F1).

Figure 1. Morphological characters and seed developmental stages of *Dendrobium officinale*. (**Left column**): symbiotic germination (**A,C,E,G,I**); (**right column**): asymbiotic germination (**B,D,F,H,J**). (**A,B**), stage 1: embryo swollen, turned light green, no germination; (**C,D**) stage 2: continued embryo enlargement, rupture of testa (germination); (**E,F**), stage 3: appearance of protomeristem (protocorm); (**G–H**) stage 4: emergence of first leaf (seedling); (**I,J**), stage 5: elongation of the first leaf. Scale bar = 0.5 mm.

Table 1. The content of five kinds of endogenous plant hormones during the seed germination of *Dendrobium officinale* ($n = 3$). Abbreviation: FW: fresh weight; U0: ungerminated seeds; SG: symbiotic germination; AG: asymbiotic germination; 2, 3, 4 means stage 2, stage 3 and stage 4 during the germination of *D. officinale* seed, respectively; S6 means the free-living mycelium on PDA of our mycorrhizal fungus; GA_3: gibberellic acids; ABA: abscisic acids. IAA: indole-3-acetic acid; ZT: *trans*-zeatin; JA: jasmonic acid.

Sample	ABA (ng/g·FW)	GA_3 (ng/g·FW)	IAA (ng/g·FW)	ZT (ng/g·FW)	JA (ng/g·FW)
U0	12.78 ± 2.87 b	0.99 ± 0.46 abc	9.00 ± 1.2 ab	0.014 ± 0.00 ab	0 a
SG2	2.89 ± 0.40 a	1.40 ± 0.27 cd	16.52 ± 7.13 b	0.016 ± 0.00 bc	16.85 ± 1.37 abc
SG3	3.21 ± 0.50 a	1.19 ± 0.35 bc	25.91 ± 9.66 c	0.052 ± 0.00 bc	31.76 ± 18.57 c
SG4	2.15 ± 0.17 a	0.66 ± 0.08 ab	15.70 ± 7.51 b	0.012 ± 0.00 ab	21.16 ± 13.95 bc
AG2	5.06 ± 1.00 a	1.90 ± 0.37 d	1.22 ± 0.40 a	0.026 ± 0.00 d	85.95 ± 14.63 d
AG3	2.85 ± 0.75 a	0.95 ± 0.31 abc	0.48 ± 0.36 a	0.014 ± 0.00 ab	12.96 ± 7.37 abc
AG4	3.35 ± 1.30 a	0.87 ± 0.36 abc	0.19 ± 0.14 a	0.021 ± 0.01 cd	9.38 ± 0.89 ab
S6	4.29 ± 2.79 a	0.44 ± 0.05 a	4.09 ± 2.19 a	0.0075 ± 0.00 a	1.63 ± 0.07 ab

Note: Different letters a, b, c, d represent significant difference ($p < 0.05$) of content of phytohormone in different development stage of *D. officinale* seeds. The analysis was performed using the Duncan method in SPSS 17.0 software.

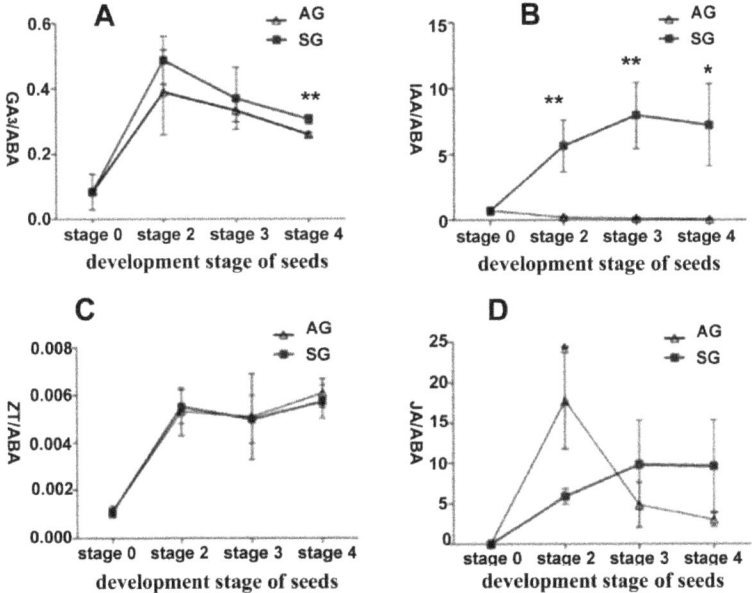

Figure 2. Endogenous hormone content change in different development stages during symbiotic and asymbiotic germination of *Dendrobium officinale* seeds. (**A**) GA_3/ABA ratio in different stage between symbiotic and asymbiotic germination; (**B**) IAA/ABA ratio; (**C**) ZT/ABA ratio; (**D**) JA/ABA ratio. Mean and SE values were calculated from at least three replicates. Asterisks indicate significant differences in same development stage between asymbiotic and symbiotic germination according to the *t*-test (* $p < 0.05$ and ** $p < 0.001$). AG, asymbiotic germination; SG, symbiotic germination.

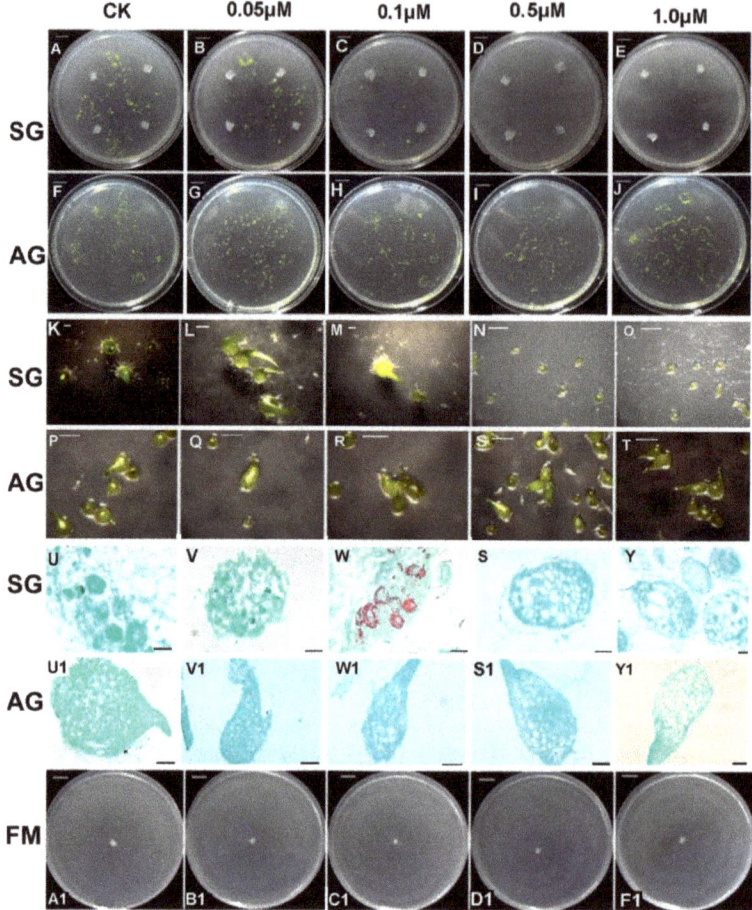

Figure 3. Effect of exogenous GA$_3$ concentration on mycorrhizal fungi colonization in symbiotic and asymbiotic germination of *Dendrobium officinale* at 4 weeks after sowing seeds. (**A–J**). seed germination when inoculated with *Tulasnella* sp., or without the fungus, at different GA$_3$ treatment concentrations. (**K–T**). morphological characters of symbiotic or asymbiotic germination at different GA$_3$ treatment concentrations under a stereomicroscope; (**U–Y1**). morphological characters of symbiotic or asymbiotic germination at different GA$_3$ treatment concentrations under a light microscope; (**A1–F1**). Colony of free-living mycelium of mycorrhizal fungus on PDA medium at different GA$_3$ treatment concentrations. Scale bars: (**A–J**) = 1 cm; (**K–T**) = 1 mm; (**J–W**), (**U1–Y1**) = 10 μm; (**S–Y**) = 5 μm; (**A1–F1**) = 1 cm; SG, symbiotic germination; AG, asymbiotic germination.

2.3. Identification of Hormone-Related Genes and Common Symbiosis Pathway-Related Genes and Their Expression Profiles in SG and AG of D. officinale Seeds

Based on previous RNA-Seq transcriptomic data of *D. officinale* seeds inoculated with *Tulasnella* sp., we screened upregulated genes involved in GA biosynthesis and signal transduction in SG, including those encoding ent–kaurene synthase (KS), ent–kaurene oxidase (KO), GA 20 oxidase (GA20ox), GA 3-beta dioxygenase (*GA3ox*), GA 2-oxidase (GA2ox), and DELLA protein (Table 2). In addition, the expression of predicted key enzyme genes involved in ABA biosynthesis (9-*cis*-epoxycarotenoid dioxygenase, NCED), metabolism (abscisic acid 8'-hydroxylase), and signal transduction (ABA responsive element binding factor) were also induced in SG of *D. officinale* seeds.

Compared with AG, the genes related to IAA biosynthesis (YUCCA family monooxygenase and SAUR family protein) had upregulated expression in the SG stage (fold-change > 2.0 and false discovery rate (FDR) < 0.001) (Table 2).

The key genes involved in CSP were upregulated in SG in our transcriptomic database, such as calmodulin-like protein (Dendrobium_GLEAN_10048053) and the calcium-dependent protein kinase (Dendrobium_GLEAN_10016982) related to Ca^{2+} signal transduction. Putative mycorrhizal-induced genes, including those encoding bidirectional sugar transporter protein, chitinase, fatty acid desaturase, and aspartic proteinase were all significantly upregulated in SG compared to AG (Table 2).

Expression levels of putative genes involved in GAs biosynthesis such as *GA2ox (DoG2ox), GA3ox (DoGA3ox), GA3ox (DoGA3ox)* and encoding UDP-glucosyl transferase (*DoSGT*), G-box-binding factor (*DoGBF*), probable inactive receptor kinase (*DoIRK*), 9-*cis*-epoxycarotenoid dioxygenase (*DoNCED*), YUCCA family monooxygenase (*DOIPM*), SAUR family protein (*DoSAUR71*), calmodulin-like protein (*DoCML19*), calcium-dependent protein kinase (*DoCDPK26*), and nodulation signaling pathway protein (*DoNSP2-1, DoNSP2-2*) were validated by real-time quantitative PCR (qPCR); as were the mycorrhiza-induced genes encoding lysosomal pro-X carboxypeptidase (*DoPRCP*), hevamine A-like (*DoHAL*), glucan endo-1,3-beta-glucosidase (*DoGGLU*), bidirectional sugar transporter (*DoSWEET14*), beta-1,3-glucanase (*DoGLU*), and aspartic proteinase CDR1-like (*DoCDR1*) (Figures 4–8). After seed germination (stage 2, stage 3, stage 4), all of these genes were usually highly expressed and upregulated in protocorm (stage 3) and seedling development stages (stage 4 and stage 5) in SG compared to AG of *D. officinale*.

2.4. Effect of Exogenous GA_3 Treatment on Genes' Expression

2.4.1. Gene Expression Related to GAs Biosynthesis

After treatment with the exogenous GA_3 concentrations, genes involved in GAs biosynthesis showed diverse expression patterns. We compared the differential expression of the above genes between asymbiotic and symbiotic conditions at a given treatment concentration GA_3 treatment (Figures 4–8). These results showed those genes related to the biosynthesis of GAs (*GA20ox, GA3ox*) were upregulated in seed germination (stage 2), protocorm formation (stage 3), and seedling (stage 4) in the SG, while *DoGA2ox* underwent significantly upregulated expression at the protocorm stage (stage 3) (Figure 4A). After applying exogenous GA_3, the expression of *GA3ox* gene in SG was 10.19, 26.42, 74.74, 109.36, and 104.15 times that in AG at 0, 0.05 µM, 0.1 µM, 0.5 µM and 1 µM exogenous GA_3 treatment concentrations, respectively (Figure 4B). In addition, the expression level of GA2ox, the key gene encoding gibberellin oxidase, which catalyzes the degradation of active GAs, was upregulated sharply at a higher GA_3 treatment concentration (0.5 µM) in SG compared to AG (246.17 fold-change). This implied a crosstalk interaction between the biosynthesis and metabolism of GAs and mycorrhizal establishment.

Table 2. Putative mycorrhizal induced genes, GA biosynthesis and other hormone gene homologs induced during symbiotic germination of *D. officinale* seeds. Abbreviation: FPKM: fragments per kilobase per million and the value represents the differential expression level; FDR: false discovery rate; A1, A2, A3 represent asymbiotic germination stage 2, stage 3, and stage 4 respectively; S1, S2, S3 represent symbiotic germination stage 2, stage 3, and stage 4, respectively. * gene expression were validated by qPCR.

Pathway	KEGG Description	Gene ID	FPKM						log2(S1/A1)	FDR	log2(S2/A2)	FDR	log2(S3/A3)	FDR
			A1	S1	A2	S2	A3	S3						
GA biosynthesis and metabolism	ent-kaurene synthase [EC:4.2.3.19] (*KS*)	Dendrobium_GLEAN_10038616	0.00	9.32	0.00	3.30	0.67	6.56	13.19	1.06×10^{-11}	11.69	5.00×10^{-3}	3.29	3.68×10^{-5}
	ent-kaurene oxidase [EC:1.14.13.78] (*KO*)	Dendrobium_GLEAN_10138923 *	0.60	30.24	0.69	40.62	0.45	15.44	5.66	1.22×10^{-49}	5.88	4.93×10^{-63}	5.10	4.57×10^{-24}
		Dendrobium_GLEAN_10138922	2.02	76.16	1.15	64.58	0.96	36.56	5.24	1.08×10^{-38}	5.81	6.06×10^{-31}	5.25	1.23×10^{-16}
	gibberellin 2-oxidase [EC:1.14.11.13] (*GA2ox*)	Dendrobium_GLEAN_10025219 *	1.72	97.72	0.00	119.45	0.28	51.62	5.83	4.63×10^{-87}	16.87	2.52×10^{-109}	7.53	3.21×10^{-47}
		Dendrobium_GLEAN_10027180	3.64	74.66	7.80	30.53	1.56	12.74	4.36	5.08×10^{-63}	1.97	3.56×10^{-11}	3.03	7.02×10^{-8}
		Dendrobium_GLEAN_10062493	23.77	78.49	20.87	54.56	0.00	31.03	1.72	2.15×10^{-22}	1.39	6.37×10^{-11}	14.92	1.11×10^{-27}
		Dendrobium_GLEAN_10062492	7.36	51.88	15.18	51.46	0.00	15.66	2.82	2.88×10^{-28}	1.76	4.23×10^{-15}	13.93	4.75×10^{-15}
		Dendrobium_GLEAN_10040612	2.73	32.22	6.21	17.34	0.00	10.67	3.56	1.90×10^{-12}	1.48	6.00×10^{-3}	13.38	2.23×10^{-5}
	gibberellin 3-beta-dioxygenase [EC:1.14.11.15] (*GA3ox*)	Dendrobium_GLEAN_10043501 *	9.70	48.02	4.69	46.72	1.95	28.86	2.31	1.32×10^{-30}	3.32	8.54×10^{-41}	3.89	1.54×10^{-30}
	gibberellin 20-oxidase [EC:1.14.11.12] (*GA20ox*)	Dendrobium_GLEAN_10090677	20.94	31.84	3.19	32.77	11.55	33.28	0.60	1.3×10^{-2}	3.36	3.27×10^{-16}	1.53	3.71×10^{-6}
		Dendrobium_GLEAN_10048964 *	19.32	67.28	7.35	40.77	7.45	23.72	1.80	6.2×10^{-22}	2.47	3.29×10^{-18}	1.67	1.83×10^{-6}
		Dendrobium_GLEAN_10048963	11.72	39.81	3.22	17.42	2.18	9.39	1.76	1.48×10^{-12}	2.44	8.50×10^{-8}	2.11	6.00×10^{-4}

Table 2. *Cont.*

Pathway	KEGG Description	Gene ID	FPKM A1	S1	A2	S2	A3	S3	log2(S1/A1)	FDR	log2(S2/A2)	FDR	log2(S3/A3)	FDR
	momilactone-A synthase [EC:1.1.1.295]	Dendrobium_GLEAN_10065156	11.57	36.43	10.14	21.53	4.64	19.74	1.65	5.60×10^{-13}	1.09	2.0×10^{-4}	2.09	9.71×10^{-9}
	isoprene synthase [EC:4.2.3.27]	Dendrobium_GLEAN_10097477	8.69	20.33	5.75	51.87	4.84	21.22	1.23	1.0×10^{-4}	3.17	2.17×10^{-29}	2.13	3.21×10^{-8}
	2-hydroxyisoflavanone dehydratase [EC:4.2.1.105]	Dendrobium_GLEAN_10030908	0.17	1.15	0.20	12.38	0.00	15.51	2.76	9.3×10^{-2}	5.95	4.03×10^{-17}	13.92	6.73×10^{-25}
GA signal transduction	DELLA protein	Dendrobium_GLEAN_10018544	0.00	13.28	0.41	21.73	0.24	36.19	13.70	2.27×10^{-30}	5.73	2.14×10^{-41}	7.24	1.76×10^{-79}
		Dendrobium_GLEAN_10067459	14.06	120.53	1.54	74.55	2.13	43.34	3.10	2.79×10^{-74}	5.60	7.66×10^{-64}	4.35	1.77×10^{-33}
		Dendrobium_GLEAN_10068560	11.66	106.49	12.73	41.88	6.48	43.20	3.19	1.0×10^{-136}	1.72	8.15×10^{-25}	2.74	4.32×10^{-47}
		Dendrobium_GLEAN_10060070	34.54	97.03	21.39	93.50	26.06	141.08	1.49	1.58×10^{-45}	2.13	5.49×10^{-67}	2.44	8.15×10^{-123}
		Dendrobium_GLEAN_10024051 *	0.00	0.59	0.00	2.23	1.05	5.65	9.20	1.15×10^{-1}	11.12	3.0×10^{-4}	2.43	1.11×10^{-5}
	zeta-carotene desaturase [EC:1.3.5.8]	Dendrobium_GLEAN_10021410	0.00	518.36	0.19	474.48	0.00	80.22	18.98	0	11.29	0	16.29	6.05×10^{-137}
ABA biosynthesis and metabolism	9-cis-epoxycarotenoid dioxygenase [EC:1.13.11.51] (*NCED*)	Dendrobium_GLEAN_10070249 *	1.74	10.23	0.50	43.83	1.17	24.04	2.56	6.99×10^{-10}	6.45	2.31×10^{-72}	4.36	3.32×10^{-35}
	abscisic acid 8'-hydroxylase [EC:1.14.13.93]	Dendrobium_GLEAN_10055771	27.91	129.63	54.03	68.91	34.40	109.62	2.22	4.28×10^{-115}	0.35	6.0×10^{-4}	1.67	1.59×10^{-64}
	momilactone-A synthase [EC:1.1.1.295]	Dendrobium_GLEAN_10065156	11.57	36.43	10.14	21.53	4.64	19.74	1.65	5.60×10^{-13}	1.09	2.0×10^{-4}	2.09	9.71×10^{-9}
ABA signal and transduction	ABA responsive element binding factor (*ABF*)	Dendrobium_GLEAN_10081660 *	11.98	141.30	12.06	88.75	5.13	54.98	3.56	1.99×10^{-112}	2.88	4.30×10^{-54}	3.42	2.13×10^{-40}
	abscisic acid receptor PYR/PYL family	Dendrobium_GLEAN_10090645	19.08	66.48	14.51	96.18	22.40	100.66	1.80	2.20×10^{-15}	2.73	6.49×10^{-32}	2.17	4.12×10^{-23}
		Dendrobium_GLEAN_10028158	28.88	74.18	36.12	106.46	20.79	66.96	1.36	9.24×10^{-17}	1.56	3.17×10^{-26}	1.69	3.42×10^{-17}
	ubiquitin-conjugating enzyme E2 H [EC:6.3.2.19] (*UBE2N*)	Dendrobium_GLEAN_10040978 *	0.20	100.23	0.07	64.50	0.00	32.99	8.97	1.12×10^{-269}	9.85	6.43×10^{-163}	15.01	1.04×10^{-89}

Table 2. Cont.

Pathway	KEGG Description	Gene ID	FPKM						log2(S1/A1)	FDR	log2(S2/A2)	FDR	log2(S3/A3)	FDR
			A1	S1	A2	S2	A3	S3						
IAA biosynthesis and metabolism	indoleacetaldoxime dehydratase [EC:4.99.1.6]	Dendrobium_GLEAN_10124247	6.48	12.52	5.33	19.60	4.10	31.70	0.95	1.00×10^{-3}	1.88	6.42×10^{-12}	2.95	2.10×10^{-33}
	YUCCA family monooxygenase [EC:1.14.13.8] (YUCCA)	Dendrobium_GLEAN_10119687	0.14	38.55	0.00	49.20	0.28	24.59	8.11	2.22×10^{-74}	15.59	9.99×10^{-91}	6.46	5.28×10^{-45}
		Dendrobium_GLEAN_10046914	0.99	37.15	0.00	51.97	0.73	37.02	5.23	3.77×10^{-21}	15.67	2.55×10^{-30}	5.66	2.96×10^{-20}
		Dendrobium_GLEAN_10072646	0.47	7.76	2.85	15.23	1.32	21.73	4.05	1.47×10^{-3}	2.42	4.14×10^{-15}	4.04	1.04×10^{-36}
		Dendrobium_GLEAN_10012601 *	0.00	29.08	0.00	39.65	0.84	20.41	14.83	2.90×10^{-35}	15.28	2.46×10^{-44}	4.60	6.04×10^{-20}
		Dendrobium_GLEAN_10061932	4.93	20.80	8.44	42.64	3.07	22.84	2.08	2.89×10^{-15}	2.34	4.02×10^{-33}	2.90	1.34×10^{-23}
		Dendrobium_GLEAN_10046916	4.01	30.78	1.65	44.97	0.34	33.09	2.94	3.33×10^{-33}	4.77	9.35×10^{-66}	6.60	1.00×10^{-59}
	SAUR family protein	Dendrobium_GLEAN_10110599	1.27	46.87	0.48	27.33	0.41	10.16	5.21	6.32×10^{-28}	5.83	1.38×10^{-15}	4.63	1.57×10^{-5}
		Dendrobium_GLEAN_10011274	2.42	65.45	0.69	24.78	0.00	9.10	4.76	3.59×10^{-27}	5.17	8.22×10^{-10}	13.15	3.00×10^{-4}
		Dendrobium_GLEAN_10033477	0.00	137.57	0.00	44.21	0.00	10.04	17.07	1.59×10^{-71}	15.43	1.72×10^{-20}	13.29	8.65×10^{-5}
IAA signal transduction		Dendrobium_GLEAN_10011273	4.55	61.11	0.52	22.76	0.00	18.76	3.75	6.23×10^{-28}	5.45	6.40×10^{-12}	14.20	4.45×10^{-11}
		Dendrobium_GLEAN_10033975 *	1.76	27.04	2.67	29.54	1.64	34.50	3.94	1.14×10^{-10}	3.47	1.70×10^{-9}	4.39	3.53×10^{-12}
		Dendrobium_GLEAN_10001422	1.25	17.82	0.00	8.76	1.74	13.04	3.83	9.05×10^{-7}	13.10	5.00×10^{-4}	2.91	1.00×10^{-3}
		Dendrobium_GLEAN_10116922	0.00	4.71	0.00	1.44	0.29	3.71	12.20	9.15×10^{-5}	0.16	7.2×10^{-1}	0.23	7.99×10^{-1}
		Dendrobium_GLEAN_10075175	0.00	6.74	0.00	0.63	0.00	0.00	12.72	4.00×10^{-4}	9.30	5.8×10^{-1}	-	-
	extracellular signal-regulated kinase [EC:2.7.11.24]	Dendrobium_GLEAN_10064705	0.00	16.86	0.40	5.95	0.00	16.86	14.04	6.55×10^{-13}	3.89	1.00×10^{-3}	14.04	6.55×10^{-13}
	calcium-binding protein CML	Dendrobium_GLEAN_10048053 *	0.45	13.61	4.33	86.28	0.45	13.61	4.92	1.18×10^{-6}	4.32	8.12×10^{-38}	4.92	1.18×10^{-6}
Ca^{2+} signal related	calcium-dependent protein kinase [EC:2.7.11.1]	Dendrobium_GLEAN_10140021	0.80	13.82	0.00	53.38	0.80	13.82	4.11	6.54×10^{-7}	15.70	8.13×10^{-34}	4.11	6.54×10^{-7}
		Dendrobium_GLEAN_10016982 *	0.00	10.17	0.62	7.70	0.00	10.17	13.31	1.78×10^{-10}	3.63	9.86×10^{-6}	13.31	1.78×10^{-10}
		Dendrobium_GLEAN_10079575	0.80	17.06	3.19	17.30	0.80	17.06	4.41	1.86×10^{-34}	2.44	3.90×10^{-19}	4.41	1.86×10^{-34}
		Dendrobium_GLEAN_10071601	1.40	14.50	2.14	23.68	1.40	14.50	3.37	1.69×10^{-16}	3.47	3.14×10^{-40}	3.37	1.69×10^{-26}
		Dendrobium_GLEAN_10022554	2.48	11.31	8.94	41.67	2.48	11.31	2.19	3.00×10^{-2}	2.22	2.41×10^{-11}	2.19	3.00×10^{-3}
		Dendrobium_GLEAN_10064211	6.40	61.24	1.65	45.27	6.40	61.24	3.26	8.29×10^{-71}	4.78	2.30×10^{-66}	3.26	8.29×10^{-71}
		Dendrobium_GLEAN_10022555	11.00	94.82	31.00	171.02	11.00	94.82	3.11	1.12×10^{-90}	2.46	2.24×10^{-122}	3.11	1.12×10^{-90}
The common symbiotic pathway	nodulation-signaling pathway 2 protein	PEQU_11738-D2 *	0.87	27.28	1.40	18.44	0.00	10.96	4.97	1.85×10^{-36}	3.72	3.32×10^{-19}	13.42	3.44×10^{-17}
		Dendrobium_GLEAN_10030409 *	2.24	138.59	2.42	80.85	1.55	68.84	5.95	1.32×10^{-295}	5.06	4.49×10^{-149}	5.47	2.04×10^{-141}

Table 2. *Cont.*

Pathway	KEGG Description	Gene ID	FPKM						log2(S1/A1)	FDR	log2(S2/A2)	FDR	log2(S3/A3)	FDR
			A1	S1	A2	S2	A3	S3						
putative mycorrhiza-induced genes	aspartic proteinase CDR1-like [EC:2.7.1.-]	Dendrobium_GLEAN_10074206	0.00	727.07	0.00	629.87	0.00	435.89	19.47	0	19.26	0	18.73	0
	glucan endo-1,3-beta-glucosidase-like isoform X1	Dendrobium_GLEAN_10113668 *	0.00	1685.91	0.00	795.80	0.00	379.47	20.69	0	19.60	0	18.53	0
	Non-specific lipid-transfer protein	Dendrobium_GLEAN_10036826	0.00	1460.16	0.70	1564.23	0.00	275.31	20.48	0	123.78	0	18.07	5.83×10^{-121}
	bidirectional sugar transporter SWEET14-like	Dendrobium_GLEAN_10125587 *	0.00	216.32	0.00	228.84	0.00	239.67	17.72	1.96×10^{-205}	17.80	1.46×10^{-199}	17.87	4.20×10^{-218}
	Aspartic proteinase nepenthesin-1 precursor	Dendrobium_GLEAN_10098789	0.13	739.91	0.00	439.57	0.00	217.37	12.47	0	18.75	0	17.73	0
	Subtilisin-like serine endopeptidase family protein	Dendrobium_GLEAN_10135421	0.00	336.63	0.00	369.14	0.00	211.92	18.36	0	18.49	0	17.69	0
	non-specific lipid-transfer protein-like protein	Dendrobium_GLEAN_10080723	-	-	0.48	583.44	0.00	193.03	-	-	105.01	0	17.56	4.93×10^{-125}
	chitinase [EC:3.2.1.14]	Dendrobium_GLEAN_10053378	0.00	124.35	0.00	198.49	0.00	183.43	16.92	9.45×10^{-96}	17.60	8.18×10^{-140}	17.48	7.11×10^{-133}
		Dendrobium_GLEAN_10042237 *	0.00	106.39	0.00	167.62	0.00	153.94	16.70	9.42×10^{-151}	17.35	1.86×10^{-22}	17.23	2.75×10^{-215}
	fatty acid desaturase [EC:3.1.4.4]	Dendrobium_GLEAN_10098792 *	0.00	493.08	0.00	423.04	0.00	147.60	18.91	0	18.69	0	17.17	1.83×10^{-281}
	beta-1,3-glucanase	Dendrobium_GLEAN_10033071 *	0.36	783.01	0.20	383.72	0.00	136.56	11.09	0	118.94	0	17.06	8.82×10^{-213}
	glucan endo-1,3-beta-glucosidase-like isoform X1	Dendrobium_GLEAN_10050850	0.00	391.39	0.00	273.17	0.00	122.44	18.58	0	18.06	0	16.90	2.09×10^{-197}
	mannose-specific lectin 3-like	Dendrobium_GLEAN_10079890	1.19	2499.75	0.27	1267.07	0.24	1020.79	11.04	0	148.75	0	12.05	0
	lysosomal Pro-X carboxypeptidase-like [EC:3.4.16.2]	Dendrobium_GLEAN_10008416 *	0.11	1471.74	0.00	1564.87	0.23	765.35	13.71	0	20.58	0	11.70	0
	mannose-specific lectin-like	Dendrobium_GLEAN_10065758	2.20	2498.33	1.01	1576.93	0.84	989.65	10.15	0	112.54	0	10.20	0
	subtilisin-like protease SDD1 [EC:3.4.21.112]	Dendrobium_GLEAN_10069506	1.14	1423.95	0.35	1729.18	0.85	950.92	10.29	0	150.56	0	10.13	0

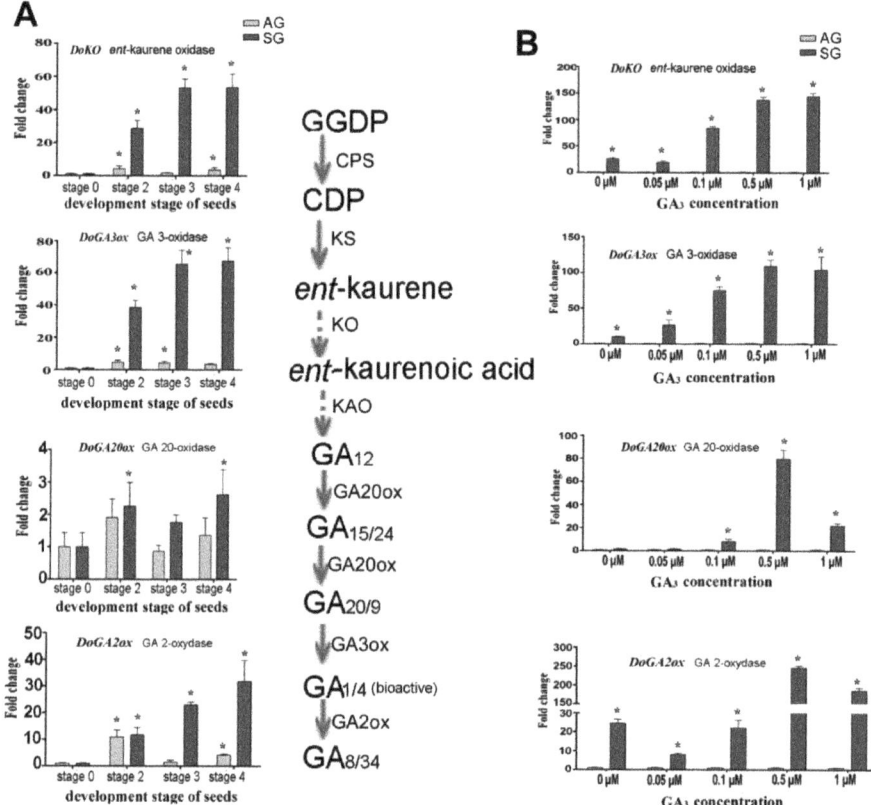

Figure 4. Expression levels of genes related to GA biosynthesis during symbiotic germination of *Dendrobium officinale* for quantitative qPCR analysis. (**A**). Genes' expression at different development stages between asymbiotic and symbiotic germination; (**B**). Genes' expression affected by GA concentrations at 4 weeks after sowing seeds. Note the fold-change values are relative to asymbiotic germination. PCR amplifications were performed for three biological replicates and two distinct technical replicates for each sample. Expression levels were calculated by the $2^{-\Delta\Delta C_T}$ method normalized against the expression of *EF1-α*, using the expression level of ungerminated seed (stage 0) (**A**) or asymbiotic germination (**B**) as control and the fold change > 2.0 was marked significant differential expression (*). AG, asymbiotic germination; SG, symbiotic germination.

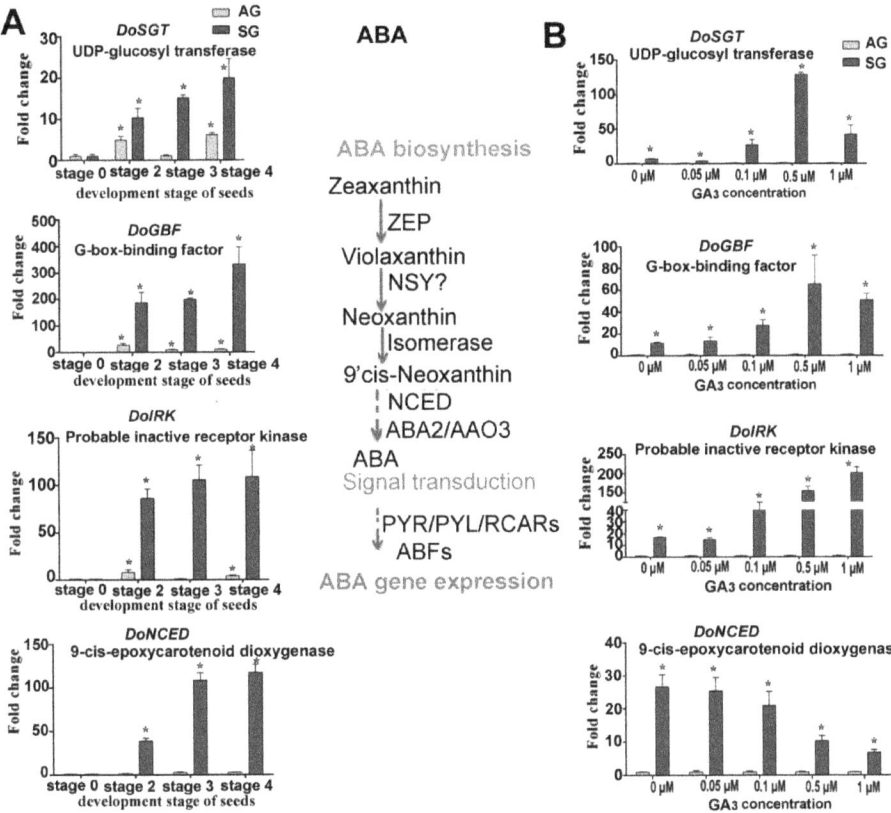

Figure 5. Expression levels of genes related to ABA biosynthesis during symbiotic germination of *Dendrobium officinale* for quantitative qPCR analysis. (**A**). Genes' expression at different development stages between asymbiotic and symbiotic germination (no GA_3 treatment); (**B**). Genes expression' affected by GA_3 concentrations at 4 weeks after sowing seeds. Expression levels were calculated by the $2^{-\Delta\Delta C_T}$ method normalized against the expression of *EF1-α*, using the expression level of ungerminated seed (stage 0) (**A**) or asymbiotic germination (**B**) as control and the fold change > 2.0 was marked significant differential expression (*). AG, asymbiotic germination; SG, symbiotic germination.

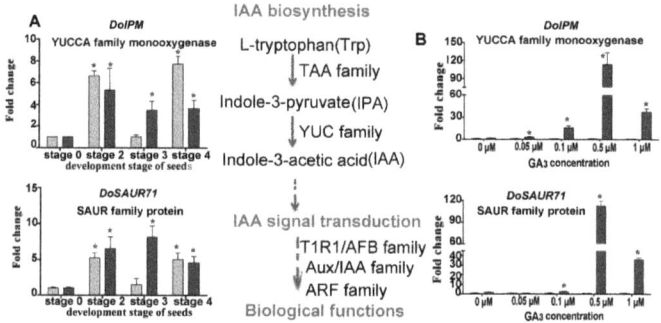

Figure 6. Expression levels of genes related to IAA biosynthesis during symbiotic germination of *Dendrobium officinale* for quantitative qPCR analysis. (**A**). Genes' expression at different development stages between asymbiotic and symbiotic germination (no GA$_3$ treatment); (**B**). Genes' expression was affected by GA$_3$ concentrations at 4 weeks after sowing seeds. Expression levels were calculated by the $2^{-\Delta\Delta C_T}$ method normalized against the expression of *EF1-α*, using the expression level of ungerminated seed (stage 0) (**A**) or asymbiotic germination (**B**) as control and the fold change > 2.0 was marked significant differential expression (*). AG, asymbiotic germination; SG, symbiotic germination.

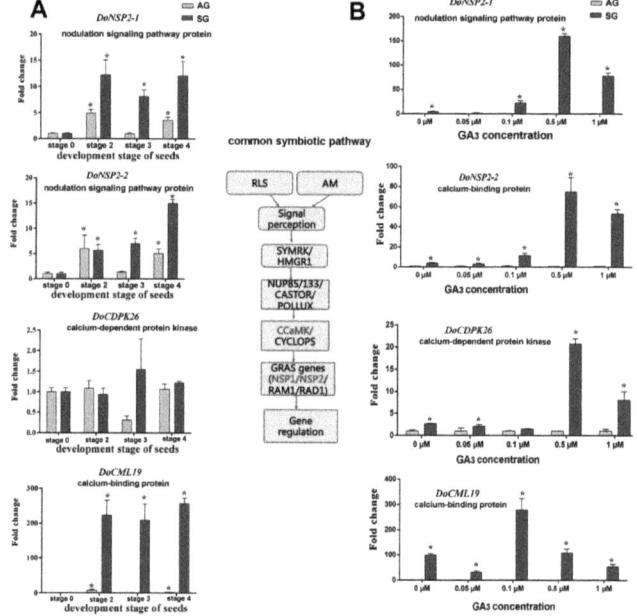

Figure 7. Expression levels of genes related to common symbiosis pathway during symbiotic germination of *Dendrobium officinale* for quantitative qPCR analysis. (**A**). Genes' expression at different development stages between asymbiotic and symbiotic germination (no GA$_3$ treatment); (**B**). Genes' expression was affected by GA$_3$ concentrations at 4 weeks after sowing seeds. Expression levels were calculated by the $2^{-\Delta\Delta C_T}$ method normalized against the expression of *EF1-α*, using the expression level of ungerminated seed (stage 0) (**A**) or asymbiotic germination (**B**) as control and the fold change > 2.0 was marked significant differential expression (*). AG, asymbiotic germination; SG, symbiotic germination.

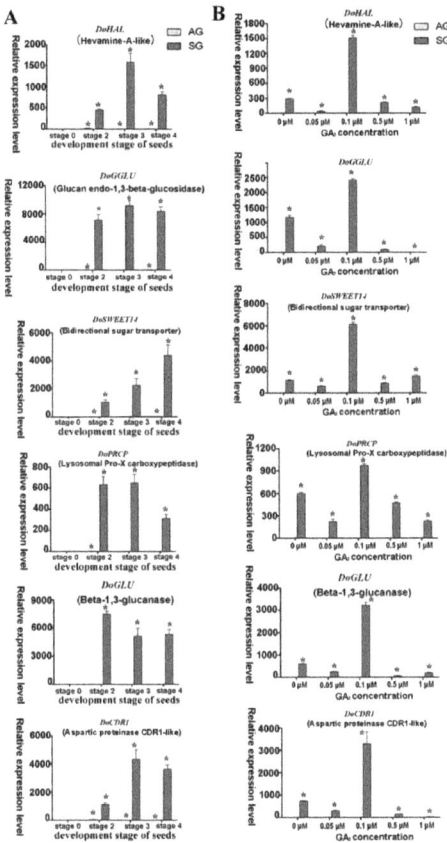

Figure 8. Expression levels of putative mycorrhiza-induced genes involved in orchid symbiotic germination of *Dendrobium officinale* for quantitative qPCR analysis. (**A**). Genes' expression at different development stages between asymbiotic and symbiotic germination (no GA$_3$ treatment); (**B**). Genes' expression was affected by GA$_3$ concentrations at 4 weeks after sowing seeds. Expression levels were calculated by the $2^{-\Delta\Delta C_T}$ method normalized against the expression of *EF1-α*, using the expression level of ungerminated seed (stage 0) (**A**) or asymbiotic germination (**B**) as control and the fold change > 2.0 was marked significant differential expression (*). AG, asymbiotic germination; SG, symbiotic germination.

2.4.2. Gene Related to ABA Biosynthesis and Signaling Transduction

Treated with exogenous GA$_3$ in the germination experiment, the genes involved in ABA biosynthesis and signaling transduction displayed diverse expression profiles (Figure 5A). For example, the gene DoNCED responsible for ABA biosynthesis was downregulated with a greater GA$_3$ concentration, while the genes involved in the signal transduction of ABA (*DoSGT, DoIRK*, and *DoGBF*) were all upregulated, implying ABA metabolism has a very active response to a changed GA$_3$ concentration (Figure 5B). The expression of genes participating in auxin biosynthesis also displayed a similar profile. Notably, DoIPM, a key gene that belongs to the YUCCA family was upregulated in SG compared to AG under the 0.5-μM GA$_3$ treatment concentration (Figure 6).

2.4.3. Expression Analysis of Putative Genes Involved in Mycorrhizal Symbiosis and Common Symbiosis Pathway

The putative symbiosis-specific expression genes, including *DoHAL*, *DoPRCP*, *DoGGLU*, *DoGLU*, *DoSWEET*, *DoCDR1*, *DoCDPK2*, and *DoNSP2* featured similar expression levels in SG after treatment with different concentrations of GA_3. In the SG group with no exogenous GA_3 treatment, the expression level of these genes increased substantially compared to AG, indicating the expression of these genes was induced by mycorrhizal fungi invasion. However, their expression underwent a similar change after imposing the exogenous GA_3 treatment; namely, genes were at first highly expressed in 0.1 µM of exogenous GA_3 but then suppressed as the GA_3 treatment concentration increased (Figures 7 and 8). The expression of *DoCDPK26* was not significantly changed in AG across the GA_3 treatment concentrations but it was significantly and highly expressed in the 0.5-µM GA_3 treatment in the SG group. Similarly, the gene *DoCML19* also was highly expressed in SG yet not significantly changed by exogenous GA_3; this implied the expression of these two genes was induced by mycorrhizal fungi but each responded differently to the exogenous GA_3 treatment.

3. Discussion

Symbiotic germination of orchid seeds involves the dual process of seed self-development and mutualistic interaction with their mycorrhizal fungi. Thus, the process is quite complex physiologically and ecologically. Orchid seeds are too tiny to perform genetic manipulations and this has inevitably limited the studies on their mechanisms of symbiotic germination, yet recent breakthroughs on arbuscular mycorrhiza have laid the foundation for investigating the SG of orchid seeds [16]. Recent studies show that the mycoheterotrophic symbiosis between orchids and mycorrhizal fungi possesses major components shared with mutualistic plant–mycorrhizal symbioses [17]. Many studies have revealed that plant hormones, especially gibberellins, are important factors affecting seed germination [10], and they are also critical for the establishment of mycorrhizal symbiosis [18,19]. In our study, the contents of five plant hormones (GA_3, ABA, IAA, ZT, and JA) was determined at four different developmental stages of seed germination of the orchid *D. officinale*. Our results revealed that the mature and ungerminated seed have the highest ABA content (12.78 ng/g·FW) but this declined further along the seed germination process, and is consistent with two other studies [20,21]. A little GA_3 was detected in the early germination stage of SG and AG group but the content is no significant difference between SG and AG group. Exogenous GA_3 negligibly affected asymbiotic germination at all concentrations used in our study, a result supporting early statements by Arditti [6] that, in general, gibberellins appear to have no effect on germinating orchid embryos, in line as well with reported findings on asymbiotic germination testing by Hadley and Harvais [22]. However, exogenous gibberellins did significantly affect symbiotic germination in our study, implying its important role in mycorrhizal establishment. In addition, although the content of GA_3 was similar between the symbiotic and asymbiotic groups, the ratio GA_3/ABA changed faster at seedling development stage in SG, indicating fungal infection probably affected the balance of endogenous GAs and ABA. Previous results indicated the gibberellin/abscisic acid balance was capable of governing the seed germination of palm and maize plants [23,24]. In tomato, the level of GAs increases as a consequence of a symbiosis-induced mechanism requiring functional arbuscules that depends on a functional ABA pathway in mycorrhizal symbiosis during the establishment of arbuscular mycorrhiza [25]. Additionally, at least 130 forms of GAs have been identified to date yet only a handful of these (GA_1, GA_3, GA_4, GA_5, and GA_7) are known to be biologically active [26]. Thus, in our next research project, we plan to quantify other active GAs molecules in *D. officinale* seeds.

The amount of IAA rose dramatically during the seed germination process, but especially during the seedling development stage of the SG group, indicating that IAA production was probably induced by mycorrhizal fungus in SG. Early research has shown that only traces of auxin occur in *Cypripedium* seed but none at all in *Dendrobium* seeds [6,27]; however, in our study, IAA was detected at relatively high content in the ungerminated stage and this content declined in the course of AG. The conflicting results are likely due the detection methods used. UHPLC is undoubtedly more sensitive for the quantification of trace amounts of plant hormones. Auxin is recognized as a secondary dormancy phytohormone, controlling seed dormancy and germination [28]. In addition, auxin metabolism and signaling also plays a crucial role in the modification of roots growth during their colonization by the ectomycorrhizal fungus *Laccria bicolor* [29]. Our result suggests IAA production was induced greatly during orchid mycorrhizal establishment, which provides a possible explanation for the faster differentiation of embryo when the seed of *D. officinale* was inoculated with the mycorrhizal fungus.

In this study, jasmonic acid (JA) content went undetected in ungerminated seeds and low JA (1.63 ng/g·FW) occurred in the free-living fungus, whereas the most JA was present in the early germination stage (stage 2) in AG (Table 1). JA is widely known to be involved in the response of plants to various stress factors, yet surprisingly little research has been carried out on JA's roles in seed germination [30]. Work by Dave et al. [31] found no massive increase in their contents during seed maturation of *Arabidopsis*, suggesting their accumulation instead occurred during early seed development. A recent study reported crosstalk between JA and ABA contributed to modulating seed germination in bread wheat and *Arabidopsis* [32]. Evidently, more research is required to unravel the molecular mechanisms by which jasmonates regulate the germination of seeds.

Besides inducing plant hormone production, the mycorrhizal fungus itself also produces hormones and this may influence its plant partners in crucial ways. In our study, all five hormones were detected in the mycorrhizal fungus *Tulasnella* sp. As for the dynamic change of hormones in symbiotic germination group of *D. officinale* seeds, whether their production arose from mycorrhizal fungi or from host plant induced by fungus is still unclear and merits further exploration in the future.

Exogenous GA_3 treatment had a dose-dependent effect on the SG of *D. officinale* seeds but did not significantly affect either the AG or free-living mycelium growth in the phenotype. Based on our initial results, we speculate the signal recognition between seed and their mycorrhizal fungi was probably impaired in some way by a higher concentration of GA_3. We did not detect fungal invasion (colonization) of the seed embryo when using either 0.5 µM and 1.0 µM exogenous GA_3. Under the microscope, we saw the seed embryo enlarged but no germination ensued at these high GA_3 concentrations in the SG group (Figure 3S–Y). Furthermore, high concentrations of GA_3 did not stop free-living mycelium from growing. A previous study has shown that GAs are phytohormones able to inhibit arbuscular mycorrhizal fungal infection by inhibiting arbuscular mycorrhizal hyphal entry into the host root where they suppressed the expression of Reduced Arbuscular Mycorrhization1 (RAM1) and RAM2 homologs that function in hyphal entry and arbuscule formation [19]. A similar scenario probably occurred in SG of *D. officinale* seeds.

Furthermore, after receiving the exogenous GA_3, plant hormone-related genes such as biosynthesis and signal transduction of GA, ABA or IAA were characterized by a similar expression profile. Namely, sharply increasing expression in response to 0.5 µM exogenous GA_3 followed by transcriptional downregulation; accordingly, we infer that exogenous GA_3 disturbed the balance of endogenous hormones and crosstalk regulation occurred between GA, IAA, and ABA during the seed germination of *D. officinale* inoculated with the *Tulasnella* sp. fungus. Normally, genes involved in GA and IAA synthesis are highly expressed in SG, especially in the protocorm and seedling stages of orchids. The symbiosis between *Cymbidium goeringii* and a *Rhizoctonia*-like mycorrhizal fungus caused the release of hormones, which were able to promote the growth of *C. goeringii* seedlings [7]. Similarly, it has been demonstrated that auxin promotes *Arabidopsis* root growth by modulating its gibberellin response [33]. We plan to quantify the endogenous hormones to further confirm the relationship between hormone content and gene expression under an exogenous GA_3 treatment during orchid seed germination.

Based on our previous RNA-seq and iTRAQ data, we found four proteins encoding genes involved in the common symbiotic signal pathway, including two genes function-annotated as nodulation signaling pathway protein (*DoNSP2-1* and *DoNSP2-2*) and two Ca^{2+} signal-related proteins, a calcium-dependent protein kinase (*DoCDPK26*), and a calmodulin-like protein (*DoCML19*). All these genes were highly expressed in SG but differed markedly. The Ca^{2+} signal is a universal second messenger, and increases in cytosolic Ca^{2+} concentration are among the earliest signaling events occurring in plants challenged with mutualistic partners or pathogens [34,35]. CDPK and CML are the two principal protein families of plant Ca^{2+} sensors [36]. The gene encoding CDPK was also identified from *D. officinale* roots infected by an orchid mycorrhizal fungus (*Mycena* sp.) by using the reverse transcription-polymerase chain reaction (RT-PCR) and rapid amplification of cDNA ends (RACE) [37]. In our study, the genes encoding CDPK (*DoCDPK26*) and CML (*DoCML19*) exhibited sharply higher expression levels in SG across the applied concentration gradient exogenous GA_3, especially under 0.5 µM (for *DoCDPK26*) and 1.0 µM (for *DoCML19*), respectively. However, this expression of *DoCML19* was similar to SG lacking exogenous GA_3 treatment, suggesting gene expression was induced by mycorrhizal fungi and only weakly related to exogenous GA_3. Conversely, the gene *DoCDPK26* showed a significant different expression in SG group with versus without exogenous GA_3 treatment, which implied that the CDPK and CML proteins probably participate in this plant–microbe interaction in different ways. Given the difficulty of genetically manipulating orchid seeds and orchid mycorrhizae, in our future research biochemical and physiological methods will be applied to confirm the mechanistic linkage between this plant hormone and Ca^{2+} signal during the SG of orchid seed, as well as changed Ca^{2+} concentrations across a gradient of exogenous GA_3 during seed germination of *D. officinale*.

We also found that the expression of genes encoding probable mycorrhizal signaling pathway proteins (*DoNSP2-1* and *DoNSP2-2*) (function-annotated as nodulation signaling pathway proteins), both of which encode GARS-family transcriptional regulators, considerably increased under 0.5 µM exogenous GA_3 treatment in SG compared to AG. This result suggests exogenous GA_3 probably affected the mycorrhizal-specific gene expression by controlling the mycorrhizal-signaling pathway. Gibberellin's ability to govern the nodulation signaling pathway in *Lotus japonicus* has been clarified by Maekawa et al. [38], who found that exogenous application of biologically active GA_3 inhibited the formation of infection threads and nodules; hence they suspected GA halted the nodulation signaling pathway downstream of cytokinin, possibly at NSP2, which is required for Nod factor-dependent NIN expression. Whether a similar situation, in which GA inhibited the downstream gene expression of the mycorrhizal signaling pathway, occurs in orchid mycorrhiza needs to be confirmed (or not) in a co-culture system of orchid seedlings with its mycorrhizal fungi.

Several typical putative mycorrhizal-fungi-induced expression genes were identified in the SG of *D. officinale* seeds based on our transcriptomic data: *DoCDR1*, *DoGGLU*, *DoGLU*, *DoPRCP*, and *DoSWEET*. For these genes, hardly any expression happened in AG but they were highly expressed in specific ways among different development stages of SG for the *D. officinale* seeds. The gene *DoCDR1* encodes an aspartic protease. Studies have found that the aspartic protease gene in rice, OsCDR1, can induce defense responses in plants and increase plant resistance to bacterial and fungal diseases [39]. *DoCDR1* was also upregulated in different germination stages of SG in the absence of the GA_3 treatment: low concentration of it did not cause this gene's expression to change, but 0.1 µM endogenous GA_3 treatment strongly elevated *DoCDR1*'s expression, suggesting that fungi induced it. Exogenous GA_3 probably affected the expression level by interfering with the balance of endogenous hormones.

DoGGLU and *DoGLU* are two genes encoding β-1,3-glucanase, belonging to the pathogenesis-related proteins class that plays an important role in biotic and abiotic stress responses of plants [40]. It has been shown that colonization by mycorrhizal fungi in orchid root does not trigger strong plant defense responses in orchid mycorrhiza of *Serapias vomeracea* with *T. calospora*, given the nonstimulated expression of the plant's defense genes [41]. However, our proteomic analysis showed that fungus invasion activated the plant defense reaction because genes encoding catalase isozyme, L-ascorbate peroxidase, and superoxide dismutase—all of which are enzymes involved in defense mechanisms—were upregulated during the SG of *D. officinale* seeds [7]. High expression levels of β-1,3-glucanase genes suggest the host plant probably produced an antifungal defense reaction, especially in the protocorm stage, via the lysis of pelotons so as to limit the extent of invasion during the SG of *D. officinale*. Finally, since the high GA_3 treatment concentrations triggered the strong expression of *DoGGLU*, *DoGLU*, this indicated the genes respond to exogenous environment stress.

SWEET family sugar exporters in arbuscula mycorrhizal symbiosis in *Medicago truncatula* are known to play a vital role in the transport of glucose across the peri-arbuscular membrane to maintain arbuscular for a healthy mutually beneficial symbiosis [42]. Genes encoding SWEET family proteins are often expressed more in the symbiotic tissues of mycorrhizal protocorms of the orchid *S. vomeracea* with *T. calospora*. In our study, evidence for a similar phenomenon was found. Mycorrhiza-induced genes were specifically expressed in SG and its expression rose sharply under the 0.1-µM exogenous GA_3 treatment; hence, these genes responded to a changed exogenous GA_3 concentration during the SG of *D. officinale* seed. Therefore, we propose that GAs is involved in the crosstalk signal pathway between GAs biosynthesis and common symbiotic signal pathway during *D. officinale* seeds' symbiotic germination and is thereby able to influence the expression of mycorrhizal-induced genes.

4. Materials and Methods

4.1. Plant Materials and Growing Conditions

Seeds of *D. officinale* were collected from a greenhouse in Jinhua County of Zhejiang Province, China, in November 2015. Mature capsules were surface sterilized, and their axenic seeds were stored at 4 °C in wax paper packets inside 1.5-mL sterilized tubes containing sterilized silica gel [7]. A mycorrhizal fungus that was a *Tulasnella* sp. (S6), isolated previously from root of *D. nobile*, was cultured in potato dextrose agar (PDA) medium. Symbiotic germination (SG) testing was carried out in oatmeal agar plates (OMA, 0.25% oat meal and 1% agar) and the asymbiotic germination (AG) testing was performed in 1/2 Murashige & Skoog (1/2 MS) medium without fungi, under a 12-h/12-h light/dark (L/D) cycle at 25 °C. In our previous work, we demonstrated this fungus is able to stimulate seed germination of *D. officinale* prior to AG, by reducing time to germination and increasing germination rate [7].

4.2. Determination of Endogenous Hormone during Seed Germination of D. officinale

Endogenous hormones, including gibberellic acid (GA_3), abscisic acid (ABA), indole-3-acetic acid (IAA), *trans*-zeatin (ZT) and jasmonic acid (JA), were examined on a total of eight samples at three different developmental stages (stage 2, stage 3, and stage 4) of AG and SG, ungerminated seed, and free-living mycelium of fungus. Each sample consisted of three biological replicates. Standards of ABA, ZT, indole-3-aceticacid, GA_3, and JA (Sigma, St. Louis, MO, USA) were used for the quantification of endogenous hormones. Hormone extraction and fractionation followed the description of Kojima et al. [43]. Briefly, 50–200 mg of fresh seeds or fungi were frozen in liquid nitrogen and homogenized with a lysis buffer (methanol:water:formic acid = 7.9:2:0.1) in a 2-mL microcentrifuge tube. The homogenate was kept at 4 °C for at least 15 h. After centrifugation at $10,000\times g$ for 15 min, the ensuing supernatant was transferred to a new collection tube. The combined eluate was evaporated and then reconstituted with 1 mL of 1 M formic acid, and then the hormone-containing fraction was passed through an MAX column. Quantitative analysis was performed using ultra-high performance liquid chromatography (UHPLC, Agilent 1290 Infinity, Agilent, Santa Clara, CA, USA) coupled with tandem mass spectrometry (MS/MS, Agilent 6490 Triple Quadrupole, Agilent, Santa Clara, CA, USA). Automatic identification and integration of each MRM transition was done under default parameter settings in Masshunter software (Agilent, Santa Clara, CA, USA), but assisted with manual inspections. The mass spectral peak area of the analyte was taken as the ordinate, and a linear regression standard curve drawn with the concentration of the analyte as the abscissa, from which the regression equation was obtained. Then, the mass spectral peak area of the analyte of a given sample was substituted into the linear equation, to calculate the content of each endogenous hormone.

4.3. Exogenous GA_3 Treatment on Symbiotic and Asymbiotic Germination of D. officinale

The concentration of exogenous GAs was selected in preliminary experiments, which spanned 0.05 µM to 1 µM, according to a previous study [19]. Seeds were sown in OMA medium (for SG with fungus) and $\frac{1}{2}$ MS medium (for AG without fungus) with four concentrations of exogenous GA_3 (0.05,

0.1, 0.5, and 1 µM; Beijing Solarbio Science & Technology Co., Ltd., Beijing, China). We designed four groups, including seeds in OMA medium (no fungus), fungus in OMA medium (no seeds) with different concentrations of exogenous GA_3, and the normal AG group (seeds in 1/2MS without fungus) and SG group (seed in OMA medium with fungus) without exogenous GA_3. Next, the petri dishes were incubated at 25 °C under a 12-h/12-h light/dark (L/D) cycle. Morphological changes during seed germination were observed daily, under a stereomicroscope and a Leica light microscope DM2500 (Leica Microsystems, Wetzlar, Germany).

4.4. Transcriptome Analysis by RNA-Seq

Transcriptome analysis of the eight samples was performed in our previous study; the samples corresponded to three different germination stages (stage 2, stage 3, and stage 4) of symbiotic and asymbiotic seeds, respectively, and to free-living mycelium of mycorrhizal fungus and ungerminated seeds [7]. The original transcriptomic data was deposited in the public NCBI and SRA database (accession No. PRJNA279934). Based on this transcriptomic data, we screened the putative genes involved in biosynthesis and signal transduction of plant hormones, the common symbiotic pathway, and specific gene expression in SG of *D. officinale*.

4.5. RNA Extraction and Quantitative Real-Time PCR

RNA extraction and quantitative RT-PCR was performed as described in our prior study [7]. Briefly, total RNA was extracted from 200 mg of seeds using the RNeasy Plant Mini Kit (Qiagen, CA, USA) and treated with an RNase-free DNase I digestion kit (Beijing Aidlab Biotech Company, Beijing, China) to remove any residual genomic DNA. Then 1 µg of RNA was reverse-transcribed to cDNA, using a reverse transcription system (Bio-Rad Laboratories, Inc., Richmond, CA, USA), and the cDNA equivalent to 25 ng of total RNA served as a template for each PCR reaction, carried out using SYBR Green supermix (Bio-Rad Laboratories, Inc., Richmond, CA, USA) with a final concentration of 1.6 mM of each primer. The primer sequences are listed in Table S1. The qRT-PCR experiments were done using a SYBR Premix Ex TaqTM (Takara Biotechnology Co., Ltd., Dalian, China) on the LightCycler 480 machine (Roche Applied, Mannheim, Germany). PCR amplifications of three biological replicates were performed, which also included three distinct technical replicates. A no-template control (i.e., RNase-free water) was included for every qPCR run. Transcript abundance was normalized using the housekeeping gene *EF-1α* and a given gene's expression level amount was calculated by the $2^{-\Delta\Delta C_T}$ method [44].

4.6. Data Analysis

The data of hormone content were analyzed with one-way ANOVA and the statistical analysis was performed using software SPSS 11.0. Data were presented as means ± SD from at least three independent experiments. *p* values < 0.05 were considered significant difference.

5. Conclusions

This study mainly explored the relationship between endogenous hormones and symbiotic germination of orchid *D. officinale* seeds and the effects upon seed germination from exogenous GA_3. Endogenous hormonal change regulated the seed germination of *D. officinale* and mycorrhizal fungi invasion can greatly stimulate its host plant's endogenous IAA accumulation. This could explain the faster differentiation of the embryo at the protocorm stage during symbiotic germination. Exogenous GA_3 has a dose-dependent effect on the establishment of the mycorrhizal relationship between the fungus and seeds, such that a high concentration of GA_3 probably acts upon the genes or proteins of the common symbiotic pathway, thereby inhibiting the recognition between orchid seeds and mycorrhizal fungi to further influence seed germination. Gene expression of the putative mycorrhizal-induced and symbiotic signal pathway responds to exogenous GA_3 concentration change, implying GA_3 contributes to the crosstalk between the hormone biosynthetic pathway and common symbiotic signal pathway.

This study lays a foundation for the further exploration of seed germination, especially the symbiotic germination mechanism of orchid seeds.

Supplementary Materials: The following figures are available online at http://www.mdpi.com/1422-0067/21/17/6104/s1. Table S1: Primer sequences used in qRT-PCR amplification. All primers were designed using Primer 3.0 software and synthesized by Genewiz Company (China).

Author Contributions: J.C. and S.G. designed the study. J.C. wrote the manuscript. B.Y. performed all experiments. Y.T., Y.X., Y.L. and D.Z. participated in data analysis and reference searching and revised the manuscript. All authors have read and agreed to the published version of the manuscript.

Funding: This research was funded by the National Natural Science Foundation of China (81573527; 81973423) and the CAMS Innovation Fund for Medical Sciences (CIFMS) (2017-I2M-3-013).

Conflicts of Interest: The authors declare no conflict of interest. The funders had no role in the design of the study; in the collection, analyses, or interpretation of data; in the writing of the manuscript, or in the decision to publish the results.

References

1. Fay, M.F. Orchid conservation: Further links. *Ann. Bot.* **2016**, *118*, 89–91. [CrossRef] [PubMed]
2. Rasmussen, H.N. *Terrestrial Orchids, from Seed to Mycotrophic Plant*; Cambridge University Press: Cambridge, UK, 1995.
3. Selosse, M.A.; Martos, F. Do chlorophyllous orchids heterotrophically use mycorrhizal fungal carbon? *Trends Plant Sci.* **2014**, *19*, 683–685. [CrossRef]
4. Arditti, J.; Ghani, A.K.A. Numerical and physical properties of orchid seeds and their biological implications. *New Phytol.* **2000**, *145*, 367–421. [CrossRef]
5. Smith, S.E.; Read, D.J. *Mycorrhizal Symbiosis*; Academic Press: San Diego, CA, USA, 2008.
6. Arditti, J. Factors affecting the germination of orchid seeds. *Bot. Rev.* **1967**, *33*, 1–97. [CrossRef]
7. Chen, J.; Liu, S.S.; Kohler, A.; Yan, B.; Luo, H.M.; Chen, X.M.; Guo, S.X. iTRAQ and RNA-Seq analyses provide new insights into regulation mechanism of symbiotic germination of *Dendrobium officinale* Seeds (Orchidaceae). *J. Proteome Res.* **2017**, *16*, 2174–2187. [CrossRef]
8. Li, Y.Y.; Chen, X.M.; Zhang, Y.; Cho, Y.H.; Wang, A.R.; Yeung, E.C.; Zeng, X.; Guo, S.X.; Lee, Y.I. Immunolocalization and changes of hydroxyproline-rich glycoproteins during symbiotic germination of *Dendrobium officinale*. *Front. Plant Sci.* **2018**, *9*, 552. [CrossRef]
9. Genre, A.; Timmers, T. The symbiotic role of the actin filament cytoskeleton. *New Phytol.* **2019**, *221*, 611–613. [CrossRef]
10. Ogawa, M.; Hanada, A.; Yamaychi, Y.; Kuwahara, A.; Kamiya, Y.; Yamaguchi, S. Gibberellin biosynthesis and response during *Arabidopsis*. *Plant Cell* **2003**, *15*, 1591–1604. [CrossRef]
11. Jin, Y.; Liu, H.; Luo, D.; Yu, N.; Dong, W.; Wang, C.; Zhang, X.; Dai, H.; Yang, J.; Wang, E. DELLA proteins are common components of symbiotic rhizobial and mycorrhizal signalling pathways. *Nat. Commun.* **2016**, *7*, 12433. [CrossRef]
12. Yu, N.; Luo, D.; Zhang, X.; Liu, J.; Wang, W.; Jin, Y.; Dong, W.; Liu, J.; Liu, H.; Yang, W.; et al. DELLA protein complex controls the arbuscular mycorrhizal symbiosis in plants. *Cell Res.* **2014**, *24*, 130–133. [CrossRef]
13. Genre, A.; Russo, G. Does a common pathway transduce symbiotic signals in plant-microbe interactions? *Front. Plant Sci.* **2016**, *7*, 96. [CrossRef] [PubMed]
14. Yan, L.; Wang, X.; Liu, H.; Tian, Y.; Lian, J.; Yang, R.; Hao, S.; Wang, X.; Yang, S.; Yang, Q.; et al. The genome of *Dendrobium officinale* Illuminates the biology of the important traditional chinese Orchid herb. *Mol. Plant* **2015**, *8*, 922–934. [CrossRef] [PubMed]
15. Kohler, A.; Kuo, A.; Nagy, L.G.; Morin, E.; Barry, K.W.; Buscot, F.; Canbäck, B.; Choi, C.; Cichocki, N.; Clim, A.; et al. Convergent losses of decay mechanisms and rapid turnover of symbiosis genes in mycorrhizal mutualists. *Nat. Genet.* **2015**, *47*, 410–415. [CrossRef] [PubMed]
16. MacLean, A.M.; Bravo, A.; Harrison, M.J. Plant signaling and metabolic pathways enabling arbuscular mycorrhizal symbiosis. *Plant Cell* **2017**, *29*, 2319–2335. [CrossRef] [PubMed]

17. Miura, C.; Yamaguchi, K.; Miyahara, R.; Yamamoto, T.; Fuji, M.; Yagame, T.; Haruko Imaizumi-Anraku, H.; Yamato, M.; Shigenobu, S.; Kaminaka, H. The Mycoheterotrophic symbiosis between orchids and mycorrhizal fungi possesses major components shared with mutualistic plant-mycorrhizal symbioses. *Mol. Plant Microbe Interact.* **2018**, *31*, 1032–1047. [CrossRef]
18. Foo, E.; Plett, J.M.; Lopez-Raez, J.A.; Reid, D. Editorial: The role of plant hormones in plant-microbe symbioses. *Front. Plant Sci.* **2019**, *10*, 1391. [CrossRef]
19. Takeda, N.; Handa, Y.; Tsuzuki, S.; Kojima, M.; Sakakibara, H.; Kawaguchi, M. Gibberellins interfere with symbiosis signaling and gene expression and alter colonization by arbuscular mycorrhizal fungi in *Lotus japonicus*. *Plant Physiol.* **2015**, *167*, 545–557. [CrossRef]
20. Lee, Y.I.; Chung, M.C.; Yeung, E.C.; Lee, N. Dynamic distribution and the role of abscisic acid during seed development of a lady's slipper orchid, *Cypripedium formosanum*. *Ann. Bot.* **2015**, *116*, 403–411. [CrossRef]
21. Yan, X.N.; Tian, M.; Liu, F.; Wang, C.X.; Zhang, Y. Hormonal and morphological changes during seed development of *Cypripedium japonicum*. *Protoplasma* **2017**, *254*, 2315–2322. [CrossRef]
22. Hadley, G.; Harvais, G. The effect of certain growth substances on asymbiotic germination and development of *Orchis purpurella*. *New Phytol.* **1968**, *67*, 441–445. [CrossRef]
23. Bicalho, E.M.; Pintó-Marijuan, M.; Morales, M.; Müller, M.; Munné-Bosch, S.; Garcia, Q.S. Control of macaw palm seed germination by the gibberellin/abscisic acid balance. *Plant Biol.* **2015**, *17*, 990–996. [CrossRef] [PubMed]
24. White, C.N.; Proebsting, W.M.; Hedden, P.; Rivin, C.J. Gibberellins and seed development in maize. I. evidence that Gibberellin/Abscisic acid balance governs germination versus maturation pathways. *Plant Physiol.* **2000**, *122*, 1081–1088. [CrossRef] [PubMed]
25. Martín-Rodríguez, J.A.; Huertas, R.; Ho-Plágaro, T.; Ocampo, J.A.; Turečková, V.; Tarkowská, D.; Ludwig-Müller, J.; García-Garrido, J.M. Gibberellin-Abscisic acid balances during arbuscular mycorrhiza formation in tomato. *Front. Plant Sci.* **2016**, *7*, 1273. [CrossRef] [PubMed]
26. Hayashi, S.; Gresshoff, P.M.; Ferguson, B.J. Mechanistic action of gibberellins in legume nodulation. *J. Integr. Plant Biol.* **2014**, *56*, 971–978. [CrossRef]
27. Poddubnaya-Arnoldi, V.A. Study of fertilization in the living material of some angiosperms. *Phytomorphology* **1960**, *10*, 185–198.
28. Liu, X.; Zhang, H.; Zhao, Y.; Feng, Z.; Li, Q.; Yang, H.Q.; Luan, S.; Li, J.; He, Z.H. Auxin controls seed dormancy through stimulation of abscisic acid signaling by inducing ARF-mediated ABI3 activation in *Arabidopsis*. *Proc. Natl. Acad. Sci. USA* **2013**, *110*, 15485–15490. [CrossRef]
29. Alice, V.; Ales, P.; Judith, F.; Annegret, K.; Karin, L.; Francis, M.; Valérie, L. Development of the *Poplar-Laccaria bicolor* ectomycorrhiza modifies root auxin metabolism, signaling, and response. *Plant Physiol.* **2015**, *169*, 890–902.
30. Linkies, A.; Leubner-Metzger, G. Beyond gibberellins and abscisic acid: How ethylene and jasmonates control seed germination. *Plant Cell Rep.* **2012**, *31*, 253–270. [CrossRef]
31. Dave, A.; Hernandez, M.L.; He, Z.; Andriotis, V.M.E.; Vaistij, F.E.; Larson, T.R.; Graham, I.A. 12-oxo-phytodienoic acid accumulation during seed development represses seed germination in *Arabidopsis*. *Plant Cell* **2011**, *23*, 583–599. [CrossRef]
32. Ju, L.; Jing, Y.; Shi, P.; Liu, J.; Chen, J.; Yan, J.; Chu, J.; Chen, K.M.; Sun, J. JAZ proteins modulate seed germination through interaction with ABI5 in bread wheat and *Arabidopsis*. *New Phytol.* **2019**, *223*, 246–260. [CrossRef]
33. Fu, X.; Harberd, N.P. Auxin promotes *Arabidopsis* root growth by modulating gibberellin response. *Nature* **2003**, *421*, 740–743. [CrossRef] [PubMed]
34. Aldon, D.; Mbengue, M.; Mazars, C.; Galaud, J.P. Calcium signalling in plant biotic interactions. *Int. J. Mol. Sci.* **2018**, *19*, 665. [CrossRef] [PubMed]
35. Seybold, H.; Boudsocq, M.; Romeis, T. CDPK activation in PRR signaling. *Methods Mol. Biol.* **2017**, *1578*, 173–183. [PubMed]
36. McCormack, E.; Tsai, Y.C.; Braam, J. Handling calcium signaling: *Arabidopsis* CaMs and CMLs. *Trends Plant Sci.* **2015**, *10*, 383–389. [CrossRef] [PubMed]
37. Zhang, G.; Zhao, M.M.; Li, B.; Song, C.; Zhang, D.W.; Guo, S.X. Cloning and expression analysis of a calcium-dependent protein kinase gene in *Dendrobium Officinale* in response to mycorrhizal fungal infection. *Acta Pharm. Sin.* **2012**, *47*, 1548–1554.

38. Maekawa, T.; Maekawa-Yoshikawa, M.; Takeda, N.; Imaizumi-Anraku, H.; Murooka, Y.; Hayashi, M. Gibberellin controls the nodulation signaling pathway in *Lotus japonicus*. *Plant J.* **2009**, *58*, 183–194. [CrossRef]
39. Prasad, B.D.; Creissen, G.; Lamb, C.; Chattoo, B.B. Overexpression of rice (*Oryza sativa* L.) OsCDR1 leads to constitutive activation of defense responses in rice and *Arabidopsis*. *Mol. Plant Microbe Interact.* **2009**, *22*, 1635–1644. [CrossRef]
40. Balasubramanian, V.; Vashisht, D.; Cletus, J.; Sakthivel, N. Plant β-1,3-glucanases their biological functions and transgenic expression aga. *Biotechnol. Lett.* **2012**, *34*, 1983–1990. [CrossRef]
41. Perotto, S.; Rodda, M.; Benetti, A.; Sillo, F.; Ercole, E.; Rodda, M.; Girlanda, M.; Murat, C.; Balestrini, R. Gene expression in mycorrhizal orchid protocorms suggests a friendly plant-fungus relationship. *Planta* **2014**, *239*, 1337–1349. [CrossRef]
42. An, J.; Zeng, T.; Ji, C.; Graaf, S.; Zheng, Z.; Xiao, T.T.; Deng, X.; Xiao, S.; Bisseling, T.; Limpens, E.; et al. A *Medicago truncatula* SWEET transporter implicated in arbuscule maintenance during arbuscular mycorrhizal symbiosis. *New Phytol.* **2019**, *224*, 396–408. [CrossRef]
43. Kojima, M.; Kamada-Nobusada, T.; Komatsu, H.; Takei, K.; Kuroha, T.; Mizutani, M.; Ashikari, M.; Ueguchi-Tanaka, M.; Matsuoka, M.; Suzuki, K.; et al. Highly sensitive and high-throughput analysis of plant hormones using MS-probe modification and liquid chromatography-tandem mass spectrometry: An application for hormone profiling in *Oryza sativa*. *Plant Cell Physiol.* **2009**, *50*, 1201–1214. [CrossRef] [PubMed]
44. Livak, K.J.; Schmittgen, T.D. Analysis of relative gene expression data using real-time quantitative PCR and the 2(-Delta Delta C(T)) Method. *Methods* **2001**, *25*, 402–408. [CrossRef] [PubMed]

© 2020 by the authors. Licensee MDPI, Basel, Switzerland. This article is an open access article distributed under the terms and conditions of the Creative Commons Attribution (CC BY) license (http://creativecommons.org/licenses/by/4.0/).

Article

Upregulation of the MYB2 Transcription Factor is Associated with Increased Accumulation of Anthocyanin in the Leaves of *Dendrobium bigibbum*

Gah-Hyun Lim [1], Se Won Kim [1,2], Jaihyunk Ryu [1], Si-Yong Kang [1], Jin-Baek Kim [1] and Sang Hoon Kim [1,*]

[1] Advanced Radiation Technology Institute, Korea Atomic Energy Research Institute (KAERI), Jeongeup 56212, Korea; kah7702@kaeri.re.kr (G.-H.L.); sewonk@korea.kr (S.W.K.); jhryu@kaeri.re.kr (J.R.); sykang@kaeri.re.kr (S.-Y.K.); jbkim74@kaeri.re.kr (J.-B.K.)

[2] National Institute of Agricultural Sciences, Rural Development Administration, Jeonju 54874, Korea

* Correspondence: shkim80@kaeri.re.kr; Tel.: +82-63-570-3318; Fax: +82-63-570-3811

Received: 27 July 2020; Accepted: 31 July 2020; Published: 6 August 2020

Abstract: Orchids with colorful leaves and flowers have significant ornamental value. Here, we used γ-irradiation-based mutagenesis to produce a *Dendrobium bigibbum* mutant that developed purple instead of the normal green leaves. RNA sequencing of the mutant plant identified 2513 differentially expressed genes, including 1870 up- and 706 downregulated genes. The purple leaf color of mutant leaves was associated with increased expression of genes that encoded key biosynthetic enzymes in the anthocyanin biosynthetic pathway. In addition, the mutant leaves also showed increased expression of several families of transcription factors including the *MYB2* gene. Transient overexpression of *D. biggibum MYB2* in *Nicotiana benthamiana* was associated with increased expression of endogenous anthocyanin biosynthesis genes. Interestingly, transient overexpression of orthologous *MYB2* genes from other orchids did not upregulate expression of endogenous anthocyanin biosynthesis genes. Together, these results suggest that the purple coloration of *D. biggibum* leaves is at least associated with increased expression of the *MYB2* gene, and the *MYB2* orthologs from orchids likely function differently, regardless of their high level of similarity.

Keywords: anthocyanin; MYB2; orchid; *Dendrobium bigibbum*; γ-irradiation

1. Introduction

In Orchidaceae, *Dendrobium* species are one of the most popular orchids known for their medicinal and commercial value in potted and cut flower industries [1]. The *Dendrobium* genus contains approximately 1800 species and are mainly distributed throughout Asia and the South Pacific [2]. *Dendrobium catenatum* (also named *Dendrobium officinale*), *Dendrobium nobile*, and *Dendrobium candidum* are used in herbal medicines in many Asian countries [3]. Moreover, the *Dendrobium* genus is known for its valuable floral traits including colors, morphologies, and scent and *Dendrobium* species are regarded as some of the important commercial cut flowers. A variety of *Dendrobium* hybrids have been created that have improved flower colors. However, limitation of genetic resources in *Dendrobium* limits the extent to which flower color can be modified.

D. bigibbum is an epiphytic or lithophytic orchid that contains cylindrical pseudobulbs, each having between three and five green or purplish leaves and arching flowering stems with up to 20 usually lilac-purple flowers. The *D. bigibbum* plants containing purple spots on their leaves are very popular in the commercial market. Although colorful leaves and flowers add significant ornamental values to orchids, our understanding of the differential pigmentation in *D. bigibbum* remains limited.

Natural agents extracted from various parts of *Dendrobium* contain bioactive substances, such as phenolic compounds, anthocyanins, and polysaccharides [4–6]. Many of these phenolic compounds and anthocyanins have well-known antioxidant activities [7] and contribute to leaf and flower coloration [8]. Anthocyanins are water-soluble, which are present in the vacuoles of plant epidermal cells and impart an orange, red, or blue color to flowers, fruits, stems, leaves, and roots [9]. Anthocyanin biosynthesis is a well-studied secondary metabolic pathway in plants that involves the conversion of phenylalanine into 4-coumaryl-CoA, followed by their conversions to flavonoid compounds. Studies in antirrhinum [10], petunia [11,12], maize [13,14], *Brassica* [15,16], and *Arabidopsis* [17,18] have identified genes that regulate anthocyanin production, and these can be broadly classified into two major groups. The first group consists of enzymes that participate in anthocyanin biosynthesis, including phenylalanine ammonia-lyase (PAL), chalcone synthetase (CHS), chalcone isomerase (CHI), flavanone 3-β-hydroxylase (F3H), dihydroflavonol 4-reductase (DFR), anthocyanin synthase (ANS), and UDP-glucose flavonoid 3-O-glucosyl transferase [19]. Loss-of-function mutations in *CHS*, *CHI*, *F3H*, *DFR*, or *ANS* abolish anthocyanin biosynthesis, and plants harboring these mutations often produce colorless tissues [20–24]. Anthocyanin accumulations in green and red leaves of *Dendrobium officinale* stems have been associated with *ANS* and UDP-glucose flavonoid 3-O-glucosyl transferase expression [25]. Moreover, Yu et al. suggested that among anthocyanins, delphinidin 3,5-O-diglucoside and cyanidin 3-O-glucoside may be responsible for the red peel color of *D. officinale* [25]. The second group contains MYBs, basic helix-loop-helixes (bHLHs), or WD40 repeat transcription factors (TFs) that regulate the expression levels of genes involved in anthocyanin biosynthesis [26,27]. Earlier studies on *Arabidopsis* and *Medicago truncatula* indicated that MYB2 acts as a transcriptional repressor of anthocyanin biosynthesis and that the overexpression of MYB2 abolishes anthocyanin biosynthesis [28,29]. However, the overexpression of orchid *MYB2* in petunia results in increased petal pigmentation [11]. Likewise, the transient overexpression of *Phalaenopsis equestris MYB2* positively regulates anthocyanin pigmentation and is associated with the increased expression of downstream genes *PeF3H5*, *PeDFR1*, and *PeANS3* [30]. Conversely, silencing of *PeMYB2* results in reduced anthocyanin accumulation [31]. Thus, depending on the plant system, MYB2 appears to serve as either a negative or positive regulator of anthocyanin biosynthesis.

In this study, we characterized an orchid mutant that was isolated on the basis of its unusual leaf color. The *D. bigibbum* mutant accumulated higher levels of anthocyanin, which in turn was associated with the increased expression of genes regulating anthocyanin biosynthesis. This also included the *MYB2* gene, which, when transiently expressed in a heterologous system, led to induction of genes associated with anthocyanin biosynthesis.

2. Results

2.1. The Purple Mutant of D. Bigibbum Accumulates Higher Levels of Anthocyanin

We used γ-irradiated *D. bigibbum* rhizomes to produce a mutant that developed purple leaves in comparison to the green leaves seen on wild type (WT) plants (Figure 1A,B) (Figure S1). This mutant, designated as RB016-S7, was propagated through four generations of tissue culturing. To determine whether the purple coloration of the mutant's leaves was associated with anthocyanin pigmentation, we used a pH-differential-based method to quantify the anthocyanin content. The anthocyanin content in the purple leaves (11.68 mg/g dry weight) was ~7.0-fold higher than in the green leaves (1.66 mg/g dry weight) (Figure 1C). Thus, the purple coloration of RB016-S7 leaves was likely associated with the increased biosynthesis of anthocyanins.

To understand the biochemical basis of the increased anthocyanin production in the RB16-S7 mutant, we analyzed genome-wide changes in gene expression. Total RNAs from WT and RB016-S7 leaves were used to construct six cDNA libraries that were sequenced using the Illumina HiSeq 2500 platform. After filtering and quality trimming the raw reads, we obtained 47–66 million high quality reads. Using Trinity, the clean reads from the six libraries were assembled into 110,104 transcripts, with an average length of 1116 bp, and these were then assembled into 32,575 unigenes, with an average length of 1048 bp (Table 1). The sequence length distribution of unigenes showed that 8373 unigenes (25.7%) ranged from 100 to 500 bp, 11,350 unigenes (34.8%) ranged from 501 to 1000 bp, and 6488 unigenes (19.91%) had lengths of more than 1500 bp (Figure 2). The 30,714 unigenes were matched with the non-redundant (nr) database, and among these, 26,851 unigenes matched sequences from *Dendrobium catenatum,* followed by *Phalaenopsis equestris* (2263) and *Apostasia shenzhenica* (209) (Figure 3A). Furthermore, this was consistent with the phylogenetic analysis carried out among native *Dendrobium spp*, *Cymbidium spp*, *P. equestris* and *A. shenzhenica* orchids, which, as expected, showed relatedness among *Dendrobium spp* (Figure 3B).

Figure 1. Images of *Dendrobium bigibbum* and the anthocyanin contents in the leaves. (**A**) Morphological phenotypes of typical wild type (WT) and RB016-S7 mutant *D. bigibbum* plants. (**B**) Relative anthocyanin contents in the WT and RB016-S7 mutant. Error bars represent standard deviations ($n = 3$). The experiment was repeated three times with similar results. (**C**) The number of up- and downregulated differentially expressed genes (DEGs) in the WT 3 versus RB016-S7 mutant comparison.

Table 1. Summary of RNA sequencing and de novo transcriptome assembly results.

Sequences	Control 1	Control 2	Control 3	RB016-S7-1	RB016-S7-2	RB016-S7-3
BEFORE TRIMMING						
Total nucleotides (bp)	4,962,911,568	5,845,449,712	5,115,158,264	4,956,765,904	5,110,844,048	4,343,315,032
Number of raw reads	65,301,468	76,913,812	67,304,714	65,220,604	67,247,948	57,148,882
AFTER TRIMMING						
Total nucleotides (bp)	4,288,037,900	4,994,465,471	4,370,988,523	4,215,819,862	4,288,137,551	3,561,476,236
Number of clean reads	56,856,166	66,205,490	57,973,956	55,934,410	56,856,134	47,287,332
GC content (%)	46.42	44.22	45.73	47.06	46.79	46.55
Q30 percentage (%)	95.67	95.86	95.56	95.43	95.68	95.22
AFTER ASSEMBLY						
Number of transcripts in the combined data	110,104					
Number of unigenes in the combined data	32,575					
Total nucleotides of transcripts (bp)	122,947,955					
Total nucleotides of unigenes (bp)	34,155,642					
Mean length of transcripts (bp)	1,116					
Mean length of unigenes (bp)	1,048					
N50 of unigenes (bp)	1350					

Q30, base call accuracy of 99.9%; N50, the sequence length of the shortest unigene at 50% of the total genome length.

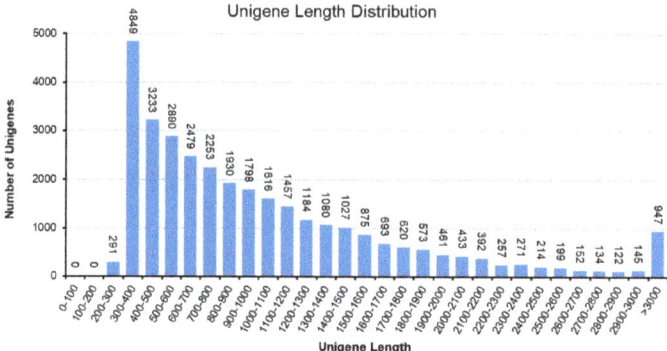

Figure 2. Sequence length distribution of the unigenes in *D. bigibbum* transcriptomes. The *x*-axis indicates unigene length intervals from 200 bp to >3000 bp. The *y*-axis indicates the number of unigenes of each given length.

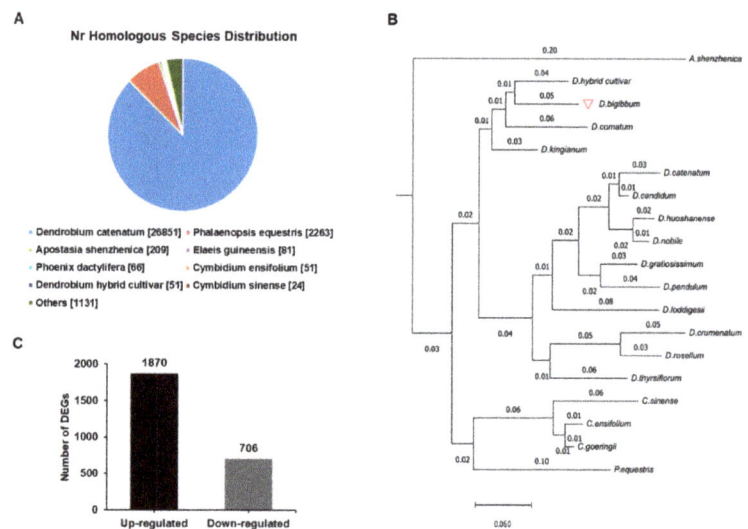

Figure 3. Species distribution of the BLAST search results in the nr database. (**A**) A cut off E-value of 10^{-5} was used. Different species are indicated by different colors. (**B**) A reference phylogenetic tree derived from rDNA ITS 2 sequences of 14 species of *Dendrobium*, 3 species of *Cymbidium*, *Apostasia shenzhenica*, and *Phalaenopsis equestris*. (**C**) The number of up- and downregulated differentially expressed genes (DEGs) in the wild type versus RB016-S7 mutant comparison.

2.2. Functional Annotation and Classification

In the Gene Ontology (GO) analysis, 17,498 unigenes (53.71%) were assigned to three GO terms and were categorized into 41 functional groups (FDR < 0.05) (Table S1). The GO assignments were divided into three categories: biological process (BP), cellular component (CC), and molecular function (MF). Among these, 10,569 unigenes (32.4%), 9195 unigenes (28.2%), and 12,401 unigenes (38%), were assigned to BP, CC, and MF, respectively. In the BP category, the predicted proteins were mainly distributed in metabolic process (30.61%) and cellular process (28.36%), followed by biological regulation (7.34%), localization (6.87%), and regulation of biological process (6.09%). Predicted proteins assigned to the CC category were mainly associated with cellular anatomical entity (55.61%), intracellular (30.19%), and protein-containing complex (12.84%). Furthermore, in the MF category, the most heavily represented groups were linked to catalytic activity (47.13%), binding (40.78%), and transporter activity (5.04%) (Figure S2).

To predict and classify the gene functions, we queried all the unigenes against the evolutionary genealogy of genes: Non-supervised Orthologous Groups (eggNOG) (v4.5) database. This database contains the functional descriptions and classifications of the orthologous proteins, including Clusters of Orthologous Groups and euKaryotic Orthologous Groups. This analysis allowed us to allocate 27,963 unigenes to 25 eggNOG classifications. Among them, the eggNOG category of functional unknown (S, 27.88%) represented the largest group, followed by signal transduction mechanisms (T, 8.58%), posttranslational modification, protein turnover, chaperones (O, 8.13%), transcription (K, 8.09%), and carbohydrate transport and metabolism (G, 5.70%) (Figure S3).

Next, we mapped the assembled unigenes to the reference anthocyanin pathways, including metabolism, genetic information processing, environmental information processing, and cellular processes, in the KEGG (http://www.kegg.jp/kegg/pathway.html). The 6314 unigenes were assigned to 394 KEGG sub-pathways (Table S2). These pathways included KEGG orthology (KO) entries for metabolism (3503 KOs), genetic information processing (950 KOs), environmental information processing (488 KOs), cellular processes (702 KOs), and organismal systems (671 KOs) (Figure S4).

2.3. Analysis of Differentially Expressed Genes (DEGs) Associated with Anthocyanin Biosynthesis

A total of 2513 DEGs (FDR < 0.05) were identified between the WT and RB016-S7 mutant. Compared with WT, 1870 and 706 genes were up- and downregulated in the RB016-S7 mutant, respectively (Figure 3C; Table S3). The top 20 significant pathways for the up- and downregulated genes were selected for further analysis. The upregulated genes were mainly enriched in ribosome biogenesis, MAPK signaling pathway, plant–pathogen interaction, plant hormone signal transduction, phenylpropanoid biosynthesis, starch and sucrose metabolism, and flavonoid biosynthesis (Table 2). The 20 significant pathways for the downregulated genes are listed in Table 3. The downregulated genes were mainly enriched in folate biosynthesis, starch and sucrose metabolism, plant hormone signal transduction, and phenylpropanoid biosynthesis.

Table 2. Top 20 enriched Kyoto Encyclopedia of Genes and Genomes (KEGG) pathways of upregulated differentially expressed genes (DEGs).

Pathway	DEG Number	Pathway ID
Ribosome	58	ko03010
MAPK signaling pathway-plant	40	ko04016
Plant-pathogen interaction	40	ko04626
Plant hormone signal transduction	37	ko04075
Phenylpropanoid biosynthesis	28	ko00940
Starch and sucrose metabolism	25	ko00500
Flavonoid biosynthesis	23	ko00941
Fluid shear stress and atherosclerosis	23	ko05418
Phenylalanine metabolism	19	ko00360
Cancer-related pathways	18	ko05200
Protein processing in the endoplasmic reticulum	18	ko04141
Cellular senescence	16	ko04218
Endocytosis	16	ko04144
Glycolysis/Gluconeogenesis	16	ko00010
β-Alanine metabolism	15	ko00410
Calcium signaling pathway	14	ko04020
Oxytocin signaling pathway	14	ko04921
Phagosome	14	ko04145
Amino sugar and nucleotide sugar metabolism	13	ko00520
Arginine and proline metabolism	13	ko00330

Table 3. Top 20 enriched KEGG pathways of downregulated DEGs.

Pathway	DEG Number	Pathway ID
Folate biosynthesis	13	ko00790
Starch and sucrose metabolism	8	ko00500
Plant hormone signal transduction	7	ko04075
Brassinosteroid biosynthesis	6	ko00905
Phenylpropanoid biosynthesis	6	ko00940
Circadian rhythm - plant	5	ko04712
Cyanoamino acid metabolism	5	ko00460
Glyoxylate and dicarboxylate metabolism	5	ko00630
Protein processing in the endoplasmic reticulum	5	ko04141
Renin-angiotensin system	5	ko04614
β-Alanine metabolism	4	ko00410
Glycine, serine and threonine metabolism	4	ko00260
Lysosome	4	ko04142
Phenylalanine metabolism	4	ko00360
Photosynthesis	4	ko00195
Photosynthesis - antenna proteins	4	ko00196
Platinum drug resistance	4	ko01524
Protein digestion and absorption	4	ko04974
Purine metabolism	4	ko00230
Tropane, piperidine, and pyridine alkaloid biosynthesis	4	ko00960

2.4. Analysis of Anthocyanin Biosynthetic Genes in Identified DEGs

The mutant showed an increased accumulation of anthocyanin; therefore, we used a KEGG functional enrichment to search for genes associated with anthocyanin biosynthesis among the 2513 DEGs. A total of 17 DEGs, encoding eight key enzymes, were identified, and they were three *PAL* genes (*PAL1*: denphalae05809, *PAL2*: denphalae05806, and *PAL4*: denphalae05808), two cinnamic acid 4-hydroxylase genes (*C4H*: denphalae10925 and denphalae10926), four 4-coumarate CoA-ligase genes (*4CL*: denphalae18583, denphalae22607, denphalae27156, and denphalae27157), four *CHS* genes (denphalae02657, denphalae02658, denphalae05188, and denphalae11910), and one gene each of *F3H* (denphalae02991), flavonoid 3′-monooxygenase (*F3′H*: denphalae11915), *DFR* (denphalae11241), and *ANS* (denphalae18276). All these DEGs were significantly upregulated in the RB016-S7 mutant compared with WT (Table 4). Among other notable genes that were upregulated in RB016-S7 were TFs that belonged to WRKY (33 genes), MYB (20 genes), bHLH (23 genes), and WD40 (1 gene) groups. Among these, *DbMYB2*, *-4*, *-30*, and *-44*, as well as *DbbHLH1*, *-62*, *-96*, *-114*, and *-148*, were highly expressed in the RB016-S7 mutant (Table S4). The expression patterns of the anthocyanin biosynthetic genes were consistent with the increased anthocyanin levels in the RB16-S7 mutant (Figure 4A,B).

Table 4. Expression profiles of anthocyanin biosynthetic genes.

Gene Name	Unigene ID	Gene Length	FPKM		Fold Change	log2Fold Change
			Wild Type	S7 Mutant		
PAL1	denphalae05809	2223	3.98	89.98	21.44	4.42
PAL2	denphalae05806	2093	12.98	196.34	14.13	3.82
PAL3	denphalae05808	2139	8.59	71.56	7.45	2.90
C4H	denphalae10925	1518	1.92	18.63	8.97	3.16
	denphalae10926 *	1518	6.09	36.67	5.50	2.46
4CL	denphalae18583 *	1731	15.16	79.17	4.78	2.26
	denphalae22607	1698	3.97	13.60	2.89	1.53
	denphalae27156	1473	0.63	3.23	5.15	2.36
	denphalae27157	1695	2.91	6.02	-	-
CHS	denphalae02657 *	1173	2.52	9.64	3.40	1.77
	denphalae02658	1170	10.50	36.49	3.15	1.65
	denphalae05188	1092	0.52	1.62	-	-
	denphalae11910	1188	102.38	659.26	5.62	2.49
F3H	denphalae02991	1137	98.69	245.15	2.07	1.05
F3'H	denphalae11915	1563	31.60	152.97	4.28	2.10
DFR	denphalae11241	1059	102.75	311.67	2.54	1.34
ANS	denphalae18276	1083	70.86	180.97	2.22	1.15

FPKM, fragments per kilobase of transcript per million mapped reads.

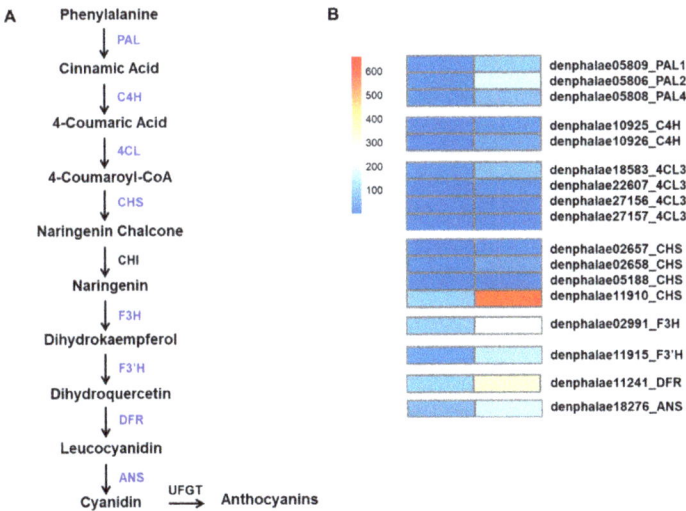

Figure 4. Flavonoid–anthocyanin biosynthetic genes in *D. bigibbum*. (**A**) The differentially expressed genes (DEGs) between the WT and RB016-S7 mutant found in leaves are highlighted in blue. Phenylalanine ammonia-lyase, PAL; cinnamic acid 4-hydroxylase, C4H; 4-coumarate CoA-ligase, 4CL; chalcone synthase, CHS; chalcone isomerase, CHI; flavanone 3-hydroxylase, F3H; flavonoid 3'-monooxygenase, F3'H; dihydroflavonol 4-reductase, DFR; and anthocyanidin synthase, ANS. (**B**) Expression profiles determined using fragments per kilobase of transcript per million mapped reads (FPKM) values obtained from RNA-Seq data. Expression values (as FRKM) were not scaled per row to allow the visualization of original FPKM values among samples. The heatmap was generated using the R package pheatmap.

2.5. Quantitative Real-Time PCR (qRT-PCR) Analysis of the Genes Involved in Anthocyanin Biosynthesis

To confirm the RNA-Seq data, we first selected 10 candidate genes associated with anthocyanin biosynthesis and analyzed their expression levels using qRT-PCR. The qRT-PCR analysis confirmed ~1.5-, 2.3-, and 2.4-fold higher levels for *PAL1*, *PAL2*, and *PAL4*, respectively, in the RB016-S7 mutant compared with the WT. In addition, the qRT-PCR analysis showed that *C4H*, *4CL*, and *CHS* were induced ~2-, 3.3-, and 10-fold in the RB016-S7 mutant compared with the WT. Similarly, *F3'H*, *F3H*, *DFR*, and *ANS* were induced ~1.2-, 7.2-, 5.0-, and 3-fold in the RB016-S7 mutant compared with the WT (Figure 5). The qRT-PCR data were consistent with results obtained from the RNA-Seq data. Thus, the purple pigmentation in RB016-S7 may be associated with the increased expression levels of genes involved in anthocyanin biosynthesis.

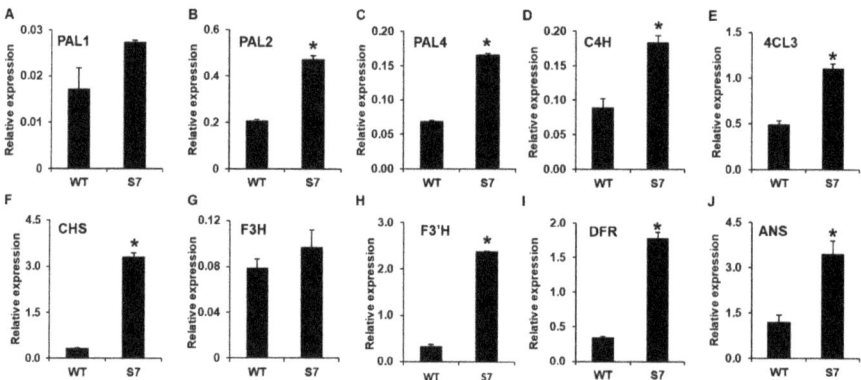

Figure 5. qRT-PCR analysis of 10 genes showing altered expression levels in the RNA sequencing (RNA-Seq) analysis. The genes were associated with anthocyanin biosynthesis. More specifically, (**A–J**) indicate the relative expression levels of *PAL1*, *PAL2*, *PAL3*, *C4H*, *4CL*, *CHS*, *F3H*, *F3'H*, *DFR*, and *ANS*, respectively. The elongation factor 1-alpha (*EF1a*) gene served as an internal control. Error bars indicate standard deviations ($n = 3$). The experiment was repeated three times with similar results. Asterisks denote a significant difference between respective WT and RB016-S7 mutant leaves samples (*t*-text, $p < 0.0001$).

Next, we analyzed the expression levels of regulatory genes associated with anthocyanin biosynthesis. The RNA-Seq dataset showed that *DbMYB2*, *-30*, and *-44* were highly upregulated in the RB016-S7 mutant compared with the WT, while the expression of *DbMYB75* was not significantly different from in WT plants. Notably, the qRT-PCR analysis was only able to confirm a ~13-fold induction in *DbMYB2*, while the expression of *DbMYB30*, *-44*, and *-75* remained at WT levels.

Comparisons of expression levels of genes encoding bHLH TFs showed that only *DbbHLH1* was expressed at higher levels in RB016-S7 than WT plants. In comparison, *DbbHLH96*, *-114*, and *-153* showed WT-like expression levels. *DbWD40*, which showed a 67.97% identity to the *Arabidopsis* ortholog *AtTTG1*, had a WT-like expression level [32]. A recent report also suggests roles for WRKY TFs in anthocyanin biosynthesis. RNA-Seq data showed that several WRKY TFs were highly expressed in the RB016-S17 mutant compared with WT. However, the qRT-PCR analysis was only able to confirm ~1.5–3-fold inductions of *DbWRKY24*, *WRKY31*, and *WRKY40* genes (Figure 6). Thus, only a select group of TFs were upregulated in the mutant plant, and these, in turn, could play roles in the regulation of genes involved in anthocyanin biosynthesis.

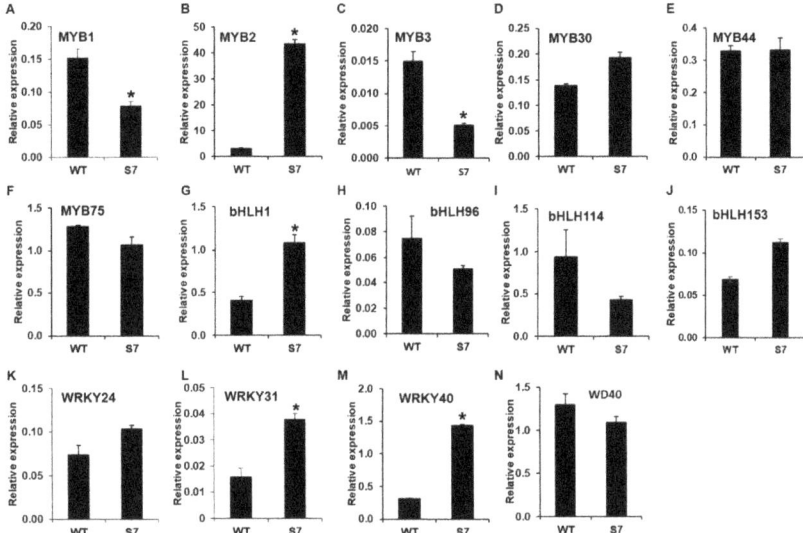

Figure 6. qRT-PCR analysis of 14 genes showing altered expression levels in the RNA-Seq analysis. The relative expression levels of transcription factor genes in the leaves. More specifically, (**A–N**) indicate the relative expression levels of *MYB1*, *MYB2*, *MYB3*, *MYB30*, *MYB44*, *MYB75*, *bHLH1*, *bHLH96*, *bHLH114*, *bHLH153*, *WRKY24*, *WRKY31*, *WRKY40*, and *WD40*. The *EF1a* gene served as an internal control. Error bars indicate standard deviations ($n = 3$). The error bars indicate SD ($n = 3$). Results are representative of two independent experiments. Asterisks denote a significant difference between respective WT and RB016-S7 mutant leaves samples (*t*-test, $p < 0.0001$).

2.6. DbMYB2 Positively Regulates Anthocyanin Biosynthesis

Increased expression of *DbMYB2* in the RB016-S17 mutant suggested that MYB2 could positively regulate expression of anthocyanin genes and thereby anthocyanin levels. This is further supported by an earlier study that showed that *Dendrobium* hybrid MYB2 positively regulated anthocyanin pigmentation in *Dendrobium* petals. Amino acid alignment of DbMYB2 with DhMYB2 BS No.3 [33] showed ~92% identity. Likewise, amino acid alignment of MYB2 orthologs from *D.* hybrid, *D. candidum*, *D. nobile*, and *Cymbidium sinense* showed ~80%, ~80%, ~62%, and 63% identity, respectively, with DbMYB2 (Figure 7A). The amino acid alignment showed that the R2R3 repeat region was highly conserved among various MYB proteins (Figure 7A). Phylogenetic analysis between these MYB proteins placed DbMYB2, DhMYB2, and DcMYB2 in the same clade (Figure 7B).

To determine whether increased expression of DbMYB2 positively regulated expression of anthocyanin genes, we expressed *MYB2* genes from *D. bigibbum*, *D. candidum*, *D. nobile*, *D.* hybrid, and *C. sinense* in *Nicotiana benthamiana* and evaluated expression of *N. benthaminana* genes *ANS*, *DFR*, and *CHS*, which are associated with anthocyanin biosynthesis. All the *MYB2* genes showed varying levels of increased expression at 36 h post-agroinfiltration (Figure 8D–H). Interestingly, however, only transient expression of *DbMYB2* was associated with increased expression of *ANS*, *DFR*, and *CHS* in *N. benthamiana* (Figure 8A–C). These results strongly suggest that DbMYB2 positively regulates expression of genes associated with anthocyanin biosynthesis, and that higher anthocyanin levels in the RB016-S17 mutant are likely due to higher expression levels of *DbMYB2*. Inability of other *MYB2* orthologs to increase expression of *ANS*, *DFR*, and *CHS* suggests that, regardless of their homology, the *MYB2* orthologs function differently.

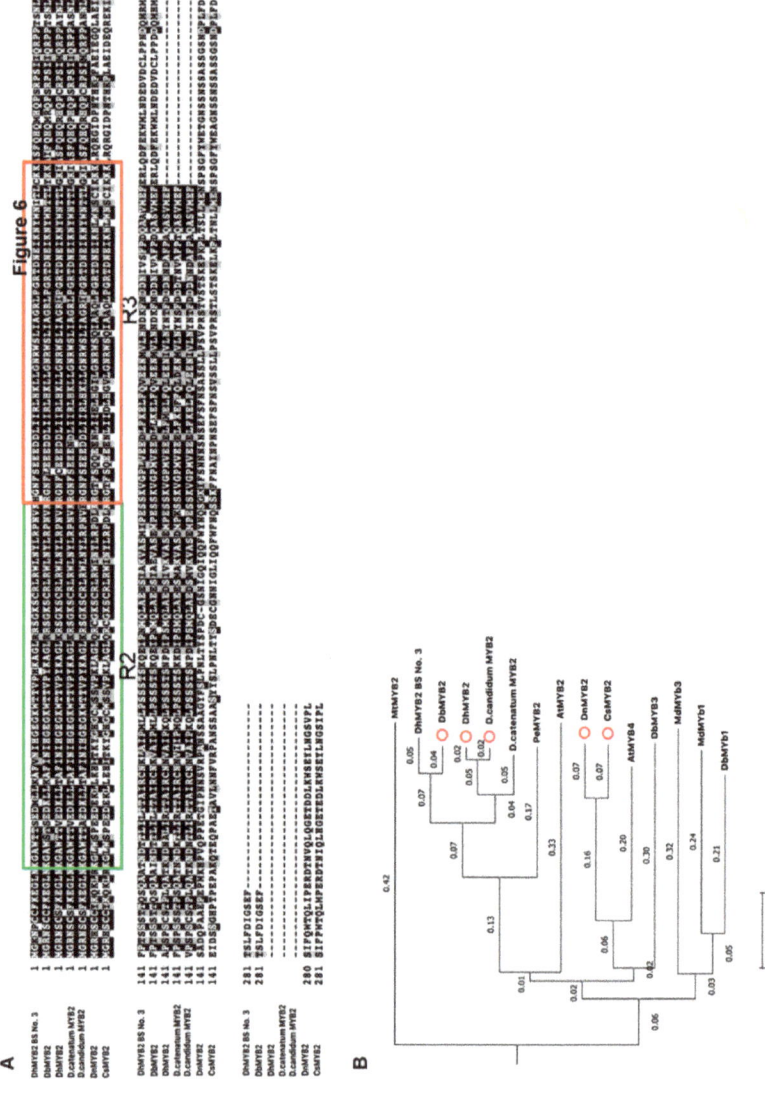

Figure 7. Sequence alignments and phylogenetic tree of DbMYB2 from orchids with known R2R3-MYBs domain. (**A**) Alignments of the full-length deduced amino acid sequences of DbMYB2 with other R2R3-MYBs present in other orchids. (**B**) Phylogenetic relationship of DbMYB2 with known anthocyanin MYB regulators from other orchid species.

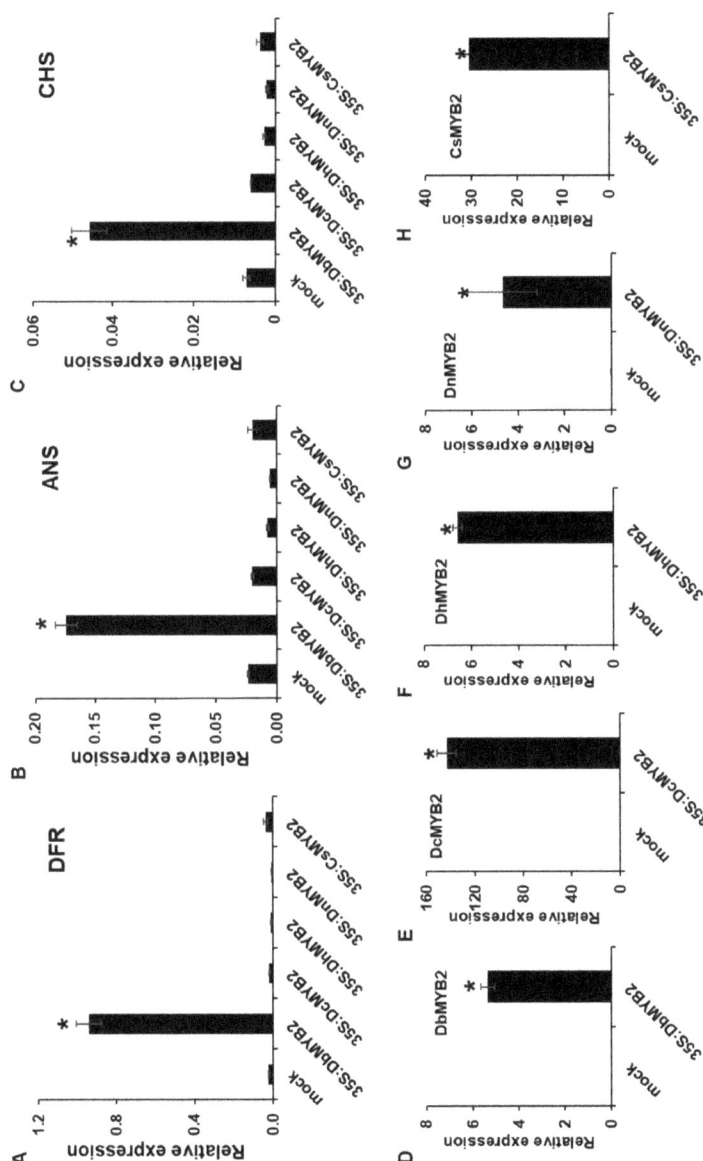

Figure 8. Transient expression of DbMYB2s were associated with increased expression of ANS, DFR, and CHS in *N. benthamiana*. (**A–C**) Real-time quantitative RT-PCR showing relative expression levels of *NbDFR*, *NbANS*, and *NbCHS* genes in *N. benthamiana* transiently overexpressing *MYB2* genes from different orchids. (**D–H**) Real-time quantitative RT-PCR showing relative expression levels of *MYB2* orthologs from different orchids. The error bars indicate SD ($n = 4$). Results are representative of three independent experiments. Asterisks denote a significant difference with mock and MYB2 overexpressed plants (*t*-test, $p < 0.0001$).

3. Discussion

Anthocyanins are pigments that confer color to various plant parts [34]. The color is determined by the composition and concentration of pigments, which vary greatly among plant species [35]. Cyanidin-3-glucoside is a major anthocyanin found in most plants [36]. Other common anthocyanin pigments present in plants include delphinidin, pelargonidin, peonidin, malvidin, and petunidin. Earlier studies on *Dendrobium* orchids primarily focused on anthocyanin profiles in flowers and stems, which contain pelargonidin, cyanidin, peonidin, delphinidin, and/or malvidin [24]. In contrast, we were only able to detect malvidin in the leaves of *D. bigibbum* (data not shown), and its levels were associated with increased purple pigmentation in the RB016-S7 mutant's leaves. Thus, anthocyanin pigments present in leaves versus flowers and stems might be associated with the specific genes expressed in these tissues. We determined that the increased anthocyanin accumulation in RB016-S7 was associated with increased expression levels of *PAL*, *CHS*, *F3′H*, and *DFR* genes that are involved in anthocyanin biosynthesis. Although the anthocyanin biosynthetic genes are well-conserved, the timing, level, and spatial distribution of anthocyanin biosynthesis are primarily determined by TFs.

A recent study offered useful insights into the functions of WRKY TFs in anthocyanin biosynthesis [37], which in turn regulates the MYB/bHLH/WD40complex [27,38–40]. Our analysis also identified 33 WRKY-, 20 MYB-, 23 bHLH- and 1WD40-encoding genes that were differentially expressed in the RB016-S7 mutant leaves. It is possible that these TFs regulate the expression of one or more genes involved in the anthocyanin pathway (Table S5). An example of complex regulation underlying anthocyanin biosynthesis includes the feed-forward loop mechanism in which TFs regulate each other and jointly regulate target genes [28]. In *Dendrobium* hybrid petals, *DhMYB2* and *DhbHLH1* TFs play regulatory roles in the anthocyanin biosynthetic pathway [33]. Consistent with this finding, we determined that the expression levels of *DbMYB2* and *DbHLH1* were significantly higher in the RB016-S7 mutant. Notably, the *A. thaliana* and *M. truncatula* MYB2 proteins act as transcriptional repressors of anthocyanin biosynthesis, and the overexpression of either MYB2 abolishes anthocyanin biosynthesis [28,29]. Likewise, heterologous overexpression of *Malus domestica* MYB3 (*MdMYB3*) in *Nicotiana tabacum* is associated with increased anthocyanin biosynthesis [41]. Thus, orthologs of MYB and possibly other genes involved in anthocyanin biosynthesis may play opposite roles in different plants. This was further evident in our analysis, which showed that increased expression of *MYB2* in RB016-S7 plants positively correlated with anthocyanin biosynthesis. This was further consistent with our result that heterologous overexpression of *D. bigibbum MYB2* in *N. benthamiana* led to increased expression of *ANS*, *DFR*, and *CHS* genes. Interestingly, transient overexpression of *MYB2* orthologs from other *Dendrobium spp.* or *C. sinense* did not alter expression of *ANS*, *DFR*, and *CHS* genes. These results strongly suggest that DbMYB2 positively regulates expression of genes associated with anthocyanin biosynthesis, even though DbMYB2 showed high levels of homology with other MYB2 orthologs. Thus, subtle changes in MYB2 sequence are likely sufficient to alter their function. It is possible that MYB2 in other orchids could regulate anthocyanin biosynthesis genes by serving as a part of the bigger complex that contains other factors like bHLHs or WD40. Deciphering the exact biochemical functions of various TFs involved in anthocyanin biosynthesis in *D. bigibbum* will require more detailed analyses of these proteins.

4. Materials and Methods

4.1. Plant Materials

Protocorm-like bodies (PLBs) of *Dendrobium bigibbum* var. compactum were cultured on pH5.3 Hyponex medium (6.5:6.0:19:0 N:P:K; Hyponex Japan Corp., Ltd, Osaka, Japan) supplemented with sucrose (3% *w/v*) and agar (0.4% *w/v*) (Duchefa Biochemie B.V., Haarlem, The Netherlands). PLBs were cultured in 220 ml glass jars containing 30ml medium, which were closed with semipermeable plastic caps. All the cultures were maintained at 22–25 °C and >60% humidity. Plants were grown under white fluorescent light (PPFD = 50 μmol/m^2/s) with a 16-h illumination and 8-h dark photoperiod.

4.2. γ-Irradiation of in Vitro Shoot Cultures

Six-month-old in vitro regenerated shoots of approximately 3 cm in length were exposed to γ-radiation using 60Co γ-irradiator (60 Gy/24 h) at the Korea Atomic Energy Research Institute, Jeongeup, Korea [42]. The first vegetative generation in which treatment was performed was referred to as M1V1. The study continued until the fourth generation (M1V4) to confirm the stability of the induced traits. The purple-colored leaf mutant RB016-S7 was obtained and its physiological traits were analyzed.

4.3. RNA Extraction and Quantitative Real-Time PCR (qRT-PCR) Analysis

Total RNA was extracted from the WT and the RB016-S7 plants using an RNeasy plant mini kit (Qiagen, Hilden, Germany), following the manufacturer's instructions. The RNA concentration and quality of each sample were determined using a Nanodrop 2000 spectrophotometer (Thermo Fisher Scientific, Waltham, MA, USA) and agarose electrophoresis, respectively. The cDNA was transcribed from 500 ng of total RNA using a ReverTra Ace-α-kit (Toyobo Co. Ltd, Osaka, Japan). The qRT-PCR was performed with a CFX96 touch real-time PCR detection system (Bio-Rad, Hercules, CA, USA) using iQ™ SYBR® Green supermix (Bio-Rad, Hercules, CA, USA). The *D. bigibbum* actin gene was used as an internal control, and the $2^{-\Delta\Delta Ct}$ method was used to analyze differential expression levels. Cycle threshold values were calculated using CFX Manager 3.1 software (Bio-Rad, Hercules, CA, USA). Gene-specific primers are listed in Table S5.

4.4. Measurement of Total Anthocyanin Content

After harvesting the leaves, the samples were freeze-dried and subjected to solvent extraction using a solution of 85% ethanol acidified with 15% 1.5 N HCl. The samples were incubated at 4 °C for 24 h. Samples were diluted in two buffer solutions: potassium chloride buffer 0.025 M (pH 1.0) and sodium acetate buffer 0.4 M (pH 4.5). Absorbance was measured via spectrophotometer at 510 and 700 nm after 15 min of incubation at room temperature, respectively. Absorbance was calculated as

$$\text{Anthocyanin pigment (cyanidin} - 3 - \text{glucoside equivalents, mg/L)} = \frac{A \times MW \times DF \times 10^3}{\varepsilon \times l} \quad (1)$$

A = (A510 nm − A700 nm) pH1.0 − (A510 nm − A700 nm) pH4.5; MW (molecular weight) = 449.2 g/mol for cyanidin-3-glucoside (cyd-3-glu); DF = dilution factor; l = pathlength in cm; ε = 26,900 molar extinction coefficient, in L·mol^{-1}·cm^{-1}, for cyd-3-glu; and 10^3 = factor for conversion from g to mg [43].

4.5. RNA-Seq Analysis, De Novo Assembly, and Unigene Generation

The cDNA libraries were prepared independently from both WT and RB016-S7 leaves. Low quality and duplicated reads, as well as adapter sequences, were removed from RNA-seq raw data using Trimmomatic with default parameters [44]. The de novo assembly was performed using Trinity (ver. 2.8.4) with default parameters [45]. Afterwards, redundant sequences were removed from the assembled transcript sequences using cd-hit-est (ver. 4.7) with a similarity threshold of 90% (i.e., removing similar sequences sharing more than a 90% identity), generating nr transcript sequences. Protein coding sequences (CDSs) were predicted and extracted from the nr transcript sequences using TransDecoder (ver. 5.5) with a parameter of selection of the longest CDS by comparison with Pfam database. The collection of extracted CDSs was designated as the unigene set and used for further analyses. The completeness of the unigene set was validated by analysis with Benchmarking Universal Single-Copy Orthologs (ver. 3.1.0) [46].

4.6. Functional Annotation of Unigenes

Sequences homologous to unigenes were identified using BLASTP analyses (cutoff e-value 1e-5) against the NCBI nr protein database. The GO terms, and eggnog (ver. 3.0) and KEGG pathways, were assigned to the unigenes based on BLASTP results using the Blast2GO program (ver. 5.2.5). Conserved domains in the unigene sequences were identified using InterProScan program (ver. 5.34-73.0) with default parameters. In addition, a KEGG pathway analysis was also performed with the KEGG Automatic Annotation KAAS Server using the single-directional best hit method and searching against representative gene sets from both eukaryotes and monocots.

4.7. Expression Profiling of Unigenes

Trimmed high-quality RNA-seq reads were mapped on the unigene sequences using BWA (ver. 0.7.17-r1188) [47] and then, RNA reads mapped on unigene sequences were counted using SAMtools (ver. 1.9) [48]. Fragments per kilobase of transcript per million mapped reads values were calculated using the number of RNA-seq reads mapped on unigene sequences and used for the expression profiling of unigenes.

4.8. Identification of DEGs between the WT and RB016-S7 Mutant

The bioconductor package DESeq (ver. 1.22.1) was used to identify DEGs between samples [49]. Genes showing over two-fold expression changes with *p*-values of less than 0.05 were considered DEGs. The GO enrichment analysis was performed for the DEGs using Fisher's exact test with an adjusted *p*-value of 0.05 in the Blast2GO program (ver. 5.2.5).

4.9. Analysis of Unigenes Involved in Anthocyanin Biosynthesis

For anthocyanin biosynthesis, unigenes assigned to the anthocyanin biosynthetic reference pathway from the KEGG pathway analysis were first selected. In addition, BLASTP searches (cutoff e-value: 1e-10) were performed using 40 *Arabidopsis* genes involved in anthocyanin biosynthesis [50] as queries, and then, unigenes with high similarity levels ($\geq$ 60% identity and $\geq$ 80% alignment length) to the query sequences were selected as candidate unigenes that could be involved in anthocyanin biosynthesis. Expression values (fragments per kilobase of transcript per million mapped reads) for the genes were retrieved from expression profiles of the unigenes set and used for generating heatmaps using the R -package pheatmap (ver. 1.0.12).

4.10. Cloning of Orchid MYB Genes

The full-length *MYB2* cDNA (denphalae23719) was PCR-amplified from *D. bigibbum* leaves and cloned into the Gateway binary vector pMDC32 vector, under the 35S CaMV promoter. The primers used for amplification of *MYB2* sequences are listed in Table S5. All the amplified products were sequenced.

4.11. Agroinfiltration of N. Benthamiana

The pMDC32-MYB2 plasmids were transformed into *A. tumefaciens* strains LBA4404 via the freeze–thaw method [51]. Agrobacteria were grown in the LB medium supplemented with 100 mg/L kanamycin and incubated at 28 °C with shaking. Bacteria were pelleted by centrifugation (14,000*g* for 5 min) and resuspended to an OD_{600} = 0.8 in a buffer containing 10 mM MES pH 5.6, 10 mM $MgCl_2$, and 200 µM acetosyringone. Cultures were then incubated for 2–4 h at room temperature. Bacteria were infiltrated into the underside of *N. benthamiana* leaves using a needleless 1 ml syringe. The agroinfiltrated plants were kept in the growth chamber maintained at 23 °C with a 16-h photoperiod.

Supplementary Materials: Supplementary materials can be found at http://www.mdpi.com/1422-0067/21/16/5653/s1. Figure S1. Images of *D. bigibbum*. Morphological phenotypes of typical wild type (WT) and the RB016-S7 mutant. Figure S2. Distribution of annotated sequences based on a Gene Ontology (GO) analysis. The GO functional classification assigned 17,498 unigenes to 41 subcategories under the three main GO categories of biological process, cellular component, and molecular function. The *x*-axis indicates the subcategories and the *y*-axis indicates the number of unigenes in each category. Figure S3. Numeric distribution of eggNOG annotations of unigenes. Letters on the *x*-axis refer to the categories on the right. The *y*-axis indicates the number of unigenes in the corresponding eggNOG category. Figure S4. Distribution of annotated sequences as assessed using a Kyoto Encyclopedia of Genes and Genomes (KEGG) pathway analysis. The *x*-axis indicates enriched KEGG pathways, and the *y*-axis represents the number of unigenes within each KEGG pathway. (A) Metabolism; (B) genetic information processing; (C) environmental information processing; (D) cellular processes; (E) organismal systems. Table S1. Significantly enriched GO terms. Table S2: Significantly enriched KEGG pathways. Table S3: Clusters of annotated GO terms in the biological process category enriched in up- and downregulated genes between WT and the RB016-S7 leaves. Table S4: Expression profiles of the regulatory genes for leaf color in *D. bigibbum*. Table S5: Gene-specific primers used for quantitative real-time PCR and *Agrobacterium* transient assay. Table S6: MYB2 gene sequence from *Dendrobium* orchids and *Cymbidium sinense* for *Agrobacterium* transient assay.

Author Contributions: Conceptualization, G.-H.L. Formal analysis, G.-H.L.; Funding acquisition, S.-Y.K. and J.-B.K.; Investigation, S.W.K.; Methodology, J.R.; Project administration, S.-Y.K., J.-B.K. and S.H.K.; Resources, S.H.K.; Writing–original draft, G.-H.L. All authors have read and agreed to the published version of the manuscript.

Funding: This work was supported by a grant from the Nuclear R&D program of the Ministry of Science and ICT and the research program of KAERI, Republic of Korea.

Conflicts of Interest: The authors declare no conflict of interest.

Abbreviations

ANS	anthocyanidin synthase
4CL	4-coumarate CoA ligase
C4H	cinnamic acid 4-hydroxylase
CHI	chalcone isomerase
CHS	chalcone synthase
DFR	dihydroflavonol 4-reductase
F3H	flavanone 3-hydroxylase
F3′H	flavonoid 3′-monooxygenase
GO	gene ontology
KEGG	Kyoto Encyclopedia of Genes and Genomes
PAL	phenylalanine ammonia-lyase
WT	wild -type

References

1. Wang, H.Z.; Feng, S.G.; Lu, J.J.; Shi, N.N.; Liu, J.J. Phylogenetic study and molecular identification of 31 *Dendrobium* species using inter-simple sequence repeat (ISSR) markers. *Sci. Hortic.* **2009**, *122*, 440–447. [CrossRef]
2. Feng, S.G.; Lu, J.J.; Gao, L.; Liu, J.J.; Wang, H.-Z. Molecular phylogeny analysis and species identification of *Dendrobium* (*Orchidaceae*) in China. *Biochem. Genet.* **2014**, *52*, 127–136. [CrossRef] [PubMed]
3. Bulpitt, C.J.; Li, Y.; Bulpitt, P.F. The use of orchids in Chinese medicine. *Genome Natl. Res. Counc. Can.* **2007**, *100*, 558–563.
4. Wang, H.; Zhang, T.; Sun, W.; Wang, Z.; Zuo, D.; Zhou, Z.; Li, S.; Xu, J.; Yin, F.; Hua, Y. Erianin induces G2/M-phase arrest, apoptosis, and autophagy via the ROS/JNK signaling pathway in human osteosarcoma cells in vitro and in vivo. *Cell Death Dis.* **2016**, *7*, e2247. [CrossRef]
5. Khoo, H.E.; Azlan, A.; Tang, S.T.; Lim, S.M. Anthocyanidins and anthocyanins: Colored pigments as food, pharmaceutical ingredients, and the potential health benefits. *Food Nutr. Res.* **2017**, *61*, 1361779. [CrossRef]
6. Muddathir, A.M.; Yamauchi, K.; Batubara, I.; Mohieldin, E.A.M.; Mitsunaga, T. Anti-tyrosinase, total phenolic content and antioxidant activity of selected Sudanese medicinal plants. *S. Afr. J. Bot.* **2017**, *109*, 9–15. [CrossRef]
7. Szymanowska, U.; Baraniak, B. Antioxidant and Potentially Anti-Inflammatory Activity of Anthocyanin Fractions from Pomace Obtained from Enzymatically Treated Raspberries. *Antioxidants* **2019**, *8*, 299. [CrossRef]
8. Jaakola, L. New insights into the regulation of anthocyanin biosynthesis in fruits. *Trends Plant. Sci.* **2013**, *18*, 477–483. [CrossRef]

9. Sadilova, E.; Stintzing, F.C.; Carle, R. Anthocyanins, colour and antioxidant properties of eggplant (Solanum melongena L.) and violet pepper (Capsicum annuum L.) peel extracts. *Zeitschrift für Naturforschung C* **2006**, *61*, 527–535. [CrossRef]
10. Goodrich, J.; Carpenter, R.; Coen, E.S. A common gene regulates pigmentation pattern in diverse plant species. *Cell* **1992**, *68*, 955–964. [CrossRef]
11. Wang, L.M.; Zhang, J.; Dong, X.Y.; Fu, Z.Z.; Jiang, H.; Zhang, H.C. Identification and functional analysis of anthocyanin biosynthesis genes in Phalaenopsis hybrids. *Biol. Plant.* **2018**, *62*, 45–54. [CrossRef]
12. Quattrocchio, F.; Wing, J.F.; Va, K.; Mol, J.N.M.; Koes, R. Analysis of bHLH and MYB domain proteins: Species-specific regulatory differences are caused by divergent evolution of target anthocyanin genes. *Plant. J.* **1998**, *13*, 475–488. [CrossRef] [PubMed]
13. Cone, K.C.; Cocciolone, S.M.; Burr, F.A.; Burr, B. Maize anthocyanin regulatory gene pl is a duplicate of c1 that functions in the plant. *Plant. Cell* **1993**, *5*, 1795–1805. [PubMed]
14. Hanson, M.A.; Gaut, B.S.; Stec, A.O.; Fuerstenberg, S.I.; Goodman, M.M.; Coe, E.H.; Doebley, J.F. Evolution of anthocyanin biosynthesis in maize kernels: The role of regulatory and enzymatic loci. *Genetics* **1996**, *143*, 1395–1407.
15. Kim, C.; Park, S.; Kikuchi, S.; Kwon, S.; Park, S.; Yoon, U.; Park, D.; Seol, Y.; Hahn, J.; Park, S. Genetic analysis of gene expression for pigmentation in Chinese cabbage (Brassica rapa). *Biochip J.* **2010**, *4*, 123–128. [CrossRef]
16. Guo, N.; Cheng, F.; Wu, J.; Liu, B.; Zheng, S.; Liang, J.; Wang, X. Anthocyanin biosynthetic genes in Brassica rapa. *BMC Genom.* **2014**, *15*, 426. [CrossRef]
17. Borevitz, J.O.; Xia, Y.; Blount, J.; Dixon, R.A.; Lamb, C. Activation tagging identifies a conserved MYB regulator of phenylpropanoid biosynthesis. *Plant. Cell* **2000**, *12*, 2383–2393. [CrossRef]
18. Payne, C.T.; Zhang, F.; Lloyd, A.M. GL3 encodes a bHLH protein that regulates trichome development in Arabidopsis through interaction with GL1 and TTG1. *Genetics* **2000**, *156*, 1349–1362.
19. Holton, T.A.; Cornish, E.C. Genetics and biochemistry of anthocyanin biosynthesis. *Plant. Cell* **1995**, *7*, 1071. [CrossRef]
20. Britsch, L.; Ruhnau-Brich, B.; Forkmann, G. Molecular cloning, sequence analysis, and in vitro expression of flavanone 3 beta-hydroxylase from *Petunia hybrida*. *J. Biol. Chem.* **1992**, *267*, 5380–5387.
21. Franken, P.; Niesbach-Klösgen, U.; Weydemann, U.; Maréchal-Drouard, L.; Saedler, H.; Wienand, U. The duplicated chalcone synthase genes C2 and Whp (white pollen) of *Zea mays* are independently regulated; evidence for translational control of Whp expression by the anthocyanin intensifying gene in. *EMBO J.* **1991**, *10*, 2605–2612. [CrossRef]
22. Inagaki, Y.; Hisatomi, Y.; Iida, S. Somatic mutations caused by excision of the transposable element, Tpn1, from the DFR gene for pigmentation in sub-epidermal layer of periclinally chimeric flowers of Japanese morning glory and their germinal transmission to their progeny. *Theor. Appl. Genet.* **1996**, *92*, 499–504. [CrossRef] [PubMed]
23. Nakatsuka, T.; Nishihara, M.; Mishiba, K.; Yamamura, S. Temporal expression of flavonoid biosynthesis-related genes regulates flower pigmentation in gentian plants. *Plant. Sci.* **2005**, *168*, 1309–1318. [CrossRef]
24. Napoli, C.A.; Fahy, D.; Wang, H.-Y.; Taylor, L.P. White anther: A petunia mutant that abolishes pollen flavonol accumulation, induces male sterility, and is complemented by a chalcone synthase transgene. *Plant Physiol.* **1999**, *120*, 615–622. [CrossRef]
25. Yu, Z.; Liao, Y.; Teixeira da Silva, J.A.; Yang, Z.; Duan, J. Differential accumulation of anthocyanins in dendrobium officinale stems with red and green peels. *Int J. Mol. Sci.* **2018**, *19*, 2857. [CrossRef] [PubMed]
26. Allan, A.C.; Hellens, R.P.; Laing, W.A. MYB transcription factors that colour our fruit. *Trends Plant Sci.* **2008**, *13*, 99–102. [CrossRef] [PubMed]
27. Schaart, J.G.; Dubos, C.; Romero De La Fuente, I.; van Houwelingen, A.M.M.L.; de Vos, R.C.H.; Jonker, H.H.; Xu, W.; Routaboul, J.M.; Lepiniec, L.; Bovy, A.G. Identification and characterization of MYB-b HLH-WD 40 regulatory complexes controlling proanthocyanidin biosynthesis in strawberry (F ragaria× ananassa) fruits. *New Phytol.* **2013**, *197*, 454–467. [CrossRef]
28. Matsui, K.; Umemura, Y.; Ohme-Takagi, M. A protein with a single MYB domain, acts as a negative regulator of anthocyanin biosynthesis in Arabidopsis. *Plant. J.* **2008**, *55*, 954–967. [CrossRef]

29. Jun, J.H.; Liu, C.; Xiao, X.; Dixon, R.A. The transcriptional repressor MYB2 regulates both spatial and temporal patterns of proanthocyandin and anthocyanin pigmentation in Medicago truncatula. *Plant Cell* **2015**, *27*, 2860–2879.
30. Hsu, C.-C.; Chen, Y.-Y.; Tsai, W.-C.; Chen, W.-H.; Chen, H.-H. Three R2R3-MYB transcription factors regulate distinct floral pigmentation patterning in Phalaenopsis spp. *Plant Physiol.* **2015**, *168*, 175–191. [CrossRef]
31. Lu, H.-C.; Hsieh, M.-H.; Chen, C.-E.; Chen, H.-H.; Wang, H.-I.; Yeh, H.-H. A high-throughput virus-induced gene-silencing vector for screening transcription factors in virus-induced plant defense response in orchid. *Mol. Plant Microbe Interact.* **2012**, *25*, 738–746. [CrossRef] [PubMed]
32. Zhang, B.; Schrader, A. TRANSPARENT TESTA GLABRA 1-dependent regulation of flavonoid biosynthesis. *Plants* **2017**, *6*, 65. [CrossRef] [PubMed]
33. Li, C.; Qiu, J.; Ding, L.; Huang, M.; Huang, S.; Yang, G.; Yin, J. Anthocyanin biosynthesis regulation of DhMYB2 and DhbHLH1 in Dendrobium hybrids petals. *Plant Physiol. Biochem.* **2017**, *112*, 335–345. [CrossRef] [PubMed]
34. Chen, C. Overview of plant pigments. In *Pigments in Fruits and Vegetables*; Springer: New York, NY, USA, 2015; pp. 1–7.
35. Veberic, R.; Slatnar, A.; Bizjak, J.; Stampar, F.; Mikulic-Petkovsek, M. Anthocyanin composition of different wild and cultivated berry species. *LWT* **2015**, *60*, 509–517. [CrossRef]
36. Noda, N.; Yoshioka, S.; Kishimoto, S.; Nakayama, M.; Douzono, M.; Tanaka, Y.; Aida, R. Generation of blue chrysanthemums by anthocyanin B-ring hydroxylation and glucosylation and its coloration mechanism. *Sci. Adv.* **2017**, *3*, e1602785. [CrossRef]
37. Lloyd, A.; Brockman, A.; Aguirre, L.; Campbell, A.; Bean, A.; Cantero, A.; Gonzalez, A. Advances in the MYB-bHLH-WD Repeat (MBW) Pigment Regulatory Model: Addition of a WRKY Factor and Co-option of an Anthocyanin MYB for Betalain Regulation. *Plant Cell Physiol.* **2017**, *58*, 1431–1441. [CrossRef]
38. Albert, N.W.; Lewis, D.H.; Zhang, H.; Schwinn, K.E.; Jameson, P.E.; Davies, K.M. Members of an R2R3-MYB transcription factor family in Petunia are developmentally and environmentally regulated to control complex floral and vegetative pigmentation patterning. *Plant J.* **2011**, *65*, 771–784. [CrossRef]
39. Gonzalez, A.; Brown, M.; Hatlestad, G.; Akhavan, N.; Smith, T.; Hembd, A.; Moore, J.; Montes, D.; Mosley, T.; Resendez, J. TTG2 controls the developmental regulation of seed coat tannins in Arabidopsis by regulating vacuolar transport steps in the proanthocyanidin pathway. *Dev. Biol.* **2016**, *419*, 54–63. [CrossRef]
40. Ramsay, N.A.; Glover, B.J. MYB-bHLH-WD40 protein complex and the evolution of cellular diversity. *Trends Plant Sci.* **2005**, *10*, 63–70. [CrossRef]
41. Vimolmangkang, S.; Han, Y.; Wei, G.; Korban, S.S. An apple myb transcription factor, MdMYB3, is involved in regulation of anthocyanin biosynthesis and flower development. *BMC Plant Biol.* **2013**, *13*, 176. [CrossRef]
42. Jo, Y.D.; Kim, J.-B. Frequency and Spectrum of Radiation-Induced Mutations Revealed by Whole-Genome Sequencing Analyses of Plants. *Quantum Beam Sci.* **2019**, *3*, 7. [CrossRef]
43. Lee, J.; Durst, R.W.; Wrolstad, R.E. Determination of total monomeric anthocyanin pigment content of fruit juices, beverages, natural colorants, and wines by the pH differential method: Collaborative study. *J. AOAC Int.* **2005**, *88*, 1269–1278. [CrossRef] [PubMed]
44. Bolger, A.M.; Lohse, M.; Usadel, B. Trimmomatic: A flexible trimmer for Illumina sequence data. *Bioinformatics* **2014**, *30*, 2114–2120. [CrossRef]
45. Grabherr, M.G.; Haas, B.J.; Yassour, M.; Levin, J.Z.; Thompson, D.A.; Amit, I.; Adiconis, X.; Fan, L.; Raychowdhury, R.; Zeng, Q. Full-length transcriptome assembly from RNA-Seq data without a reference genome. *Nat. Biotechnol.* **2011**, *29*, 644. [CrossRef] [PubMed]
46. Simão, F.A.; Waterhouse, R.M.; Ioannidis, P.; Kriventseva, E.V.; Zdobnov, E.M. BUSCO: Assessing genome assembly and annotation completeness with single-copy orthologs. *Bioinformatics* **2015**, *31*, 3210–3212. [CrossRef]
47. Li, H.; Durbin, R. Fast and accurate short read alignment with Burrows–Wheeler transform. *Bioinformatics* **2009**, *25*, 1754–1760. [CrossRef]
48. Li, H.; Handsaker, B.; Wysoker, A.; Fennell, T.; Ruan, J.; Homer, N.; Marth, G.; Abecasis, G.; Durbin, R. The sequence alignment/map format and SAMtools. *Bioinformatics* **2009**, *25*, 2078–2079. [CrossRef]
49. Anders, S.; Huber, W. Differential expression analysis for sequence count data. *Genome. Bio.* **2010**, *11*, R106–R110. [CrossRef]

50. Ren, M.; Wang, Z.; Xue, M.; Wang, X.; Zhang, F.; Zhang, Y.; Zhang, W.; Wang, M. Constitutive expression of an A-5 subgroup member in the DREB transcription factor subfamily from *Ammopiptanthus mongolicus* enhanced abiotic stress tolerance and anthocyanin accumulation in transgenic Arabidopsis. *PLoS ONE* **2019**, *14*, e0224296.
51. Hellens, R.; Mullineaux, P.; Klee, H. Technical focus: A guide to *Agrobacterium* binary Ti vectors. *Trends Plant Sci.* **2000**, *5*, 446–451. [CrossRef]

© 2020 by the authors. Licensee MDPI, Basel, Switzerland. This article is an open access article distributed under the terms and conditions of the Creative Commons Attribution (CC BY) license (http://creativecommons.org/licenses/by/4.0/).

Article

Genome-Wide Identification and Expression Profile of *TPS* Gene Family in *Dendrobium officinale* and the Role of *DoTPS10* in Linalool Biosynthesis

Zhenming Yu [1,2,†], Conghui Zhao [1,3,†], Guihua Zhang [1], Jaime A. Teixeira da Silva [4] and Jun Duan [1,2,*]

1. Guangdong Provincial Key Laboratory of Applied Botany & Key Laboratory of South China Agricultural Plant Molecular Analysis and Genetic Improvement, South China Botanical Garden, Chinese Academy of Sciences, Guangzhou 510650, China; zhenming311@scbg.ac.cn (Z.Y.); zhaoconghui@scbg.ac.cn (C.Z.); zhanggh@scbg.ac.cn (G.Z.)
2. Center of Economic Botany, Core Botanical Gardens, Chinese Academy of Sciences, Guangzhou 510650, China
3. University of Chinese Academy of Sciences, No. 19A Yuquan Road, Beijing 100049, China
4. Independent Researcher, P.O. Box 7, Miki-cho Post Office, Ikenobe 3011-2, Kagawa-ken 761-0799, Japan; jaimetex@yahoo.com
* Correspondence: duanj@scib.ac.cn; Tel.: +86-020-37252978
† These authors contributed equally to this work.

Received: 25 June 2020; Accepted: 28 July 2020; Published: 30 July 2020

Abstract: Terpene synthase (TPS) is a critical enzyme responsible for the biosynthesis of terpenes, which possess diverse roles in plant growth and development. Although many terpenes have been reported in orchids, limited information is available regarding the genome-wide identification and characterization of the TPS family in the orchid, *Dendrobium officinale*. By integrating the *D. officinale* genome and transcriptional data, 34 *TPS* genes were found in *D. officinale*. These were divided into four subfamilies (TPS-a, TPS-b, TPS-c, and TPS-e/f). Distinct tempospatial expression profiles of *DoTPS* genes were observed in 10 organs of *D. officinale*. Most *DoTPS* genes were predominantly expressed in flowers, followed by roots and stems. Expression of the majority of *DoTPS* genes was enhanced following exposure to cold and osmotic stresses. Recombinant DoTPS10 protein, located in chloroplasts, uniquely converted geranyl diphosphate to linalool in vitro. The *DoTPS10* gene, which resulted in linalool formation, was highly expressed during all flower developmental stages. Methyl jasmonate significantly up-regulated *DoTPS10* expression and linalool accumulation. These results simultaneously provide valuable insight into understanding the roles of the TPS family and lay a basis for further studies on the regulation of terpenoid biosynthesis by *DoTPS* in *D. officinale*.

Keywords: terpene synthase; terpenes; methyl jasmonate; abiotic stress; orchids

1. Introduction

Terpenes, which are derived biosynthetically from two isomeric 5-carbon building blocks, dimethylallyl diphosphate (DMAPP) and isopentenyl diphosphate (IPP), are the largest family of plant secondary metabolites [1]. Plant terpenes play vital roles in attracting insect pollinators [2], plant defense response [1,3], plant–plant interactions [4], and the mediation of interactions with various ecological habitats [5]. The high volatility of terpene compounds promotes the scent in orchids. For instance, geraniol and linalool are major floral scent compounds in *Phalaenopsis bellina* [6,7]. Orchid floral volatiles, as well as flower color, shape, and fragrance are key horticultural ornamental traits in orchids, and also serve to attract pollinators in various ecological habitats [6].

The biosynthetic pathway of volatile terpenes is well characterized in plants (Figure 1). Generally, the C5 precursors DMAPP and IPP are formed, and the direct precursors farnesyl diphosphate (FPP), geranyl diphosphate (GPP), and geranylgeranyl diphosphate (GGPP) are generated. Subsequently, plant terpenes are biosynthesized by terpene synthase (TPS), which converts FPP to sesquiterpene in the cytosol via the mevalonic acid (MVA) pathway, and GPP and GGPP to monoterpenes and diterpenes, respectively in plastids by the methylerythritol phosphate (MEP) pathway [1,8]. TPS is positioned at the branch point of the isoprenoid pathway, and is a key enzyme for terpenoid synthesis.

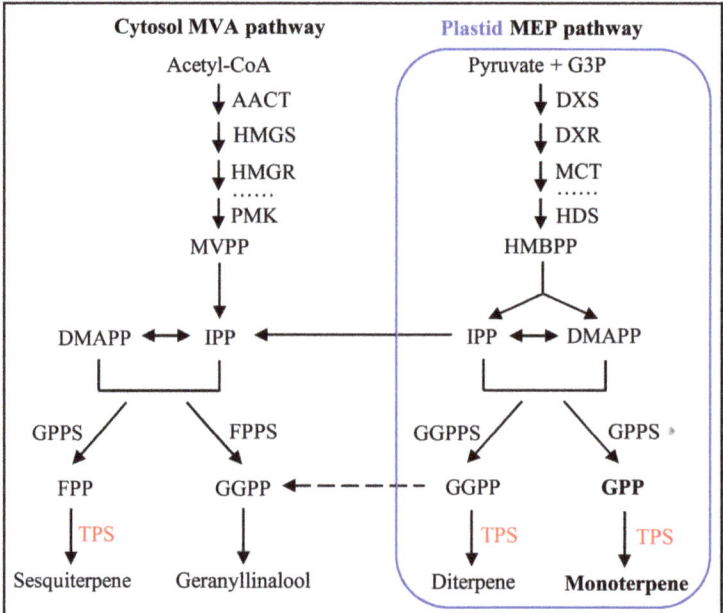

Figure 1. The pathway of terpene synthase genes responsible for the formation of terpenes in planta [3]. Terpenes are biosynthesized by the cytosol mevalonic acid (MVA) and the plastid methylerythritol phosphate (MEP) pathways, the former giving rise to sesquiterpenes and geranyllinalool, and the latter to monoterpenes and diterpenes. AACT, acetyl-CoA acetyltransferase; DMAPP, dimethylallyl pyrophosphate; DXS, 1-deoxy-d-xylulose 5-phosphate synthase; DXR, 1-deoxy-d-xylulose 5-phosphate reductoisomerase; FPP, farnesyl pyrophosphate; FPPS, FPP synthase; G3P, glyceraldehyde 3-phosphate; GGPP, geranylgeranyl pyrophosphate; GGPPS, GGPP synthase; GPP, geranyl pyrophosphate; GPPS, GPP synthase; HDS, 4-hydroxy-3-methylbut-2-en-1-yl diphosphate synthase; HMBPP, (E)-4-hydroxy-3-methylbut-2-en-1-yl diphosphate; HMGR, hydroxymethylglutaryl-CoA reductase; HMGS, hydroxymethylglutaryl-CoA synthase; IPP, isopentenyl pyrophosphate; MCT, 2-C-methyl-d-erythritol 4-phosphate cytidylyltransferase; MVPP, mevalonate 5-pyrophosphate; PMK, phosphomevalonate kinase; TPS, terpene synthase.

Each TPS is characterized by two conserved domains, PF03936 (C-terminal) and PF01397 (N-terminal) [9], as indicated in the Pfam (http://pfam.xfam.org/) database. The TPS family is phylogenetically classified into seven subfamilies (TPS-a, TPS-b, TPS-c, TPS-d, TPS-e/f, TPS-g, and TPS-h) [1]. Among them, TPS-a encodes sesquiterpene synthase that is found in both dicotyledonous and monocotyledonous plants. The angiosperm-specific TPS-b encodes monoterpene synthase with a R(R)X_8W motif that catalyzes the isomerization cyclization reaction. TPS-c is deemed to belong to the ancestral clade and catalyzes copalyl diphospate synthase. The gymnosperm-specific TPS-d performs several functions, as diterpene, monoterpene, and sesquiterpene synthases. TPS-e/f encodes copalyl diphosphate/kaurene synthases, which are critical enzymes for the production of gibberellic acid. Another angiosperm-specific TPS-g encodes monoterpene synthase without the R(R)X$_8$W motif. TPS-h is only observed in *Selaginella moellendorffii* [1,9–11]. In addition, TPS harbors conserved structural features such as DDxxD, NSE/DTE, and R(R)X$_8$W motifs [1].

To date, TPS gene families have been identified at the genome-wide level in various plant species, including *Arabidopsis thaliana* [12], *Camellia sinensis* [13], *Daucus carota* [14], *Eucalyptus globulus* and *E. grandis* [15], *Malus domestica* [16], *Solanum lycopersicum* [17], *Selaginella moellendorffii* [18], and *Vitis vinifera* [19]. Orchids form one of the largest families of flowering plants, and their metabolic profile contains various terpenes [7]. Only a few *TPS* genes have been identified thus far in orchids. *PbTPS5* and *PbTPS10* might be involved in monoterpene biosynthesis in *Phalaenopsis bellina* [20]. *FhTPS1* catalyzes the formation of linalool, while *FhTPS4*, *FhTPS6*, and *FhTPS7* are bifunctional enzymes that can simultaneously recognize FPP and GPP as substrates [21]. However, no comprehensive study about *TPS* genes in *Dendrobium officinale* exists.

D. officinale is an endangered orchid native to South and Southeast Asia, and is used for medicinal purposes in Chinese culture [22]. Moreover, *D. officinale* is a unique orchid because it grows on rocks, trees, or even cliffs. In order to adapt to harsh growth conditions, terpene compounds are synthesized [6,22–25]. Therefore, it is necessary to characterize the *TPS* gene family and study the roles of TPS in *D. officinale*. These findings will provide a valuable reference about the terpene biosynthetic pathway in orchids.

2. Results

2.1. Genome-Wide Identification and Features of TPS Proteins in D. officinale

To systematically identify the *TPS* genes in *D. officinale*, a hidden Markov model (HMM) profile of the conserved C-terminal (PF03936) and N-terminal (PF01397) domains in the TPS protein was used as a BLAST query against the *D. officinale* genome database [23]. After the removal of redundant sequences, 34 TPS genes were obtained (Table 1). The open reading frame (ORF) of *DoTPS* ranged from 378 (*DoTPS12*) to 2571 bp (*DoTPS4*), the deduced length of the amino acids ranged from 125 (*DoTPS12*) to 856 aa (*DoTPS4*), and molecular weight (Mw) ranged from 14.98 (*DoTPS12*) to 100.05 kDa (*DoTPS4*). The theoretical isoelectric point (pI) values of DoTPS proteins ranged from 4.94 (*DoTPS19*) to 7.18 (*DoTPS4*). The calculated grand average of hydrophobicity (GRAVY) values, ranging from −0.429 (*DoTPS4*) to 0.013 (*DoTPS19*), indicated most DoTPS proteins were hydrophilic, except for DoTPS19 with a GRAVY value > 0. In addition, the aliphatic index (AI) of DoTPS proteins ranged from 80.35 (*DoTPS4*) to 110.46 (*DoTPS4*), and the instability index (II) of these proteins ranged from 33.82 (*DoTPS20*) to 51.56 (*DoTPS31*). According to three widely used predictors (AtSubP [26], Plant-mPLoc [27], and pLoc-mPlant [28], all having good accuracy with greater than 70%), 14/34 DoTPS proteins were targeted to the chloroplast, other 20 DoTPS proteins were targeted to chloroplast or cytoplasm (Table 1, Tables S1–S3), suggesting that different predictors produce different results, and it was better to verify by experimental results. The prediction of secondary structures demonstrated that α-helixes and random coils were dominant in all DoTPS proteins, followed by extended strands and β-turns, accounting for on average 68.68, 23.82, 4.33, and 3.17%, respectively (Table S4).

Table 1. Information of the plant TPS gene family in *D. officinale*.

Name	Gene ID [1]	ORF [2] (bp)	AA [3] (aa)	pI [4]	Mw [5] (kDa)	AI [6]	II [7]	GRAVY [8]	Localization [9]
DoTPS1	Dca014928	960	319	6.31	36.82	90.78	48.81	−0.342	Chloroplast [a,b,c]
DoTPS2	Dca000724	1902	633	6.47	74.02	88.44	43.11	−0.260	Chloroplast [a,b,c]
DoTPS3	Dca000725	1827	608	5.46	70.52	89.67	41.24	−0.231	Chloroplast [a,b,c]
DoTPS4	Dca022838	2571	856	7.18	100.05	80.35	47.78	−0.429	Chloroplast [a,b,c]
DoTPS5	Dca003141	1692	563	5.11	65.84	98.86	42.13	−0.228	Chloroplast [a,b]/Cytoplasm [c]
DoTPS6	Dca019411	1521	506	5.13	59.48	93.68	43.71	−0.133	Chloroplast [a,b]/Cytoplasm [c]
DoTPS7	Dca003139	1692	563	5.67	65.72	92.49	39.06	−0.266	Chloroplast [a,b]/Cytoplasm [c]
DoTPS8	Dca028160	579	192	6.83	22.63	100.62	45.31	−0.121	Chloroplast [a]/Unknown [b]/Cytoplasm [c]
DoTPS9	Dca019412	1665	554	5.03	64.93	92.94	39.76	−0.189	Chloroplast [a,b]/Cytoplasm [c]

Table 1. Cont.

Name	Gene ID [1]	ORF [2] (bp)	AA [3] (aa)	pI [4]	Mw [5] (kDa)	AI [6]	II [7]	GRAVY [8]	Localization [9]
DoTPS10	Dca007746	1797	598	5.73	69.73	93.61	47.41	−0.242	Chloroplast [a,b]/Cytoplasm [c]
DoTPS11	Dca022749	696	231	5.13	27.40	104.20	49.98	−0.045	Chloroplast [a,b]/Cytoplasm [c]
DoTPS12	Dca024936	378	125	5.64	14.98	98.32	40.18	−0.326	Chloroplast [a]/Cytoplasm [b,c]
DoTPS13	Dca026570	1659	552	5.59	64.94	96.97	38.16	−0.266	Chloroplast [a,b,c]
DoTPS14	Dca005188	2550	849	6.71	98.69	86.21	46.66	−0.352	Chloroplast [a,b,c]
DoTPS15	Dca025698	1659	552	5.31	64.82	89.60	44.73	−0.284	Chloroplast [a,b,c]
DoTPS16	Dca016979	1650	549	5.62	64.23	91.62	36.94	−0.374	Chloroplast [a,b]/Cytoplasm [c]
DoTPS17	Dca008309	1653	550	5.42	64.89	97.13	47.01	−0.233	Chloroplast [a,b,c]
DoTPS18	Dca011215	1674	557	5.36	64.80	95.10	44.24	−0.302	Chloroplast [a,b,c]
DoTPS19	Dca010855	1446	481	4.94	55.95	110.46	34.63	0.013	Chloroplast [a,b,c]
DoTPS20	Dca026890	1749	582	5.20	68.07	92.84	33.82	−0.295	Chloroplast [a,b,c]
DoTPS21	Dca007747	1797	598	5.62	69.60	94.92	48.15	−0.224	Chloroplast [a,b]/Cytoplasm [c]
DoTPS22	Dca003142	1692	563	5.24	65.61	95.26	43.73	−0.245	Chloroplast [a,b]/Cytoplasm [c]
DoTPS23	Dca011214	1674	557	5.22	65.03	90.36	37.30	−0.331	Chloroplast [a,b]/Cytoplasm [c]
DoTPS24	Dca000728	1386	461	6.38	53.78	93.08	38.16	−0.180	Chloroplast [a,b]/Cytoplasm [c]
DoTPS25	Dca013782	1794	597	5.31	69.67	94.61	44.76	−0.274	Chloroplast [a,b]/Cytoplasm [c]
DoTPS26	Dca026369	1650	549	5.42	64.61	90.73	42.39	−0.438	Chloroplast [a,b]/Cytoplasm [c]
DoTPS27	Dca000723	1938	645	5.89	74.89	91.74	48.68	−0.240	Chloroplast [a,b]/Cytoplasm [c]
DoTPS28	Dca003295	1863	620	5.91	72.57	92.35	47.89	−0.262	Chloroplast [a,b]/Cytoplasm [c]
DoTPS29	Dca018407	1653	550	5.57	64.68	95.89	50.10	−0.251	Chloroplast [a,b]/Cytoplasm [c]
DoTPS30	Dca013784	1377	458	5.07	53.35	93.52	38.64	−0.254	Chloroplast [a,b]
DoTPS31	Dca016966	1089	362	7.07	41.65	99.70	51.56	−0.193	Chloroplast [a,b,c]
DoTPS32	Dca018946	2433	810	5.75	91.13	88.99	46.05	−0.169	Chloroplast [a,b,c]
DoTPS33	Dca017971	1536	511	5.57	58.70	97.18	42.66	−0.109	Chloroplast [a,b]/Cytoplasm [c]
DoTPS34	Dca020940	1797	598	5.19	69.59	91.52	45.72	−0.280	Chloroplast [a,b,c]

[1] Gene ID, it is annotated in *D. officinale* genome [23]; [2] ORF, open reading frame; [3] AA, amino acid; [4] pI, theoretical isoelectric point; [5] Mw, molecular weight; [6] AI, aliphatic index; [7] II, instability index; [8] GRAVY, grand average of hydrophobicity; [9] Localization is predicted by Plant-mPLoc [27] (Table S1, http://www.csbio.sjtu.edu.cn/bioinf/plant-multi/), AtSubP [26] (Table S2, http://bioinfo3.noble.org/AtSubP/), and pLoc-mPlant [28] (Table S3, http://www.jci-bioinfo.cn/pLoc-mPlant/) tools. [a, b, c] indicates the result of Plant-mPLoc, AtSubP, and pLoc-mPlant, respectively.

2.2. Analysis of Conserved Motifs and Gene Structure

Since analysis of gene structure will facilitate an understanding of gene evolution and possible roles, the structure of *DoTPS* genes in *D. officinale* was investigated (Figure 2A). The amount of exons ranged from 2 to 14, with an average of 6.6 for all *DoTPS* genes. *DoTPS32* contained the most exons (14), whereas *DoTPS8* and *DoTPS12* harbored the fewest exons (2). The majority of *DoTPS* genes (52.9%) had seven exons. Apart from *DoTPS7, -20, -24, -27*, and *-28*, most of the genes that clustered in the same group generally possessed a similar exon–intron structure, especially in terms of intron number and exon length (Figure 2A). This conserved exon–intron structure within each cluster was in agreement with the classification of *DoTPS* genes in a neighbor-joining (NJ) phylogenetic tree based on DoTPS sequences.

To further elucidate the structural and functional features of DoTPS, 20 conserved motifs of the DoTPS proteins were identified using MEME software (Figure 2B). The lengths of these motifs ranged from 15 to 47 amino acids (Figure S1; Table S5). DoTPS3 contained the most motifs (18/20) while DoTPS31 had only two motifs. Motif 6 was found in all DoTPS proteins, except DoTPS12. Motifs 5 and 10 were the second most common DoTPS proteins (32/34), followed by motifs 1 and 2 (31/34). DoTPS1, -8, and -31 did not contain the DDxxD motif (motif 1), and DoTPS1, -12, and -31 did not contain the R(R)X$_8$W (motif 2) motif. Intriguingly, motif 14 was found in the cluster containing DoTPS5, -6, -7, -9, 13, -15, -16, -17, -22, -26, and -29. Motifs 17 and 20 were particularly abundant in the group containing DoTPS2, -3, -10, -18, -19, -20, -21, -23, -25, -27, -28, -30, -32, and -34. Motif 20 only existed in the cluster that included DoTPS4, -14, and -32. Motif 18 was only observed in a small branch that harbored DoTPS2, -3, -24, and -27 (Figure 2B). Despite the different types of motifs among clusters, DoTPS proteins within the same cluster generally possessed similar motifs. The diversity of DoTPS phylogenetic grouping patterns was likely influenced by the gene structure and the location of motifs.

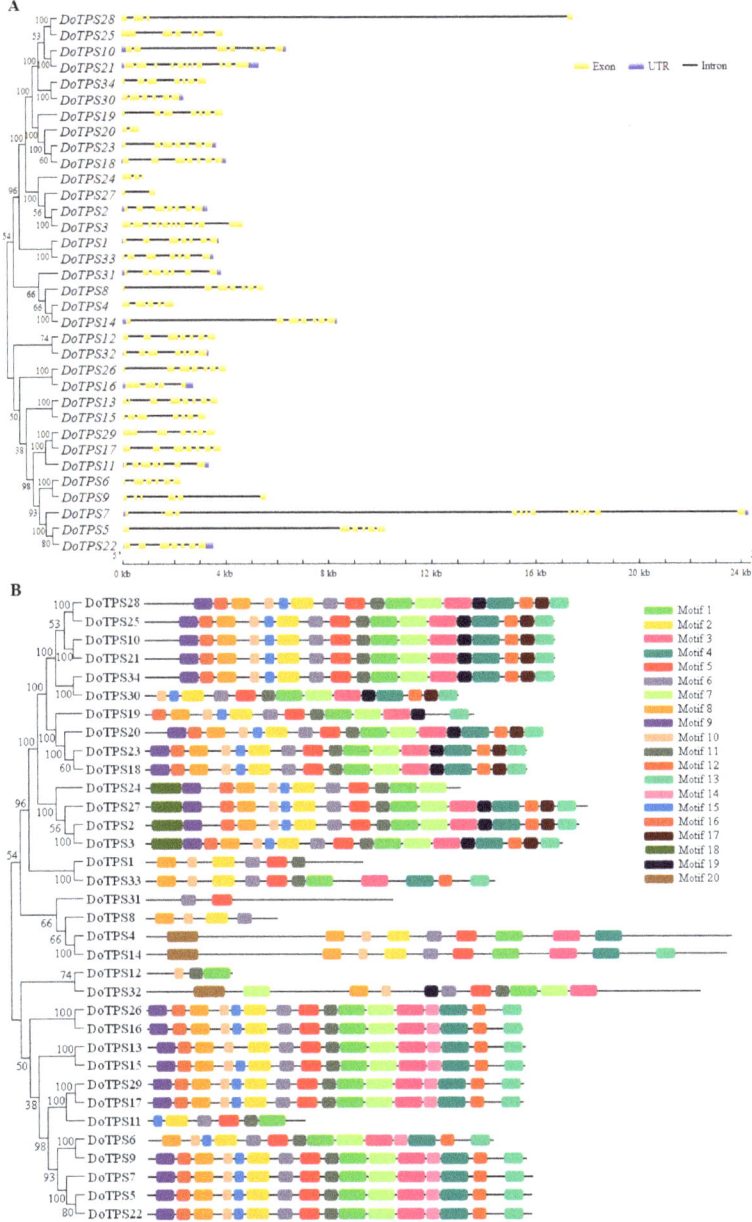

Figure 2. Phylogenetic relationships, exon–intron structure, and protein domain analysis of *DoTPS* genes in *D. officinale*. (**A**) Phylogenetic relationships and exon–intron structure of *DoTPS* genes. Exon–intron distribution was performed using GSDS 2.0 server (http://gsds.cbi.pku.edu.cn/). Yellow boxes indicate exons, black lines indicate introns. Blue boxes represent upstream/downstream-untranslated regions. (**B**) Phylogenetic relationships and motif structures of *DoTPS* genes. Phylogenetic tree was generated with MEGA 7.0 using the NJ method. Twenty classical motifs in DoTPS proteins were analyzed by the MEME tool. The width of each motif ranged from 15 to 47 amino acids. Different color blocks represented different motifs.

2.3. Phylogenetic Analysis of DoTPS Genes in D. officinale

To gain further insight into the evolutionary relationships among the TPS subfamilies, an unrooted phylogenetic tree was constructed using the neighbor-joining (NJ) method implemented in MEGA 7.0 with the Jones–Taylor–Thornton (JTT) model based on multiple sequence alignment of TPS members from *Abies grandis*, *A. thaliana*, *Apostasia shenzhenica*, *D. officinale*, *Oryza sativa*, *Phalaenopsis equestris*, *Populus trichocarpa*, *Selaginella moellendorffii*, *Solanum lycopersicum*, and *Sorghum bicolor* (Figure 3, Table S6). The phylogenetic tree demonstrates that TPS proteins were clustered into seven subfamilies (TPS-a, TPS-b, TPS-c, TPS-d, TPS-e/f, TPS-g, and TPS-h) according to the published report [1,15]. Thirty-four DoTPS proteins appeared in only four groups (TPS-a, TPS-b, TPS-c, and TPS-e/f, 14, 16, 1, and 3, respectively), and 88.2% of them belonged to TPS-a or TPS-b subgroups. Similarly, there were 20 and 9 TPS proteins present in *P. equestris* and *A. shenzhenica*, the amount of TPS-a, and TPS-b accounted for 12/20, and 7/9, respectively (Figure 3). This phenomenon was similar to that in *A. thaliana* and *O. sativa* [1,10]. Remarkably, in the TPS-a group, dicotyledonous and monocotyledonous plants formed distinct subgroups, which were observed previously [1,15] and termed them as TPS-a1 (dicots) and TPS-a2 (monocots), suggesting that the *TPS-a* genes evolved independently. Similar to the TPS-a group, the TPS-c group was further divided into dicot and monocot subclades. Consistent with a previous report, the TPS-d subfamily was specific to gymnosperms, and the TPS-h subfamily was only observed in *S. moellendorffii* [1,11,12,17], inferring that they might play a particular role in these species.

Multiple sequence alignment of DoTPS proteins was further analyzed. As illustrated in Figure 4, the arginine-tryptophan motif, $R(R)X_8W$, was found in all the DoTPS-a and DoTPS-b proteins, except DoTPS8, at the N-terminus. It plays a role in initiation of the isomerization cyclization reaction [1,10]. However, the arginine-tryptophan motif, $R(R)X_8W$, varied or was even absent in TPS-c and TPS-e/f proteins. Two aspartate-rich motifs, DDxxD and NSE/DTE, are essential for cleaving prenyl diphosphate substrate by chelating a trio of Mg^{2+} or Mn^{2+} at the C-terminus [9,11]. The DDxxD motif was conserved in almost all the DoTPS proteins, except three DoTPS proteins (DoTPS1, DoTPS8 of the TPS-a group, and DoTPS31 of the TPS-c group). DoTPS31 was the only TPS-c member in *D. officinale*. The TPS-c subfamily is mainly found in land plants and its prenyl diphosphate unit is not cleaved [1]. The NSE/DTE motif was absent in DoTPS8, -11, and -12 of the TPS-a group, DoTPS1, -19, and -24 of the TPS-b group, and DoTPS32 of the TPS-e/f group (Figure 4). Taken together, gene structure and amino acid alignment were consistent with the phylogenetic analysis. The functions of DoTPS proteins in the same group could be inferred from known TPS proteins, according to their phylogenetic relationships.

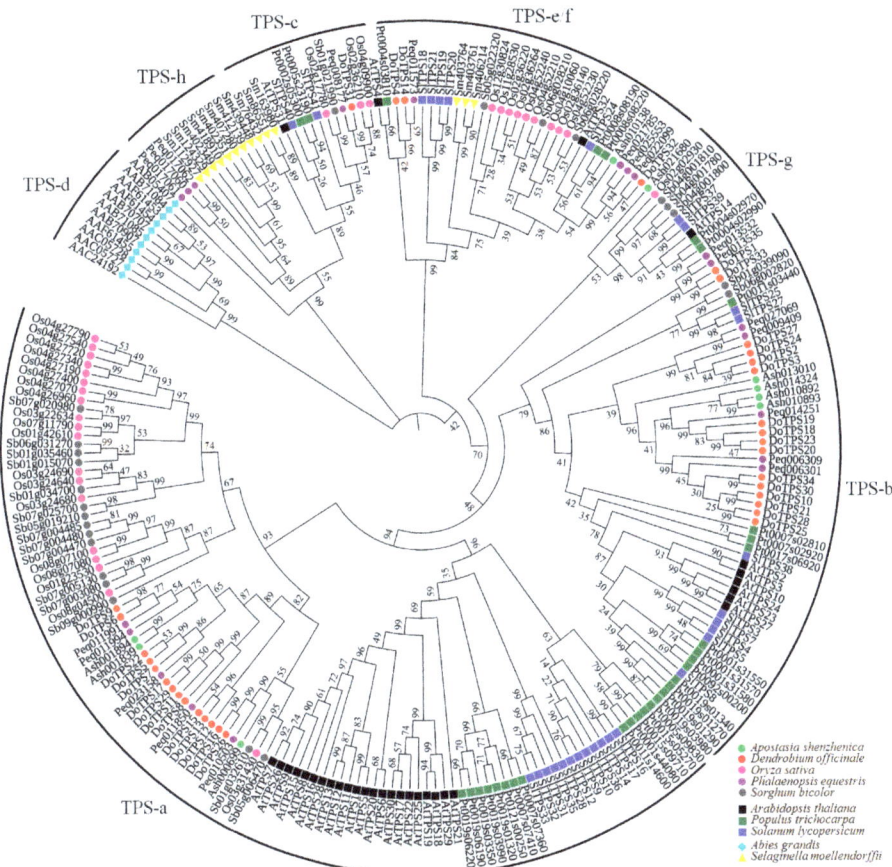

Figure 3. Phylogenetic analysis of TPS proteins in ten higher plant species. A phylogenetic tree was constructed by the neighbor-joining method with the Jones–Taylor–Thornton model and pairwise deletion option using MEGA 7.0 with 1000 bootstrap replicates. Tree visualization and labeling was performed on FigTree v1.4.4 (http://tree.bio.ed.ac.uk/software/figtree/). The TPS family was divided into seven subfamilies as previously reported [1,15]: TPS-a, TPS-b, TPS-c, TPS-d, TPS-e/f, TPS-g, and TPS-h. Circles represented monocotyledonous plants, squares represented dicotyledonous plants, the cyan diamond indicates *Abies grandis*, and yellow triangle indicates *Selaginella moellendorffii*.

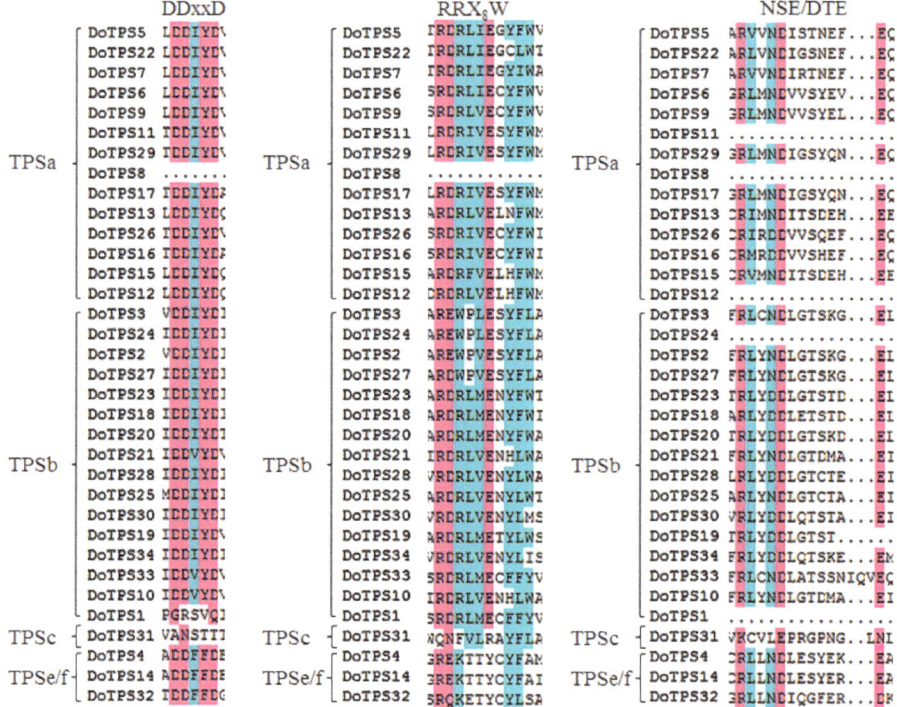

Figure 4. Comparison of DDxxD, R(R)X$_8$W, and NSE/DTE motifs in *D. officinale* DoTPS proteins.

2.4. Identification of Cis-Acting Elements in the Promoter Region of DoTPS Genes

To ascertain the potential biological roles of *DoTPS* genes in *D. officinale*, a 2000-bp upstream region of the initiation code (ATG) was identified using the PlantCARE tool. The *cis*-acting elements in the promoter regions of *DoTPS* genes were classified into three categories of *cis*-elements linked to plant growth and development, phytohormone responsiveness, and stress responsiveness (Figure 5). In the plant growth and development category (159/759), 10 *cis*-elements involved in circadian rhythms, endosperm expression (AAGAA-motif and GCN4-motif), flowering (AT-rich element, CCAAT-box, and MRE), shoot and root meristem expression (CAT-box), seed expression (RY-element), shoot expression (As-2 element), and zein metabolism (O$_2$ site) were found, the highest proportion being the As-2 element (30%). In the stress responsiveness category (221/759), various elements related to anaerobic induction (ARE, 19%), defense and stress (TC-rich repeats, 8%), dehydration (DRE, 6%), drought-inducibility (MBS, 16%), low temperature (LTR, 5%), stress (STRE, 27%), and wounding (WRE3 and WUN-motif, 9% and 10%, respectively) responsiveness were detected. Most of *cis*-elements (379/759) were related to the phytohormone responsiveness category, and were responsive to abscisic acid (ABRE), auxin (TGA-element), ethylene (ERE), gibberellin (GARE-motif, P-box, and TATC-box), MeJA (TGACG-motif and MYC), and salicylic acid (TCA-element). Notably, the largest number of *cis*-elements was the TGACG-motif and MYC associated with MeJA-responsiveness, accounting for 12% and 29% of the hormone-related *cis*-elements, respectively (Figure 5). These results suggest that *DoTPS* genes might be MeJA-induced and/or -repressed genes, and that they respond to multiple abiotic stresses.

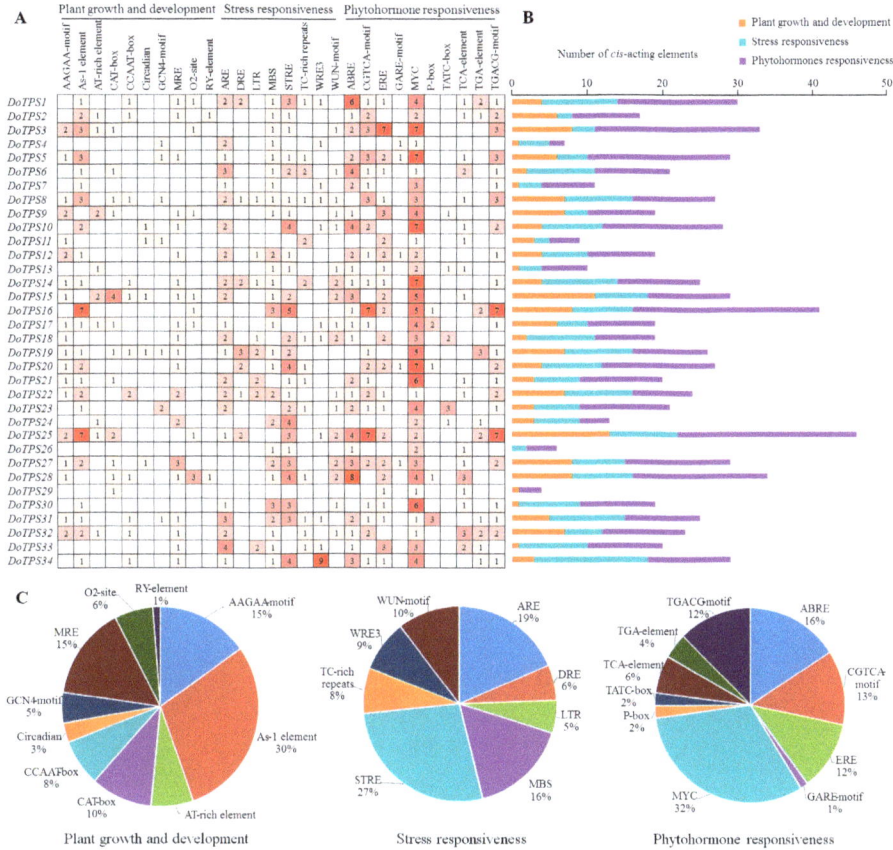

Figure 5. Information of *cis*-acting elements in *DoTPS* genes of *D. officinale*. (**A**) The gradient red colors and numbers in the grid indicate the number of different *cis*-acting elements in *DoTPS* genes. (**B**) The different colored histogram indicates the number of *cis*-acting elements in each category. (**C**) The ratio of different *cis*-acting elements in each category is shown as pie charts.

2.5. Tempospatial Expression Patterns of DoTPS Genes in Different D. officinale Organs

To obtain clues about the role of *DoTPS* genes in *D. officinale* development, an RNA-sequencing transcriptome database of flower buds, green root tips, gynostemium (column), labellum (lip), leaves, pollinia, sepals, stems, roots, and white part of roots was established (Figure 6A). Overall, *DoTPS* genes exhibited distinct organ-specific expression profiles, possibly suggesting the functional divergence of *DoTPS* genes in different *D. officinale* tissues during growth and development. *DoTPS11*, *-17*, and *-19* were highly expressed in stems. Thirteen *DoTPS* genes exhibited a high level of expression in root tissues, including five (*DoTPS4*, *-6*, *-12*, *-15*, and *-29*), five (*DoTPS13*, *-14*, *-18*, *-25*, and *-32*), and three (*DoTPS9*, *-30*, and *-34*) genes in green root tips, roots, and white part roots, respectively. Notably, 52.9% of *DoTPS* genes displayed the highest transcript abundance in floral organs. Among them, *DoTPS28* in flower buds, *DoTPS8* in gynostemium, *DoTPS5*, *-7*, *-10*, *-20*, *-21*, *-22*, and *-23* in labellum, *DoTPS16*, *-26*, and *-31* in pollinium, and *DoTPS1*, *-2*, *-3*, *-24*, *-27*, and *-33* in sepals, indicating that the preferentially expressed *DoTPS* genes might be indirectly or directly involved in the development of reproductive organs. Our data indicates that the organ-specific expression of *DoTPS* genes might be important in *D. officinale* flower growth and development.

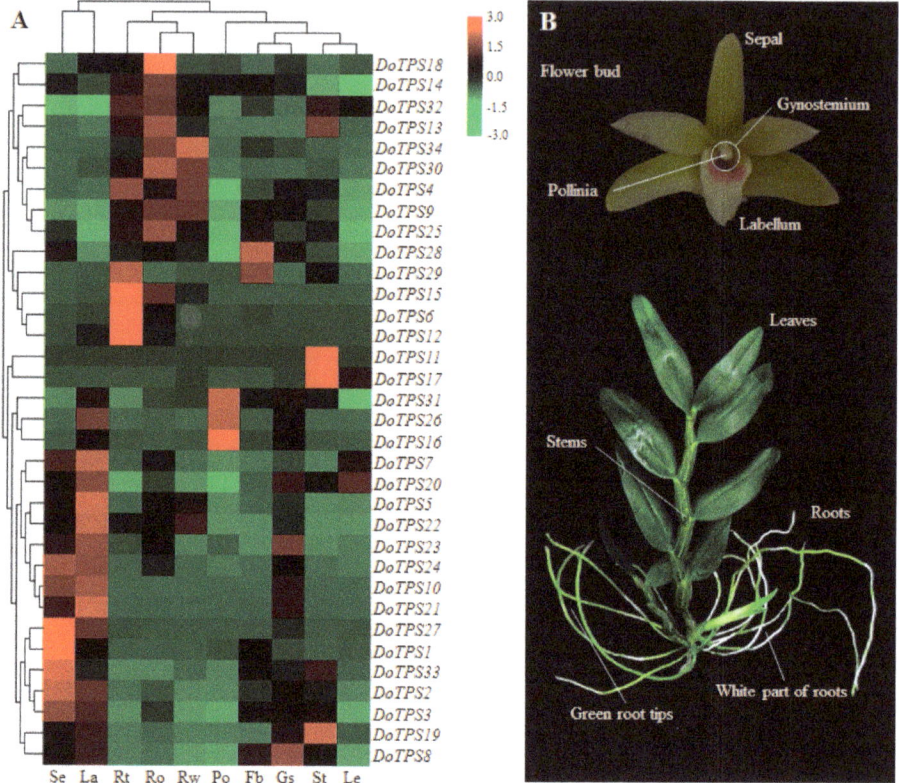

Figure 6. Tissue-specific expression profiles of *DoTPS* genes in different *D. officinale* organs. (**A**) The transcription levels of *DoTPS* genes in different tissues. The different tissues were sepals (Se), labellum (la), green root tips (Rt), roots (Ro), white part of roots (Rw), pollinia (Po), flower buds (Fb), gynostemium (Gs), stems (St), and leaves (Le) in two-year-old *D. officinale* adult plants. Heatmap was generated using the TBtools server (https://github.com/CJ-Chen/TBtools), and gradient color from green to red was expressed as the log2-transformed expression levels of each *DoTPS* gene that was normalized to the internal reference gene *DoEF-1α*, GenBank accession no. JF825419. (**B**) *D. officinale* "Zhongke 5" used in this study. All fragments per kilobase of transcript per million fragments mapped (FPKM) values that were used were downloaded from NCBI under BioProject PRJNA262478 [25], and are listed in Table S7.

2.6. Expression Patterns of DoTPS Genes under Abiotic Stress

To better understand the role of *DoTPS* genes in response to cold and osmotic stresses, transcriptome data combined with the RT-qPCR assay were employed to investigate the expression levels of *DoTPS* genes in *D. officinale* under cold (0 °C) or osmotic (mannitol) treatment. Results showed that *DoTPS* genes exhibited distinct expression patterns under osmotic treatment, showing two trends, an upward trend and a downward trend (Figure 7A). Half of the *DoTPS* genes, including *DoTPS1*, *-3*, *-6*, *-8*, *-11*, *-16*, *-18*, *-19*, *-21*, *-22*, *-23*, *-24*, *-25*, *-26*, *-27*, *-31*, and *-32*, were obviously suppressed (1.2–48.7-fold) by mannitol-induced osmotic stress. The other half of *DoTPS* genes exhibited an increasing trend, but the highest expression level was either at 12 h (*DoTPS9*, *-10*, *-14*, and *-28*), 24 h (*DoTPS2*, *-4*, *-7*, *-12*, *-13*, *-15*, *-20*, *-29*, and *-30*), or 48 h (*DoTPS5*, *-17*, *-33*, and *-34*) in response to mannitol treatment (Figure 7A).

After cold acclimation (0 °C) for 20 h, 26 *DoTPS* genes were upregulated, more than the number of suppressed genes (8; Figure 7B). Compared to the non-acclimated controls, the transcription levels of *DoTPS4*, *-8*, *-13*, *-15*, *-23*, *-26*, and *-33* were downregulated between 1.6- and 14.3-fold. In contrast, the expression levels of most other *DoTPS* genes (27/34) were clearly upregulated between 1.9- and 103.1-fold, except for the tiny variation of *DoTPS20* and *DoTPS32*. These results suggested that *DoTPS* genes might be involved in cold and osmotic stress responses in *D. officinale*.

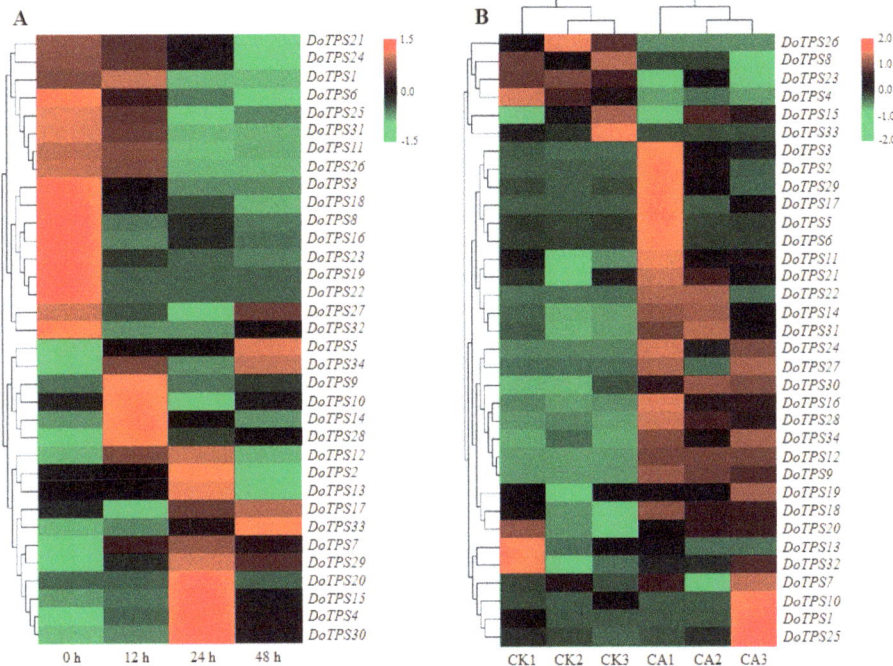

Figure 7. Transcription levels of *DoTPS* genes in *D. officinale* under cold and osmotic stresses. (**A**) Expression profiles of *DoTPS* genes in response to 200 mM mannitol treatment for 48 h. (**B**) Expression profiles of *DoTPS* genes in response to cold treatment (0 °C) for 20 h. Heatmap was drawn using TBtools software (https://github.com/CJ-Chen/TBtools). Color gradient from green to red was expressed as the log2-transformed expression level of each *DoTPS* gene. CA, cold acclimation; CK, control (non-acclimation). The expression values of *DoTPS* genes in response to mannitol treatment are listed in Table S8. The FPKM values of *DoTPS* genes exposed to cold treatment that were downloaded from a transcriptome database [29], are listed in Table S9.

2.7. Expression Patterns of DoTPS Genes Subjected to MeJA Treatment

MeJA is a signaling molecule that promotes the formation of secondary metabolic products [26]. We determined the effect of MeJA at the level of transcription of *DoTPS* genes in *D. officinale*. MeJA treatment differentially regulated *DoTPS* gene expression. Compared to the non-treated control, *DoTPS28*, *-31*, and *-32* were suppressed between 1.5- and 2.4-fold at 12 h after MeJA treatment. The suppressed genes returned to their control level at 48 h after MeJA treatment. In contrast, 31 *DoTPS* genes were upregulated between 1.2- and 45.1-fold, with the highest expression at 12 h (21/31), 24 h (8/31), or 48 h (2/31) after MeJA treatment (Figure 8A), but others showed reduced gene expression. Terpenes are dominant floral volatiles of orchids, especially geraniol and linalool [7,24]. Moreover, after treatment with MeJA, the amount of geraniol and linalool was significantly enhanced (Figure 8B). *DoTPS* genes were inducible by MeJA, which might be related to the *cis*-acting elements present in their promoters, resulting in the increased formation of terpenes in *D. officinale*.

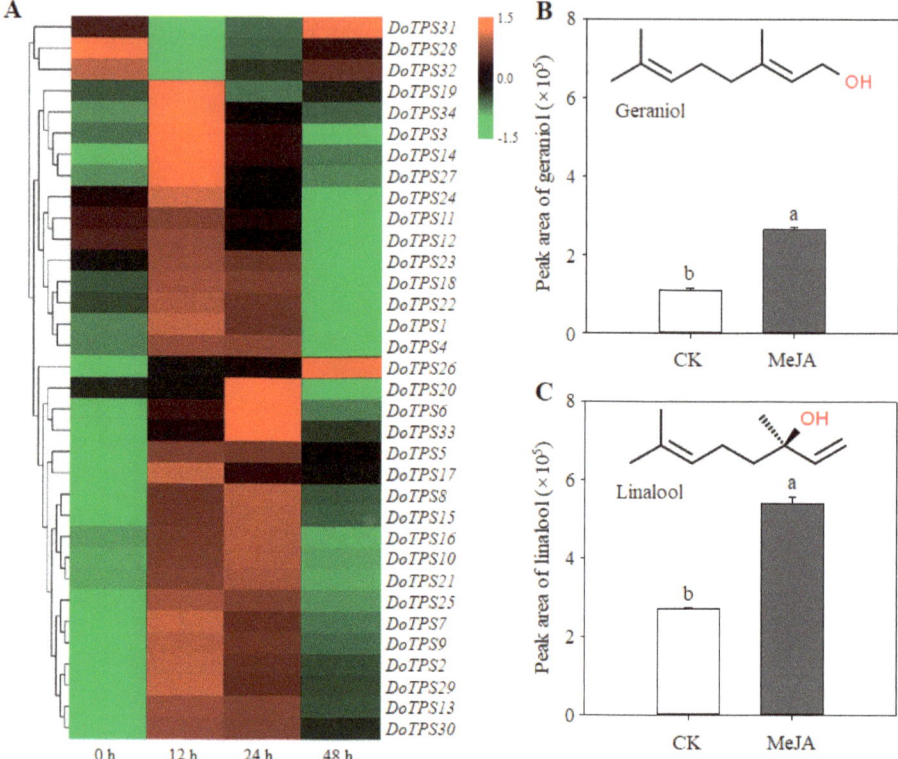

Figure 8. The transcription levels of *DoTPS* genes and synthesis of geraniol and linalool in *D. officinale* after MeJA treatment. (**A**) Effect of MeJA treatment on the expression of *DoTPS* genes. (**B**) Effect of MeJA treatment on the synthesis of geraniol. (**C**) Effect of MeJA treatment on the synthesis of linalool. Ten 10-month-old *D. officinale* seedlings exposed to 1 mM MeJA for 0, 12, 24, and 48 h were harvested. Heatmap was created using the TBtools software (https://github.com/CJ-Chen/TBtools), and gradient color from green to red was expressed as the log2-transformed expression levels of each *DoTPS* gene. Each bar represents the mean (±standard error, $n = 3$) of three independent biological replicates. Different letters above the bars indicated significant differences ($p < 0.05$, Duncan's multiple range test). CK, control treatment without MeJA. MeJA, methyl jasmonate. The expression values of *DoTPS* genes in response to MeJA treatment are listed in Table S10.

2.8. Transcription Abundance of DoTPS Genes at Budding and Flowering Stages

Since the majority of *DoTPS* genes were highly expressed in floral organs (Figure 6), we further investigated their spatial expression patterns at three floral developmental stages. *DoTPS6, -7, -9, -11, -16, -18, -21, -22, -23, -26, -28, -30, -31, -32*, and *-34* were mainly expressed in the floral budding stage. *DoTPS4, -5, -8, -10, -15, -19, -20*, and *-25* were prominently expressed during the semi-flowering stage. The remaining genes (*DoTPS1, -2, -3, -12, -13, -14, -17, -24, -27, -29*, and *-33*) displayed greatest expression levels at the full flowering stage (Figure 9). *DoTPS* genes responsible for floral fragrance showed significant differences among the three floral developmental stages. Notably, DoTPS10 had the highest level of transcription during the floral developmental stages (Figure 9), suggesting that DoTPS10 may be responsible for the biosynthesis of geraniol or linalool. Further functional characterization of the *DoTPS10* gene could be helpful.

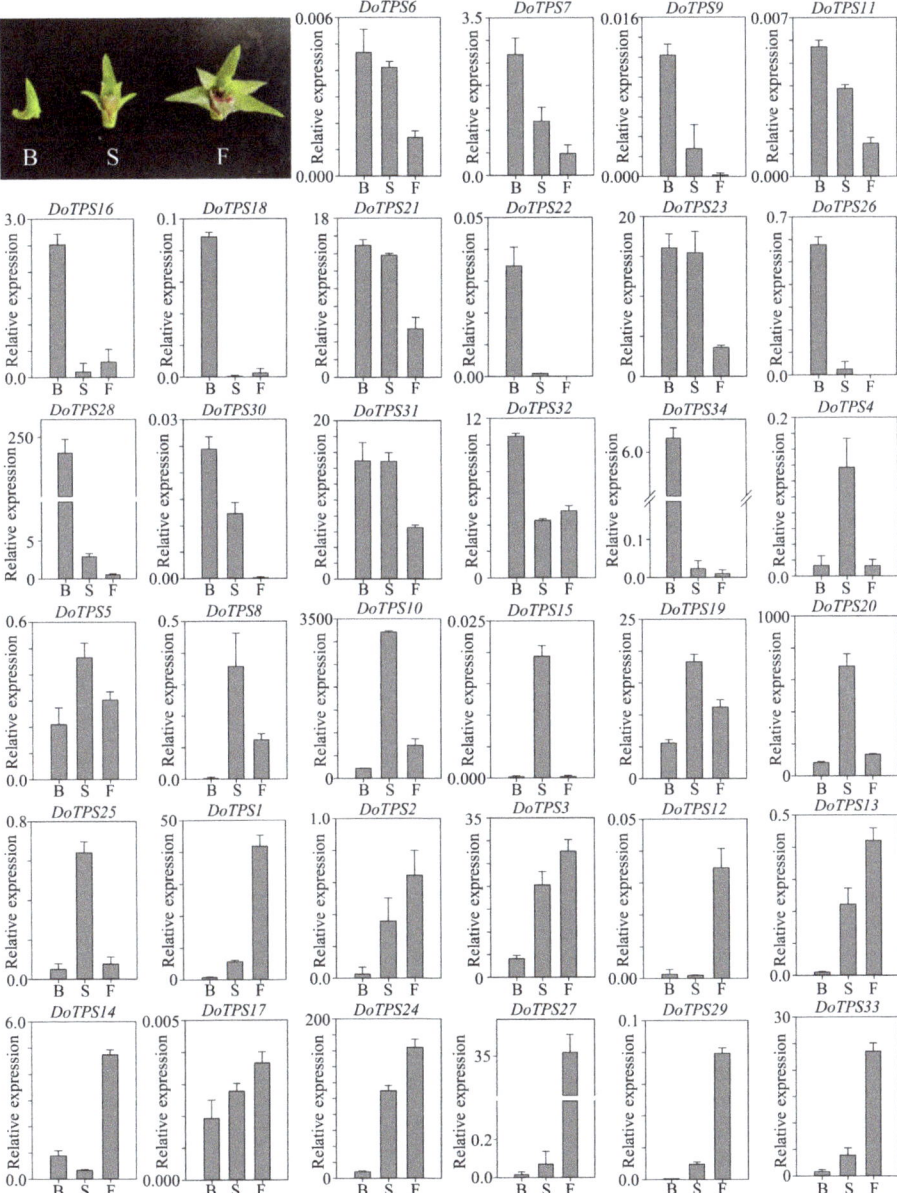

Figure 9. Transcription levels of *DoTPS* genes in *D. officinale* at three flowering stages: budding (B), semi-flowering (S) and full flowering (F). The levels of transcription were calculated by the $2^{-\Delta\Delta CT}$ method and normalized to the C_T value of *DoEF-1α*. Each bar represents the mean (±standard error, $n = 3$) of three independent biological replicates. Different letters above bars indicate significant differences ($p < 0.05$, Duncan's multiple range test). The expression values of *DoTPS* genes at three flowering stages of *D. officinale* are listed in Table S11.

2.9. Subcellular Localization of DoTPS10 in Heterologous Plants

DoTPS10 was assigned to the chloroplast with a 46 aa transit peptide. To validate the prediction, YFP-tagged DoTPS10 fusions were transiently expressed in *A. thaliana* protoplasts. In vivo YFP fluorescence signals from DoTPS10 were observed in chloroplasts (Figure 10A–D), which were consistent with a previous report of FhTPS1, FhTPS2, FhTPS4, and FhTPS5 in a *Freesia* hybrid [21].

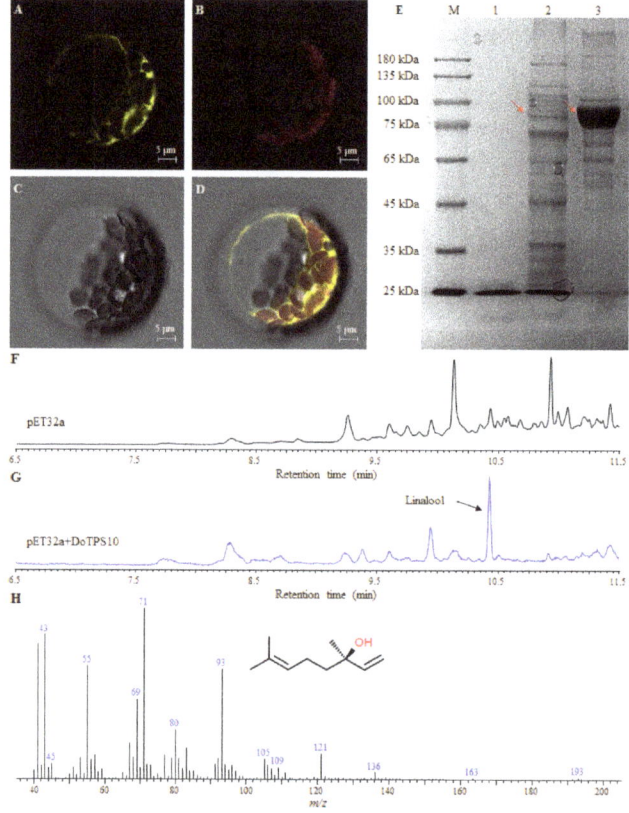

Figure 10. Functional characterization of *DoTPS10*. (**A–D**) Subcellular localization of DoTPS10 fused with yellow fluorescent protein (YFP) and transiently expressed in *A. thaliana* protoplasts. Bars = 5 μm. (**E**) SDS-PAGE analysis of DoTPS10 recombinant protein expressed in *Escherichia coli* BL21. Lanes M, 1, 2, and 3 indicate marker, pET32a, crude DoTPS10 protein, and purified DoTPS10 protein, respectively. Red arrow indicates target protein. (**F–G**) Gas chromatograms of products yielded by DoTPS10 using GPP as a substrate. (**H**) Mass spectrum of linalool was identical to the mass spectrum of the linalool standard.

2.10. Functional Characterization of DoTPS10 Involved in the Formation of Linalool

Generally, monoterpene volatiles are the main terpenes in orchids. These are produced by TPS proteins using GPP as a substrate [6,7]. To further confirm the role of DoTPS10 in the synthesis of monoterpenes, a His-tagged vector (pET32a) was used to produce the recombinant DoTPS10 protein. The vector was successfully expressed as a soluble protein in *E. coli* BL21. DoTPS10 protein was purified using a His-trap Ni-sepharose high performance column. His-tagged vector (pET32a) had an Mw of 23.8 kDa, containing 4.8 kDa of six His-tags, and recombinant DoTPS10 protein exhibited an approximate Mw of 88.3 kDa on SDS-PAGE (Figure 10E). After recombinant DoTPS10 protein was incubated with GPP, gas chromatography–mass spectrometry (GC-MS) analysis showed that empty pET32a could not produce linalool while the recombinant DoTPS10 protein singly converted the substrate GPP to the corresponding product linalool (Figure S3, Figure 9G,H).

3. Discussion

D. officinale is widely grown in subtropical and temperate regions and used as a health food, but also has a high ornamental and medicinal value [22,30]. Epiphytic or lithophytic herbs commonly suffer from adverse environmental conditions such as chilling, drought, and water deficit [22,23,29–32]. Plant volatile terpenes play critical roles not only in the formation of orchid floral scents, but also in response to environmental stresses [1,2,5–7]. TPS is the primary enzyme responsible for catalyzing the formation of monoterpenes (C_{10}), sesquiterpenes (C_{15}), or diterpenes (C_{20}) from the substrates GPP, FPP, or GGPP, respectively (Figure 1). Therefore, studies on floral scents have mainly focused on the identification and analysis of *TPS* genes responsible for the biosynthesis of terpenes [7,24].

Herein, 34 *DoTPS* genes were identified in the *D. officinale* genome according to the conserved C-terminal and N-terminal domain of TPS, followed by manual verification (Table 1; Figure 3). TPS subfamilies belong to a medium-sized family, with various gene numbers (approximately 20–150) among different plant species [1,10]. For example, 32 *AtTPS* genes were functionally discovered in *A. thaliana* [11]. A total of 14 *SmTPS*, 33 *OsTPS*, and 152 *VvTPS* genes were found in *S. moellendorffii*, *O. sativa*, and *V. vinifera*, respectively [1,10,17,18]. Furthermore, *TPS* occupied 0.26 genes/M in the *A. thaliana* genome (125 M) [33], 0.13 genes/M in *S. moellendorffii* (106 M) [34], 0.08 genes/M in the *O. sativa* genome (389 M) [35], 0.31 genes/M in *V. vinifera* (487 Mb) [36], and 0.02 genes/M in *D. officinale* (1.35 G) [22–24]. It is possible that tandem duplication may have occurred during evolution of the *D. officinale* genome, mainly in the TPS-a and TPS-b subfamilies.

Phylogenetic analysis showed that DoTPS proteins fall into four known angiosperm TPS subfamilies (TPS-a, TPS-b, TPS-c, and TPS-e/f), with the exception of the gymnosperm-specific TPS-d and *S. moellendorffii*-specific TPS-h (Figure 3). TPS-b was the largest subfamily among the DoTPS proteins, followed by TPS-a, which was consistent with *Daucus carota* [14], but in contrast to other species such as *A. thaliana* (22 of the 32 *TPS* genes were *TPS-a* genes) [12] and *S. lycopersicum* (12 of the 29 *TPS* genes were *TPS-a* genes) [17]. As illustrated in Table S12, 16 DoTPS proteins were annotated as monoterpene synthase, Fourteen DoTPS proteins were annotated as sesquiterpene synthase, and four DoTPS proteins were annotated as diterpene synthase. These four putative diterpene synthases harbored one TPS-c and three TPS-e/f proteins. TPS-e/f proteins can produce mono-, sesqui-, and di-terpenes [1]. All 14 sesquiterpene synthases and 16 monoterpene synthases were assigned to TPS-a and TPS-b, respectively (Figure 4; Table S12). For mono- and sesquiterpenes, subcellular location and availability of substrate are more important for the characterization of typical products produced in vivo. FhTPS6 was localized in the cytosol, and was deemed to be associated with the formation of a sesquiterpene (nerolidol) and several monoterpenes (myrcene, limonene, *cis*-ocimene, *trans*-ocimene, and terpinolene), suggesting that GPP and FPP might move from plastids to the cytosol [21]. TPS-a can produce monoterpenes in vitro, while TPS-b can produce hemi-, mono-, and sesquiterpenes in vitro [1]. TPS-a is an angiosperm-specific clade that is responsible for sesquiterpene or diterpene synthases, and can be further divided into monocotyledonous- and dicotyledonous-specific subgroups [1,9,10]. In *A. thaliana*, four *TPS-a* genes encode cytosolic sesquiterpene synthases, while the other nine *TPS-a* genes harbored transit peptides presumably encoding diterpene synthases, although their functions have not yet been fully investigated [12]. In *S. lycopersicum*, all *TPS-a* genes only encode sesquiterpene synthases, and 11 of 12 are cytosolic and not chloroplastic [17]. Similarly, most proteins previously functionally identified from the angiosperm-specific TPS-b clade are monoterpene synthases. For example, six *A. thaliana* TPS-b proteins clustered in the same branch that harbored AtTPS10, a monoterpene synthase that produces myrcene or ocimene [12]. These findings indicate that the TPS members share a similar functional feature within the same subfamily, however, further functional characterization is required.

The expression analysis showed that *DoTPS* genes were mainly expressed in floral organs of *D. officinale*, followed by root organs and stems (Figure 6), inferring a strict regulation of terpenoid production. Interestingly, 11 of 18 *DoTPS-b* genes with high transcript levels in floral organs were monoterpene synthase genes (Figure 6). Four monoterpene synthase genes (*DoTPS10, -19, -20,* and *-25*) were highly expressed at the semi-flowering stage (Figure 9), in agreement with the content of geraniol and linalool (Figure S2). Among them, *DoTPS10* showed the highest transcript level in floral organs (Figures 6 and 9), suggesting that it may be responsible for the biosynthesis of monoterpenes. In the present study, DoTPS10 was shown to be a single-product enzyme that could covert GPP to linalool (Figure 10), the predominant component of floral scents in orchids. TPS proteins that produce the same single product have also been found in a *Freesia* hybrid [21], *Malus domestica* [16] and *Vitis vinifera* [19]. Furthermore, the majority of *DoTPS* genes could be induced by MeJA treatment, resulting in the increased production of monoterpene volatiles such as geraniol and linalool (Figure 8). The reason why exogenous MeJA resulted in the upregulation of these *DoTPS* genes may be due to the presence of G-boxes in their promoters (Table S13), which can interact with the existing CGTCA or MYC motif of the jasmonic acid signaling pathway [37], but it needs to be further explored. Therefore, activated expression of *DoTPS10* by MeJA treatment offers a critical cue for further exploring the mechanism of linalool biosynthesis in *D. officinale*.

RT-qPCR data showed that distinct tempospatial expression profiles of *DoTPS* genes could be affected when *D. officinale* was exposed to cold or osmotic treatment (Figure 7). Similarly, previous work has emphasized the importance of terpenes in defensive response to biotic attack and abiotic stresses [1–5,7,9,10]. It is possible that the crassulacean acid metabolism plant *D. officinale*, which adheres tightly to the surface of tree bark or rocks in locations with limited soil, thus requires TPS proteins to quickly biosynthesize terpenes to circumvent adverse environments [23,25]. To better understand the mechanisms of differential terpenoid production in *D. officinale*, more efforts should be made to integrate studies on *DoTPS* expression patterns with those on profiling of terpenes.

4. Materials and Methods

4.1. Plant Materials

D. officinale "Zhongke 5" (http://www.cas.cn/syky/201811/t20181109_4669776.shtml, Figure 6B) with better adaptability to adverse habitats was cultivated in a greenhouse and in the open air at the South China Botanical Garden, Chinese Academy of Sciences (Guangzhou, Guangdong Province, China) under natural light and controlled temperatures, between 25 and 30 °C. Flowers, leaves, roots, and stems from 14-month-old adult plants of full-sib *D. officinale* were sampled at the flowering stage. For osmotic treatment, 10 independent 10-month-old *D. officinale* seedlings were transferred to fresh half-strength Murashige and Skoog (MS) medium [38] supplemented with 200 mM mannitol. Control seedlings were transferred in same way without additional mannitol. For the MeJA treatment, the same seedlings were transferred to fresh half-strength MS medium supplemented with 1 mM MeJA, and MeJA-free medium was used as the control. The leaves from osmotic and MeJA treatments were collected after treatment at 0, 12, 24, and 48 h. All samples were frozen immediately in liquid nitrogen, and stored at −80 °C until use.

4.2. Identification of TPS Family Members in D. officinale

The recently released *D. officinale* genome [23] was used in this study. Two specific TPS domains, PF03936 and PF01397, which respectively indicate the C-terminal and N-terminal domain of TPS from the Pfam database (http://pfam.xfam.org/), were used to build the corresponding Hidden Markov Model (HMM) file. HMMER v3.3 (http://www.hmmer.org/) was used to search the *D. officinale* protein database with the PF03936 and PF01397 domains model data as queries. Significant hits (e-value < 10^{-3}) were retrieved as candidate *D. officinale* TPS proteins. To verify the sequences, BLASTp (http://blast.ncbi.nlm.nih.gov) and SMART (http://smart.embl-heidelberg.de/) searches of the retrieved TPS proteins were carried out. Non-redundant sequences that did not contain the terpene synthase C-terminal domain and terpene synthase N-terminal domain were removed. The grand average of hydrophobicity (GRAVY), molecular weight (Mw), and isoelectric points (pI) of the TPS proteins were predicted from the ExPASy database (http://expasy.org/). The aliphatic index (AI), and instability index (II) were calculated by EMBOSS Pepstats tool (https://www.ebi.ac.uk/Tools/). Plant-mPLoc (www.csbio.sjtu.edu.cn/bioinf/plant-multi/), AtSubP (http://bioinfo3.noble.org/AtSubP/), and pLoc-mPlant (www.jci-bioinfo.cn/pLoc-mPlant/) were used to predict the subcellular localization of TPS proteins. The secondary structure of TPS proteins was determined using the SOPMA program (http://npsa-pbil.ibcp.fr/). In addition, TPS proteins from *P. equestris* and *A. shenzhenica* were obtained from their reported genome database [25]. The other TPS proteins from *A. grandis*, *A. thaliana*, *O. sativa*, *P. trichocarpa*, *S. moellendorffii*, *S. lycopersicum*, and *S. bicolor* were downloaded from the Phytozome version 12.1 database (https://www.phytozome.net).

4.3. Conserved Motifs, Gene Structure, and Phylogenetic Analysis

Conserved motifs of TPS proteins were analyzed with MEME software (http://meme-suite.org/) with default parameters. The exon–intron structure of TPS proteins was aligned with the Gene Structure Display Server (GSDS 2.0, http://gsds.cbi.pku.edu.cn/). Multiple sequence alignment was performed using TPS proteins from *A. grandis*, *A. shenzhenica*, *A. thaliana*, *D. officinale*, *O. sativa*, *P. equestris*, *P. trichocarpa*, *S. moellendorffii*, *S. lycopersicum*, and *S. bicolor* with ClustalX 2.1 software (www.clustal.org/). The alignments were manually adjusted and truncated with a focus on diagnostically conserved regions such as the DDxxD, NSE/DTE, and RRX_8W motifs, based on a reported protocol [15]. A phylogenetic tree was constructed using the neighbor-joining (NJ) method [39] under the Jones–Taylor–Thornton (JTT) model with 1000 bootstrap replicates in MEGA version 7.0 [40]. The generated graph was redrawn and annotated by Figtree version 1.4.4 (http://tree.bio.ed.ac.uk/software/figtree/). The sequences of TPS proteins used in this study can be found in Table S6.

4.4. Total RNA Isolation, cDNA Reverse Transcription, and RT-qPCR Analysis

Total RNA from the flowers and leaves of 14-month-old *D. officinale* "Zhongke 5" at the flowering stage were extracted using the Quick RNA Isolation Kit (Huayueyang, Beijing, China) according to the instruction manual. Genomic DNA contamination was eliminated with RNase-free DNase I (TaKaRa, Dalian, China). First-strand cDNA was synthesized by reverse transcription with the help of the PrimeScript™ RT Reagent Kit (Takara) according to the manufacturer's protocol. SYBR® Premix Ex Taq™ (TaKaRa) was applied for RT-qPCR analysis on a LightCycler 480 instrument (Roche Diagnostics, Mannheim, Germany) as described previously [41]. *D. officinale* elongation factor 1-α (*DoEF-1α*, GenBank accession no. JF825419) was selected as the internal reference gene [42]. At least three biological replicates were carried out, and relative mRNA expression data were quantified by the $2^{-\Delta\Delta CT}$ method [43]. The RT-qPCR primers of *TPS* genes listed in Table S14 were acquired by the PrimerQuest tool (http://www.idtdna.com/Primerquest/Home/Index).

4.5. Cis-Acting Elements Analysis of TPS Genes in D. officinale

The promoter sequences, 2000 bp upstream of the translational start site (ATG), of TPS genes in *D. officinale* were obtained from the *D. officinale* genome [23]. Afterwards, the online software PlantCARE (http://bioinformatics.psb.ugent.be/webtools/plantcare/html/) was employed to investigate putative *cis*-regulatory elements in the promoter region of *DoTPS* genes in *D. officinale*.

4.6. Gene Expression Analysis Based on Transcriptome Data

To gain insight into the tissue-specific transcription levels of *DoTPS* family genes, raw data from the RNA-sequencing of 10 different tissues (i.e., flower buds, green root tips, gynostemium (column), labellum (lip), leaves, pollinia, sepals, stems, roots, and the white part of roots) in two-year-old *D. officinale* adult plants was downloaded from NCBI under BioProject PRJNA262478 [25]. To study the effects of cold acclimation (0 °C for 20 h, CA) and non-acclimation (20 °C for 20 h, CK) on *DoTPS* gene expression, the raw RNA-sequencing reads were retrieved from a reported transcriptome database [29]. Fragments per kilobase of transcript per million mapped reads (FPKM) values of *DoTPS* genes in tested samples were used to evaluate transcription abundance. *DoEF-1α* was selected as the internal reference gene for normalizing each expression value. The heat maps of the *DoTPS* genes' expression patterns were illustrated using the TBtools software with default settings (https://github.com/CJ-Chen/TBtools), and the gradient color from green to red is expressed as the log2-transformed expression levels of each *DoTPS* gene.

4.7. Gas Chromatography–Mass Spectrometry Analysis of Geraniol and Linalool in Flowers of D. officinale

The frozen flowers of *D. officinale* (500 mg) were ground to a fine powder in liquid nitrogen, and then blended with precooled dichloromethane (3 mL) by vortexing for 2 min, followed by shaking at 25 °C for 8 h in the dark. The supernatant was collected by 13,000× g centrifugation, and concentrated to 200 µL using a stream of N_2 before analysis by gas chromatography–mass spectrometry (GC-MS, Shimadzu Co., Kyoto, Japan) equipped with a 30-m Supelcowax-10 column (0.25 mm diameter × 0.25 µm film thickness). The temperature program was isothermal at 60 °C for 3 min, then increased at a rate of 4 °C min^{-1} to 240 °C, and maintained at 240 °C for 20 min. MS analyses were performed in full-scan mode with a mass range from *m/z* 40 to 200. Geraniol and linalool were identified against the NIST 2008 mass spectra library (https://chemdata.nist.gov/) as described previously [44].

4.8. Prokaryotic Expression and DoTPS10 Enzyme Assay in Escherichia coli

The full-length *DoTPS10* was amplified from first-strand cDNA, as published previously [41]. The obtained PCR product was purified and inserted into the pMD-18T vector (TaKaRa) for sequencing. The gene-specific primers used for *DoTPS10* are indicated in Table S14.

A 1797-bp ORF without a stop codon (TAA) of *DoTPS10* was cloned into the pET32a vector with *Sal*I and *Xho*I restriction sites. *DoTPS10* expression in *E. coli* BL21 (DE3) cells and purification using a His-trap Ni-sepharose high performance column (GE Healthcare, Fairfield, CT, USA) were described in our previous study [45]. The purified pET32a-DoTPS10 protein was fractionated by 10% sodium dodecyl sulfate polyacrylamide gel electrophoresis (SDS-PAGE).

In vitro DoTPS10 enzyme assays were performed in screw-cap 5-mL glass vials containing 1 mL of 2-hydroxy-3-morpholinopropanesulfonic acid (MOPSO) buffer (10 mM, pH 7.0, containing 5 mM dithiothreitol, 10 mM $MgCl_2$, and 10 mM GPP as substrate) and 100 µg of DoTPS10 protein. The reactions were overlaid with 200 µL of *n*-hexane and incubated at 30 °C for 1 h. The mixtures were mixed vigorously for 1 min to obtain the enzymatic products. The organic phase was removed, and 1 µL was detected using GC-MS as described above. For comparison, His-tagged protein (empty pET32a) was used as the blank control.

4.9. Subcellular Localization of DoTPS10 in A. thaliana Mesophyll Protoplasts

To determine the localization of DoTPS10, a 1797-bp ORF without a stop codon (TAA) of *DoTPS10* was introduced into the pSAT6-EYFP-N1 vector with an *Nco*I restriction site, and was transiently transformed into mesophyll protoplasts from four-week-old *A. thaliana* leaves. After transformation in darkness at 22 °C for 20 h, YFP signals were evaluated using a Zeiss LSM 510 Meta confocal microscope (Carl Zeiss, Jena, Germany) with an excitation wavelength of 514 nm.

4.10. Statistical Analysis

IBM SPSS statistics software version 22.0 for Windows (IBM Corp., Armonk, NY, USA) was used to carry out one-way analysis of variance (ANOVA) among different samples using three replications. Duncan's multiple range test (DMRT) was used to determine significant differences ($p < 0.05$).

5. Conclusions

In this study, we reported on the identification of 34 *DoTPS* genes in *D. officinale*. Their conserved motifs, exon–intron distribution, and phylogenetic analysis was assessed. Differential expression patterns of *DoTPS* genes exposed to ten different organs and three flowering stages, highlights their involvement in regulating the biosynthesis of floral monoterpenes, as well as the responses of plants to exogenous MeJA treatment, cold, and osmotic stress. One monoterpene synthase (DoTPS10), which was targeted to chloroplasts, could specifically convert GPP into linalool in vitro. Our findings show that transcript accumulation of multiple *TPS* genes is mainly responsible for the formation of floral terpenes, and provides a foundation for further studies on orchid floral scent research through the regulation of *DoTPS* genes.

Supplementary Materials: Supplementary materials can be found at http://www.mdpi.com/1422-0067/21/15/5419/s1.

Author Contributions: Conceptualization, Z.Y. and J.A.T.d.S.; methodology, Z.Y.; validation, Z.Y., C.Z., G.Z.; formal analysis, G.Z.; investigation, Z.Y.; resources, J.D.; data curation, Z.Y. and C.Z.; writing—original draft preparation, Z.Y., C.Z. and J.A.T.d.S.; writing—review and editing, J.A.T.d.S and Z.Y.; visualization, Z.Y. and J.A.T.d.S.; supervision, J.D. and J.A.T.d.S.; project administration, J.D.; funding acquisition, J.D. All authors have read and agreed to the published version of the manuscript.

Funding: This research was funded by the National Key Research and Development Program of China, grant number 2018YFD1000400 and National Natural Science Foundation of China, grant number 31871547.

Acknowledgments: The authors are grateful to Yongxia Jia and Yangyang Xiao for assistance with GC-MS analysis of terpene compounds. We thank all the colleagues in our laboratory for help and advice. We would like to thank both the editor and reviewers for evaluating and providing constructive comments on this manuscript.

Conflicts of Interest: The authors declare no conflicts of interest.

Abbreviations

AI	Aliphatic index
bp	Base pair
CA	Cold acclimation
DMAPP	Dimethylallyl diphosphate
FPP	Farnesyl diphosphate
FPKM	Fragments per kilobase of transcript per million fragments mapped
GC–MS	Gas chromatography-mass spectrometry
GGPP	Geranylgeranyl diphosphate
GPP	Geranyl diphosphate
GRAVY	Grand average of hydrophobicity
II	Instability index
IPP	Isopentenyl diphosphate
MeJA	Methyl jasmonate
MEP	Methylerythritol phosphate
MS	Murashige and Skoog medium
Mw	Molecular weight
MVA	Mevalonic acid
NCBI	National Center for Biotechnology Information
NJ	Neighbor-joining
ORF	Open reading frame
pI	Isoelectric point
RT-qPCR	Real-time reverse transcription quantitative polymerase chain reaction
SDS-PAGE	Sodium dodecyl sulfate polyacrylamide gel electrophoresis
TPS	Terpene synthase
YFP	Yellow fluorescent protein

References

1. Chen, F.; Tholl, D.; Bohlmann, J.; Pichersky, E. The family of terpene synthases in plants: A mid-size family of genes for specialized metabolism that is highly diversified throughout the kingdom. *Plant J.* **2011**, *66*, 212–229. [CrossRef] [PubMed]
2. Byers, K.J.; Bradshaw, H.D., Jr.; Riffell, J.A. Three floral volatiles contribute to differential pollinator attraction in monkeyflowers (*Mimulus*). *J. Exp. Biol.* **2014**, *217*, 614–623. [CrossRef] [PubMed]
3. Dudareva, N.; Klempien, A.; Muhlemann, J.K.; Kaplan, I. Biosynthesis, function and metabolic engineering of plant volatile organic compounds. *New Phytol.* **2013**, *198*, 16–32. [CrossRef] [PubMed]
4. Huang, W.; Gfeller, V.; Erb, M. Root volatiles in plant-plant interactions II: Root volatiles alter root chemistry and plant-herbivore interactions of neighbouring plants. *Plant Cell Environ.* **2019**, *42*, 1964–1973. [CrossRef]
5. Campbell, D.R.; Sosenski, P.; Raguso, R.A. Phenotypic plasticity of floral volatiles in response to increasing drought stress. *Ann. Bot.* **2019**, *123*, 601–610. [CrossRef]
6. Hsiao, Y.Y.; Pan, Z.J.; Hsu, C.C.; Yang, Y.P.; Hsu, Y.C.; Chuang, Y.C.; Shih, H.H.; Chen, W.H.; Tsai, W.C.; Chen, H.H. Research on orchid biology and biotechnology. *Plant Cell Physiol.* **2011**, *52*, 1467–1486. [CrossRef]
7. Ramya, M.; Jang, S.; An, H.R.; Lee, S.Y.; Park, P.M.; Park, P.H. Volatile organic compounds from orchids: From synthesis and function to gene regulation. *Int. J. Mol. Sci.* **2020**, *21*, 1160. [CrossRef]
8. Vranová, E.; Coman, D.; Gruissem, W. Network analysis of the MVA and MEP pathways for isoprenoid synthesis. *Annu. Rev. Plant Biol.* **2013**, *64*, 665–700. [CrossRef]
9. Tholl, D. Biosynthesis and biological functions of terpenoids in plants. *Adv. Biochem. Eng. Biotechnol.* **2015**, *148*, 63–106.
10. Bohlmann, J.; Meyer-Gauen, G.; Croteau, R. Plant terpenoid synthases: Molecular biology and phylogenetic analysis. *Proc. Natl. Acad. Sci. USA* **1998**, *95*, 4126–4133. [CrossRef]
11. Jiang, S.Y.; Jin, J.; Sarojam, R.; Ramachandran, S.A. Comprehensive survey on the terpene synthase gene family provides new insight into its evolutionary patterns. *Genome Biol. Evol.* **2019**, *11*, 2078–2098. [CrossRef] [PubMed]
12. Aubourg, S.; Lecharny, A.; Bohlmann, J. Genomic analysis of the terpenoid synthase (*AtTPS*) gene family of *Arabidopsis thaliana*. *Mol. Genet. Genom.* **2002**, *267*, 730–745. [CrossRef] [PubMed]
13. Zhou, H.C.; Shamala, L.F.; Yi, X.K.; Yan, Z.; Wei, S. Analysis of terpene synthase family genes in *Camellia sinensis* with an emphasis on abiotic stress conditions. *Sci. Rep.* **2020**, *10*, 933. [CrossRef] [PubMed]
14. Keilwagen, J.; Lehnert, H.; Berner, T.; Budahn, H.; Nothnagel, T.; Ulrich, D.; Dunemann, F. The terpene synthase gene family of carrot (*Daucus carota* L.): Identification of QTLs and candidate genes associated with terpenoid volatile compounds. *Front. Plant Sci.* **2017**, *8*, 1930. [CrossRef]
15. Külheim, C.; Padovan, A.; Hefer, C.; Krause, S.T.; Köllner, T.G.; Myburg, A.A.; Degenhardt, J.; Foley, W.J. The *Eucalyptus* terpene synthase gene family. *BMC Genom.* **2015**, *16*, 450. [CrossRef]
16. Nieuwenhuizen, N.J.; Green, S.A.; Chen, X.; Bailleul, E.J.D.; Matich, A.J.; Wang, M.Y.; Atkinson, R.G. Functional genomics reveals that a compact terpene synthase gene family can account for terpene volatile production in apple. *Plant Physiol.* **2013**, *161*, 787–804. [CrossRef]
17. Falara, V.; Akhtar, T.A.; Nguyen, T.T.H.; Spyropoulou, E.A.; Bleeker, P.M.; Schauvinhold, I.; Matsuba, Y.; Bonini, M.E.; Schilmiller, A.L.; Last, R.L.; et al. The tomato terpene synthase gene family. *Plant Physiol.* **2011**, *157*, 770–789. [CrossRef]
18. Li, G.; Köllner, T.G.; Yin, Y.; Jiang, Y.; Chen, H.; Xu, Y.; Gershenzon, J.; Pichersky, E.; Chen, F. Nonseed plant *Selaginella moellendorffii* has both seed plant and microbial types of terpene synthases. *Proc. Natl. Acad. Sci. USA* **2012**, *109*, 14711–14715. [CrossRef]
19. Martin, D.M.; Aubourg, S.; Schouwey, M.B.; Daviet, L.; Schalk, M.; Toub, O.; Lund, S.T.; Bohlmann, J. Functional annotation, genome organization and phylogeny of the grapevine (*Vitis vinifera*) terpene synthase gene family based on genome assembly, FLcDNA cloning, and enzyme assays. *BMC Plant Biol.* **2010**, *10*, 226. [CrossRef]
20. Chuang, Y.C.; Hung, Y.C.; Tsai, W.C.; Chen, W.H.; Chen, H.H. PbbHLH4 regulates floral monoterpene biosynthesis in *Phalaenopsis* orchids. *J. Exp. Bot.* **2018**, *69*, 4363–4377. [CrossRef]
21. Gao, F.; Liu, B.; Li, M.; Gao, X.; Fang, Q.; Liu, C.; Ding, H.; Wang, L.; Gao, X. Identification and characterization of terpene synthase genes accounting for volatile terpene emissions in flowers of *Freesia × hybrida*. *J. Exp. Bot.* **2018**, *69*, 4249–4265. [CrossRef] [PubMed]

22. Yan, L.; Wang, X.; Liu, H.; Tian, Y.; Lian, J.; Yang, R.; Hao, S.; Wang, X.; Yang, S.; Li, Q.; et al. The genome of *Dendrobium officinale* illuminates the biology of the important traditional Chinese orchid herb. *Mol. Plant* **2015**, *8*, 922–934. [CrossRef] [PubMed]
23. Zhang, G.; Xu, Q.; Bian, C.; Tsai, W.C.; Yeh, C.M.; Liu, K.; Yoshida, K.; Zhang, L.; Chang, S.; Chen, F.; et al. The *Dendrobium catenatum* Lindl genome sequence provides insights into polysaccharide synthase, floral development and adaptive evolution. *Sci. Rep.* **2016**, *6*, 1–10. [CrossRef] [PubMed]
24. Tsai, W.C.; Dievart, A.; Hsu, C.C.; Hsiao, Y.Y.; Chiou, S.Y.; Huang, H.; Chen, H.H. Post genomics era for orchid research. *Bot. Stud.* **2017**, *58*, 61. [CrossRef] [PubMed]
25. Zhang, G.Q.; Liu, K.W.; Li, Z.; Lohaus, R.; Hsiao, Y.Y.; Niu, S.C.; Wang, J.Y.; Lin, Y.C.; Xu, Q.; Chen, L.J.; et al. The *Apostasia* genome and the evolution of orchids. *Nature* **2017**, *549*, 379–383. [CrossRef]
26. Kaundal, R.; Saini, R.; Zhao, P.X. Combining machine learning and homology-based approaches to accurately predict subcellular localization in *Arabidopsis*. *Plant Physiol.* **2010**, *154*, 36–54. [CrossRef]
27. Chou, K.C.; Shen, H.B. Plant-mPLoc: A top-down strategy to augment the power for predicting plant protein subcellular localization. *PLoS ONE.* **2010**, *5*, 11335. [CrossRef]
28. Cheng, X.; Xiao, X.; Chou, K.C. PLoc-mPlant: Predict subcellular localization of multi-location plant proteins by incorporating the optimal GO information into general PseAAC. *Mol. Biosyst.* **2017**, *13*, 1722–1727. [CrossRef]
29. Wu, Z.; Jiang, W.; Chen, S.; Mantri, N.; Tao, Z.; Jiang, C. Insights from the cold transcriptome and metabolome of *Dendrobium officinale*: Global reprogramming of metabolic and gene regulation networks during cold acclimation. *Front. Plant Sci.* **2016**, *7*, 1653. [CrossRef]
30. Teixeira da Silva, J.A.; Ng, T.B. The medicinal and pharmaceutical importance of *Dendrobium* species. *Appl. Microbiol. Biotechnol.* **2017**, *101*, 2227–2239. [CrossRef]
31. Yu, Z.; Yang, Z.; Teixeira da Silva, J.A.; Luo, J.; Duan, J. Influence of low temperature on physiology and bioactivity of postharvest *Dendrobium officinale* stems. *Postharvest Biol. Tech.* **2019**, *148*, 97–106. [CrossRef]
32. Wan, X.; Zou, L.; Zheng, B.; Tian, Y.; Wang, Y. Transcriptomic profiling for prolonged drought in *Dendrobium catenatum*. *Sci. Data* **2018**, *5*, 180233. [CrossRef] [PubMed]
33. Kaul, S.; Koo, H.L.; Jenkins, J.; Rizzo, M.; Rooney, T.; Tallon, L.J.; Feldblyum, T.; Nierman, W.; Benito, M.I.; Town, C.D.; et al. The Arabidopsis Genome Initiative. Analysis of the genome sequence of the flowering plant *Arabidopsis thaliana*. *Nature* **2000**, *408*, 796–815.
34. Banks, J.A.; Nishiyama, T.; Hasebe, M.; Bowman, J.L.; Gribskov, M.; de Pamphilis, C.; Albert, V.A.; Aono, N.; Aoyama, T.; Ambrose, B.A.; et al. The *Selaginella* genome identifies genetic changes associated with the evolution of vascular plants. *Science* **2011**, *332*, 960–963. [CrossRef]
35. Huang, X.; Kurata, N.; Wei, X.; Wang, Z.X.; Wang, A.; Zhao, Q.; Zhao, Y.; Liu, K.; Lu, H.; Li, W.; et al. A map of rice genome variation reveals the origin of cultivated rice. *Nature* **2012**, *490*, 497–501. [CrossRef]
36. Jaillon, O.; Aury, J.M.; Noel, B.; Policriti, A.; Clepet, C.; Casagrande, A.; Choisne, N.; Aubourg, S.; Vitulo, N.; Jubin, C.; et al. French-Italian Public Consortium for Grapevine Genome Characterization. The grapevine genome sequence suggests ancestral hexaploidization in major angiosperm phyla. *Nature* **2007**, *449*, 463–467.
37. Hong, G.J.; Xue, X.Y.; Mao, Y.B.; Wang, L.J.; Chen, X.Y. *Arabidopsis* MYC2 interacts with DELLA proteins in regulating sesquiterpene synthase gene expression. *Plant Cell* **2012**, *24*, 2635–2648. [CrossRef]
38. Murashige, T.; Skoog, F. A revised medium for rapid growth and bioassays with tobacco tissue cultures. *Physiol. Plantarum* **1962**, *15*, 473–497. [CrossRef]
39. Saitou, N.; Nei, M. The neighbor-joining method: A new method for reconstructing phylogenetic trees. *Mol. Biol. Evol.* **1987**, *4*, 406–425.
40. Kumar, S.; Stecher, G.; Tamura, K. MEGA7: Molecular evolutionary genetics analysis version 7.0 for bigger datasets. *Mol. Biol. Evol.* **2016**, *33*, 1870–1874. [CrossRef]
41. Yu, Z.; He, C.; Teixeira da Silva, J.A.; Luo, J.; Yang, Z.; Duan, J. The GDP-mannose transporter gene (*DoGMT*) from *Dendrobium officinale* is critical for mannan biosynthesis in plant growth and development. *Plant Sci.* **2018**, *277*, 43–54. [CrossRef]
42. An, H.; Zhu, Q.; Pei, W.; Fan, J.; Liang, Y.; Cui, Y.; Lv, N.; Wang, W. Whole-transcriptome selection and evaluation of internal reference genes for expression analysis in protocorm development of *Dendrobium officinale* Kimura et Migo. *PLoS ONE* **2016**, *11*, 163478. [CrossRef] [PubMed]
43. Livak, K.J.; Schmittgen, T.D. Analysis of relative gene expression data using real-time quantitative PCR and the $2^{-\Delta\Delta CT}$ method. *Methods* **2001**, *25*, 402–408. [CrossRef] [PubMed]

44. Yu, Z.; Zhang, G.; Teixeira da Silva, J.A.; Yang, Z.; Duan, J. The β-1,3-galactosetransferase gene *DoGALT2* is essential for stigmatic mucilage production in *Dendrobium officinale*. *Plant Sci.* **2019**, *287*, 110179. [CrossRef] [PubMed]
45. Yu, Z.; Liao, Y.; Zeng, L.; Dong, F.; Watanabe, N.; Yang, Z. Transformation of catechins into theaflavins by upregulation of *CsPPO3* in preharvest tea (*Camellia sinensis*) leaves exposed to shading treatment. *Food Res. Int.* **2020**, *129*, 108842. [CrossRef] [PubMed]

 © 2020 by the authors. Licensee MDPI, Basel, Switzerland. This article is an open access article distributed under the terms and conditions of the Creative Commons Attribution (CC BY) license (http://creativecommons.org/licenses/by/4.0/).

MDPI
St. Alban-Anlage 66
4052 Basel
Switzerland
Tel. +41 61 683 77 34
Fax +41 61 302 89 18
www.mdpi.com

International Journal of Molecular Sciences Editorial Office
E-mail: ijms@mdpi.com
www.mdpi.com/journal/ijms

www.ingramcontent.com/pod-product-compliance
Lightning Source LLC
LaVergne TN
LVHW070412100526
838202LV00014B/1446